T0137823

Studies in Computational Intelligence 482

Editor-in-Chief

Prof. Janusz Kacprzyk
Systems Research Institute
Polish Academy of Sciences
ul. Newelska 6
01-447 Warsaw
Poland
E-mail: kacprzyk@ibspan.waw.pl

For further volumes:
http://www.springer.com/series/7092

El-Ghazali Talbi (Ed.)

Metaheuristics for Bi-level Optimization

Editor

Prof. El-Ghazali Talbi
University of Lille 1
Villeneuve d'Ascq
France

ISSN 1860-949X ISSN 1860-9503 (electronic)
ISBN 978-3-642-44225-4 ISBN 978-3-642-37838-6 (eBook)
DOI 10.1007/978-3-642-37838-6
Springer Heidelberg New York Dordrecht London

© Springer-Verlag Berlin Heidelberg 2013
Softcover re-print of the Hardcover 1st edition 2013
This work is subject to copyright. All rights are reserved by the Publisher, whether the whole or part of the material is concerned, specifically the rights of translation, reprinting, reuse of illustrations, recitation, broadcasting, reproduction on microfilms or in any other physical way, and transmission or information storage and retrieval, electronic adaptation, computer software, or by similar or dissimilar methodology now known or hereafter developed. Exempted from this legal reservation are brief excerpts in connection with reviews or scholarly analysis or material supplied specifically for the purpose of being entered and executed on a computer system, for exclusive use by the purchaser of the work. Duplication of this publication or parts thereof is permitted only under the provisions of the Copyright Law of the Publisher's location, in its current version, and permission for use must always be obtained from Springer. Permissions for use may be obtained through RightsLink at the Copyright Clearance Center. Violations are liable to prosecution under the respective Copyright Law.
The use of general descriptive names, registered names, trademarks, service marks, etc. in this publication does not imply, even in the absence of a specific statement, that such names are exempt from the relevant protective laws and regulations and therefore free for general use.
While the advice and information in this book are believed to be true and accurate at the date of publication, neither the authors nor the editors nor the publisher can accept any legal responsibility for any errors or omissions that may be made. The publisher makes no warranty, express or implied, with respect to the material contained herein.

Printed on acid-free paper

Springer is part of Springer Science+Business Media (www.springer.com)

To my dear wife Keltoum. I would like to thank her for tolerating my absence and numerous trips. I am playing with her a cooperative game.

To my two sons Anis and Chahine always playing a Stackelberg game. Every time that I saw your smile it lights me up inside.

To my two daughters Besma and Lea, you are my sweet girls.

To my mother Zehour for her infinite sacrifice, and my father Ammar who continue to support me in my academic research. God bless all of you.

Preface

Importance of This Book

Applications of bi-level optimization is countless. Many challenging applications in science and industry (logistics, transportation, finance, engineering design, management, security, yield management) can be formulated as bi-level optimization problems.

A large number of real-life bi-level optimization problems in science, engineering, economics and business are complex and difficult to solve. They cannot be solved in an exact manner within a reasonable amount of time. Using metaheuristics, hybrid metaheuristics with efficient exact algorithms is the main alternative to solve this class of problems.

Purpose of This Book

The main goal of this book is to provide a state of the art of metaheuristics and hybrid metaheuristics with exact methods in solving bi-level optimization problems. The book provides a complete background that enables readers to design and implement metaheuristics to solve complex bi-level optimization problems in a diverse range of application domains. Numerous real-world examples of problems and solutions demonstrate how metaheuristics are applied in such fields as logistics and transportation, network design, security, location problems, etc.

Audience

One of the main audience of this book is **advanced undergraduate and graduate students** in computer science, operations research, applied mathematics, control, business and management, engineering, etc. Many undergraduate courses on optimization throughout the world would be interested in the contents.

In addition, the **postgraduate** courses related to optimization and complex problem solving will be a direct target of the book. Metaheuristics and bi-level optimization are present in more and more postgraduate studies (computer science, business and management, mathematical programming, engineering, control, etc).

The intended audience is also **researchers** in different disciplines. Researchers in computer science and operations research are developing new optimization algorithms. Many researchers in different application domains are also concerned by the use of metaheuristics and hybrid methods to solve bi-level optimization problems.

Many **engineers** are also dealing with bi-level optimization in their problem solving. The purpose of the book is to help engineers to use metaheuristics for solving real-world bi-level optimization problems in various domains of application. The application part of the book will deal with many important and strategic domains such as network design, transportation and logistics, pricing problems, etc.

Outline

The book is organized following different chapters dealing with :

- a taxonomy of metaheuristics to solve bi-level optimization problems.
- different metaheuristics to solve bi-level optimization problems: genetic algorithms, particle swarm optimization, co-evolutionary algorithms.
- different matheuristics combining exact algorithms and metaheuristics to solve bi-level optimization problems.
- exact algorithms to solve linear and mixed integer bi-level optimization problems.
- evolutionary algorithms to solve multi-objective bi-level optimization problems.

Lille,
March 2013

Prof. Dr. El-Ghazali Talbi
University of Lille 1, CNRS, INRIA, France

Acknowledgements

Thanks to all contributors of this book for their cooperation in bringing this book to completion.

Thanks to all the members of my teaching department GIS (Génie Informatique et Statistique) of Polytech'Lille for the excellent atmosphere.

Thanks to all the members of my research laboratories LIFL laboratory (UMR CNRS 8022) and INRIA Lille Europe, and the DOLPHIN research team. I feel there like in my family.

Finally I should like to thank the team at Springer who gave me excellent support throughout this project, and especially for their patience.

Contents

List of Contributors

Ekaterina Alekseeva
Sobolev Institute of Mathematics, 4 pr. Akademika Koptuga, Novosibirsk, Russia
e-mail: ekaterina2@math.nsc.ru

Deniz Aksen
College of Admin. Sciences and Economics, Koç University, İstanbul, Turkey
e-mail: daksen@ku.edu.tr

Necati Aras
Dept. of Industrial Engineering, Boğaziçi University, İstanbul, Turkey
e-mail: arasn@boun.edu.tr

José M. Arroyo
Universidad de Castilla – La Mancha, ETSI Industriales, Campus Universitario s/n, 13071, Ciudad Real, Spain
e-mail: JoseManuel.Arroyo@uclm.es

Antonio J. Conejo
Department of Electrical Engineering Univ. Castilla - La Mancha, Spain
e-mail: antonio.conejo@uclm.es

Kalyanmoy Deb
Indian Institute of Technology Kanpur, PIN 208016, India, and Aalto University School of Economics, PO Box 1210, FIN-101, Helsinki, Finland
e-mail: deb@iitk.ac.in

Francisco J. Fernández
Universidad de Castilla – La Mancha, ETSI Industriales, Campus Universitario s/n, 13071, Ciudad Real, Spain
e-mail: FcoJ.Fdez11@alu.uclm.es

Yury Kochetov
Institute of Mathematics, 4 pr. Akademika Koptuga, Novosibirsk, Russia
e-mail: jkochet@math.nsc.ru

Andrew Koh
Institute for Transport Studies, University of Leeds, Leeds, LS2 9JT, United Kingdom
e-mail: a.koh@its.leeds.ac.uk

George Kozanidis
Systems Optimization Laboratory Department of Mechanical Engineering, University of Thessaly, Volos, Greece
e-mail: gkoz@mie.uth.gr

François Legillon
Tasker and INRIA Lille Nord Europe, France
e-mail: francois.legillon@inria.fr

Arnaud Liefooghe
University of Lille 1, CNRS, INRIA, France
e-mail: arnaud.liefooghe@univ-lille1.fr

Magdalene Marinaki
Industrial Systems Control Laboratory, Department of Production Engineering and Management,Technical University of Crete, 73100 Chania, Crete, Greece
e-mail: magda@dssl.tuc.gr

Yannis Marinakis
Decision Support Systems Laboratory, Department of Production Engineering and Management, Technical University of Crete, 73100 Chania, Crete, Greece
e-mail: marinakis@ergasya.tuc.gr

Georgios K.D. Saharidis
Department of Mechanical Engineering, University of Thessaly, Volos, Greece
Kathikas Institute of Research and Technology Paphos, Cyprus
e-mail: saharidis@gmail.com

Ankur Sinha
Aalto University School of Economics, PO Box 1210, FIN-101, Helsinki, Finland
e-mail: ankur.sinha@aalto.fi

El-Ghazali Talbi
University of Lille 1, CNRS, INRIA, Lille-France
e-mail: talbi@lifl.fr

Chapter 1
A Taxonomy of Metaheuristics for Bi-level Optimization

El-Ghazali Talbi

Abstract. In recent years, the application of metaheuristic techniques to solve multi-level and particularly bi-level optimization problems (BOPs) has become an active research area. BOPs constitute a very important class of problems with various applications in different domains. A wide variety of metaheuristics have been proposed in the literature to solve such hierarchical optimization problems. In this paper, a taxonomy of metaheuristics to solve BOPs is presented in an attempt to provide a common terminology and classification mechanisms. The taxonomy, while presented in terms of metaheuristics, is also applicable to most types of heuristics and exact optimization algorithms.

1.1 Introduction

Multi-level and bi-level optimization are important research areas of mathematical programming [6] [13] [16]. This type of problems has emerged as an important area for progress in handling many real-life problems in different domains. The first formulation of bi-level programming was proposed in 1973 by J. Bracken and J. McGill [7]. W. Candler and R. Norton are the first authors which use the designation of bi-level and multi-level programming [10]. Since the eighties bi-level and multi-level programming receive the attention they deserved. The reader can refer to [17] [60] for extended bibliography reviews.

Multi-level optimization problems (MLOP) have been developed for distributed planning problems in a hierarchical organization with many decision makers. The decisions are taken in a sequential way and without any cooperation. These MLOPs are characterized by a hierarchy of planners; each planner is independently controlling a subset of decision variables, disjoint from the others.

El-Ghazali Talbi
University of Lille 1, CNRS, INRIA
e-mail: talbi@lifl.fr

E.-G. Talbi (Ed.): *Metaheuristics for Bi-level Optimization*, SCI 482, pp. 1–39.
DOI: 10.1007/978-3-642-37838-6_1 © Springer-Verlag Berlin Heidelberg 2013

A bi-level optimization problem (BOP) can be seen as a multi-level problem with two levels. A BOP is a hierarchy of two optimization problems (upper-level or leader, and lower-level or follower problems). The leader-follower game play is sequential and cooperation is not allowed. Each decision maker optimizes its own objective without considering the objective function of the other party, but the decision made by each party affects the objective space of the other party as well as the decision space. A subset of variables at the upper-level optimization problem is constrained to be the optimal solution of the lower-level optimization problem parameterized by the remaining variables.

Compared to single-level optimization, the difficulty in solving bi-level optimization problems lies in the following general facts:

- **Evaluation of the solutions at the upper-level problem:** It is not easy to evaluate the upper-level objective function of a BOP. The objective function at the upper-level has no explicit formulation, since it is compounded by the lower-level optimization problem. In other words, the upper-level decision maker cannot optimize his objective without regards to the reactions of the lower-level decision maker.
- **Complex interaction between the upper-level and the lower-level optimization problems:** The lower-level can be seen as a non-linear constraint and the whole problem is intrinsically a non-convex programming problem. Even if the objective function and the constraints of the upper-level and lower-level optimization problems of a BOP are all linear, the BOP is neither continuous everywhere nor convex for the objective function of the upper-level problem.

Computing optimal solutions is computationally intractable for many BOPs. In practice, we are usually satisfied with "good" solutions, which are generally obtained by metaheuristics. In addition to single-solution based metaheuristics such as descent local search (LS) [79], simulated annealing (SA) [80], tabu search (TS) [81], there is a growth interest in population-based metaheuristics. Those metaheuristics include evolutionary algorithms (EA: genetic algorithms (GA) [82], evolution strategies (ES) [83], genetic programming [85], etc.), ant colonies (AC) [87], scatter search (SS) [86], particle swarm optimization, and so on. We refer the reader to [84] for good overviews of metaheuristics.

Over the last years, interest in metaheuristics in solving BOPs has risen considerably among researchers in mathematical programming. The best results found for many practical or academic optimization problems are obtained by metaheuristics. In this paper, a taxonomy of metaheuristics in solving BOPs is presented in an attempt to provide a common terminology and classification mechanisms. The goal of the general taxonomy given here is to provide a mechanism to allow comparison of metaheuristics in a qualitative way. In addition, it is hoped the categories and their relationships to each other have been chosen carefully enough to indicate areas in need of future work as well as to help classify future work. In fact, the taxonomy could usefully be employed to classify any optimization algorithm (specific heuristics, exact algorithms) and any BOP (mono-objective BOP, multi-objective BOP,

BOP under uncertainty). However, we shall focus our attention on metaheuristics since they are general heuristics applicable to a wide class of BOPs.

The paper is organized as follows. First, section 1.2 presents bi-level optimization concepts. Sections 1.2.1 and 1.2.2 describe the different classes of BOPs, their complexity and some optimality conditions. Then, section 1.3 shows the relationships of BOPs with other classes of optimization problems (e.g multi-objective optimization, Stackelberg game). In section 1.4, some real-life applications of BOPs in different domains are outlined. An overview and unified view of metaheuristics are described in section 1.5. Section 1.6 details a taxonomy that tries to encompass all published work to date in the field of application of metaheuristics to BOPs. A focus is made on the different classes of metaheuristics for BOPs. Section 1.7 addresses an important question related to performance assessment of bi-level metaheuristics. Finally, section 1.8 summarizes the main conclusions and perspectives of this work.

1.2 Bi-level Optimization Concepts

This section covers the main concepts of bi-level optimization such as the upper-level problem, the lower-level problem, the feasible and optimal solution, the low-level reaction set and the induced region. In these definitions it is assumed, without loss of generality, the minimization of all the objectives.

In bi-level optimization problems, a hierarchical structure arises with an optimization problem on the upper level and another optimization problem on the lower level. Due to this hierarchical structure, bi-level optimization problems are closely related to Stackelberg games in game theory [55].

Example 1.1 (**Illustrative bi-level manufacturer-retailer problem).** Let us illustrate a bi-level optimization problem with a manufacturer-retailer problem [25]. The retailer orders articles from the manufacturer and sells them. Suppose tthat the involved articles are newspapers. The manufacturer is the leader and the retailer is the follower. Both the retailer and the manufacturer wish to maximize their profit as much as possible from the articles sale. The manufacturer's profit (F) and the retailer's profit (f) can be formulated as:

$$\begin{cases} F = (C-D).Q \\ f = \begin{cases} (A-C)\xi, & Q < \xi \\ (A-C)\xi - C(Q-\xi), & Q \geq \xi \end{cases} \end{cases}$$

where D is the cost of manufacturing, C is the wholesale price per article, Q is the quantity of articles ordered by the retailer, ξ is the quantity of articles sold by the retailer, and A is the retail price. The manufacturer wish the maximal wholesale price and the largest quantity of articles to be ordered. The order quantity is fixed by the retailer, and the manufacturer can control the wholesale price only. When the wholesale price increases, the retailer probably decreases the order quantity to avoid profit loss. So, increasing the wholesale price by the manufacturer does not mean increasing the profit. The problem for the manufacturer is to fix the value of the wholesale price to maximize the profit.

Definition 1.1 (Bi-level optimization problem). A bi-level optimization problem (BOP) may be defined as:

$$\begin{cases} \underset{x \in R^n, y \in R^m}{\text{Min}} \ \ \text{F}(x,y) \\ \text{subject to } G(x,y) \leq 0 \\ \qquad \begin{cases} \underset{y \in R^m}{Min} \qquad f(x,y) \\ \text{subject to } g(x,y) \leq 0 \end{cases} \end{cases}$$

where $F, f : R^n \times R^m \longrightarrow R$ are respectively the upper-level and the lower-level objective functions, $G : R^n \times R^m \longrightarrow R^p$ the constraint set of the upper-level problem, and $g : R^n \times R^m \longrightarrow R^q$ the constraint set of the lower-level problem.

Definition 1.2 (Constraint set). The constraint set of a BOP is:

$$\Omega = \{(x,y)/G(x,y) \leq 0 \text{ and } g(x,y) \leq 0\}$$

For each value of the upper-level vector x, the lower-level constraints $g(x) \leq 0$ defines the feasible set $\Omega(x)$ of the low-level problem for each x.

Definition 1.3 (Feasible set of the follower). The feasible set of the follower $\Omega(x)$ is:

$$\Omega(x) = \{y : g(x,y) \leq 0\}$$

Definition 1.4 (Rational reaction set). The rational reaction set $M(x)$ may be defined as:

$$M(x) \in \Omega(x) := \underset{y}{Argmin}\{f(x,y) : g(x,y) \leq 0\}$$

The set of variables is partitioned between two vectors: x and y. Given the vector x, the vector y is to be chosen as an optimal solution $y = M(x)$ of the lower-level optimization problem (or follower problem) parameterized by x. The solution $M(x)$ may be seen a the rational reaction of the follower on the leader's choice x. According to those definitions, a BOP can be redefined as follows:

$$\begin{cases} \underset{x,y}{\text{Min}} \qquad \text{F}(x,y) \\ \text{subject to} \quad G(x,y) \leq 0 \\ \qquad\qquad\quad y \in M(x) \end{cases}$$

Definition 1.5 (Induced region). The induced region IR of a bi-level optimization problem, which represents the feasible region at the upper-level optimization problem, is defined by the set:

$$IR = \{(x,y) \in R^n \times R^m : G(x) \leq 0, y \in M(x)\}$$

Hence the set of solutions of the lower-level problem $M(x)$ is found by minimizing the lower-level objective function f:

$$M(x) = \{y : y \in Argmin\{f(x,y) : y \in \Omega(x)\}$$

A more general definition for continuous and discrete BOP can be:

$$BOP(\Omega, F, f) = \begin{cases} \underset{x,y}{Min} F(x,y) \\ (x,y) \in \Omega \\ \text{subject to } y \in M(x) \end{cases}$$

or

$$\underset{x,y}{Min}\{F(x,y) : (x,y) \in IR\}$$

Definition 1.6 (Feasible solution). A solution (x,y) is feasible if $(x,y) \in IR$.

Definition 1.7 (Optimal solution). A solution $(x^*, y^*) \in IR$ is an optimal solution if $\forall (x,y) \in IR, F(x^*, y^*) \leq F(x,y)$

Example 1.2. **A bi-level optimization problem:** let us consider the following BOP problem:

$$\begin{cases} \underset{x,y}{\text{Min}} \quad \text{F}(x,y) = x - 2y \\ \text{subject to} \\ \quad G(x,y) = -x + 3y - 4 \leq 0 \\ \quad \begin{cases} \underset{y}{\text{Min}}\ \text{f}(x,y) = x + y \\ \text{subject to} \\ g_1(x) = -x - y \leq 0 \\ g_2(x) = x - y \leq 0 \end{cases} \end{cases}$$

For this given BOP problem:

$$\Omega(x) = \{y : y \geq |x|\}$$

and

$$M(x) = |x|$$

The induced region may be represented by:

$$\{(x,y) : -x + 3y - 4 \leq 0, y \in M(x)\} =$$
$$\{(x,y) : y = -x, -1 \leq x \leq 0\} \bigcup$$
$$\{(x,y) : y = x, 0 \leq x \leq 2\}$$

The induced set is a non-convex set but it is a connected one. If the upper-level constraints $G(x)$ are modified to:

$$G_1(x) = -x + 3y - 4 \leq 0$$
$$G_2(x) = -y + \frac{1}{2} \leq 0$$

The induced region will be:

$$I = \{(x,y) : y = -x, -1 \leq x \leq -\frac{1}{2}\} \bigcup$$
$$\{(x,y) : y = x, \frac{1}{2} \leq x \leq 2\}$$

This new induced set is disconnected and compact. For the two BOP problems, there are two local optima solutions (x,y): $(-1,1)$ and $(2,2)$ and one global optimal solution: $(2,2)$.

Optimistic versus pessemistic bi-level optimization: in the formulation of a BOP, there is a certain ambiguity in the case of the existence of multiple lower-level optimal solutions (i.e. the reaction set $M(x)$ is composed of multiple solutions). In the optimistic case , the leader selects the couple (x,y) where the reaction has to be optimal for the upper-level objective function (e.g. the leader has the last word). Then, an optimistic solution comes from a cooperative behavior of the follower. In the pessimistic case[1] the follower will select the worst case scenario in the set of rational solutions [42].

1.2.1 Bi-level Optimization Problems

As in single-level optimization, BOPs can be divided into two categories: those whose solutions are encoded with *real-valued* variables, also known as *continuous BOPs* , and those where the solutions are encoded using *discrete* variables such as *combinatorial BOPs*. Those BOPs combining continuous and discrete decision variables are considered as mixed BOPs .

In the last 30 years, the majority of the works concerned the continuous linear BOPs (particularly in the follower problem). The principal reasons of this interest are on the one hand the development of the linear programming in mathematical programming, and the relative facility to deal with such problems[2], and on the other hand the abundance of the practical cases which can be formulated in linear form. In the class of continuous BOP, this chapter deals with complex continuous MOPs (e.g. non linear[3]) for which exact algorithms cannot be applied in a reasonable amount of time.

---

[1] In this paper, the traditional optimistic case is treated.

[2] Facility to solve exactly the follower problem.

[3] The BOP formulation contains at least one non-linear problem.

Definition 1.8 (Linear Bi-level Optimization Problem). (LBOP) may be defined as:

$$\underset{x \in R^n, y \in R^m}{\text{Max}} \; F(x,y) = c_1^T x + c_2^T y$$

where y solves the following problem:

$$\begin{cases} Max & f(x,y) = d_1^T x + d_2^T y \\ \text{subject to} & \\ & g(x,y) = a^T x + b^T y \leq 0 \\ & x,y \geq 0 \end{cases}$$

where $F, f : R^n \times R^m \longrightarrow R$ are respectively the upper-level and lower-level linear objective functions, $g : R^n \times R^m \longrightarrow R^q$ the linear constraints, c_1, d_1, $a \in R^n$, and c_2, d_2, $b \in R^m$.

Non-linear continuous BOP represent also a well known class of BOPs. The non-linearity may concern the upper-level, the lower-level or both optimization problems (Tab. 1.1).

Table 1.1 Popular continuous BOPs

Upper-level problem	Lower-level problem
Linear	Linear
Non-linear	Linear
Quadratic	Quadratic
Non-linear	Non-Linear

Example 1.3. **Quadratic bi-level optimization problem:** A quadratic bi-level optimization problem (QBOP) is an optimization model formulated as follows:

$$\underset{x \in R^n, y \in R^m}{Max} F(x,y) = c_1^T x + c_2^T y + (x^T, y^T) R (x^T, y^T)^T$$

where y solves the following problem:

$$\begin{cases} Max & f(x,y) = d_1^T x + d_2^T y + (x^T, y^T) Q (x^T, y^T)^T \\ \text{subject to} & g(x,y) = a^T x + b^T y \leq 0 \\ & x,y \geq 0 \end{cases}$$

where $F, f : R^n \times R^m \longrightarrow R$ are respectively the upper-level and lower-level linear objective functions, $g : R^n \times R^m \longrightarrow R^q$ the linear constraints, $R, Q \in R^{(n+m).(n+m)}$ are symmetric matrices, c_1, d_1, $a \in R^n$, and c_2, d_2, $b \in R^m$.

The model below shows an example of a quadratic BOP [48]:

$$\begin{cases} Min & y_1^2 + y_2^2 + x^2 - 4x \\ \text{subject to} & 0 \leq x \leq 2 \end{cases}$$

where $y = (y_1, y_2)$

$$\begin{cases} Min & y_1^2 + 0.5y_2^2 + y_1.y_2 + (1-3x)y_1 + (1+x)y_2 \\ \text{subject to} & 2y_1 + y_2 - 2x \leq 1 \\ & y_1, y_2 \geq 0 \end{cases}$$

If the lower and/or upper level of a BOP is *discrete*, the problem is more difficult and the number of references in the literature solving combinatorial BOPs are rather small compared to continuous BOPs [17].

Example 1.4. **Toll-setting problem:** let us illustrate a discrete BOP by an example related to the toll-setting problem taken from [13]. The problem consists in maximizing the revenue raised from the tolls of a transportation network. The tolls are set on some links of the network. Given a network and a setting of the tolls, the users wish to minimize their travel costs. If the tolls levels are not to high, the users are detereed from using the toll arcs. Once the leader (i.e. network manager) schedule the tolls, the followers (i.e. traveler) react to this setting and choose his itinerary to minimize the total travel cost (i.e. standard cost such as time and distance plus tolls).

Given A the set of links of the transportation network and $\overline{A}$ the subset of toll links. At the upper-level, the BOP can be defined as:

$$\begin{array}{ll} \underset{T,x}{Max} & \sum_{a \in \overline{A}} T_a x_a \\ \text{subject to } l_a \leq T_a \leq u_a, \forall a \in \overline{A} & \end{array}$$

where T_a and x_a represent the toll and the flow on link a respectively, and l_a (respectively u_a) is a lower (respectively upper) bound on the toll.

From the users point of view, they are assigned to paths of minimum cost according to the current state of the transportation network. Let us consider a simple case in which thre is a congestion-free environment. In that case, the equilibrium will coincide with the flow assignement that minimizes the total travel cost. So, the path-flow vector f, and the link-flow vector x, is represented by the solution of the following lower-level discrete linear problem:

$$\begin{array}{ll} \underset{f,x}{Min} & \sum_{a \in A} c_a x_a + \sum_{a \in \overline{A}} T_a x_a \\ \text{subject to} & \sum_{p \in P_{rs}} f_p^{rs} = d_{rs}, \forall (r,s) \in \Theta \\ & x_a = \sum_{(r,s) \in \Theta} \sum_{p \in P_{rs}} \delta_{a,p}^{rs} f_p^{rs}, \forall a \in A \\ & f_p^{rs} \geq 0, \forall p \in P_{rs}, \forall (r,s) \in \Theta \end{array}$$

The objective of the follower problem is the sum of tolls T_a $(a \in \overline{A})$ and other costs (e.g. time, distance), aggregated in a measure c_a for each link. The first constraint expresses demand satisfaction in the sense that, for a given origin-destination pair (r,s) (the set of all such pairs is denoted by Θ), the sum of the flows f_p^{rs} on all paths p connecting r to s (these paths being regrouped in P_{rs}) equals the travel demand, d_{rs}. The next cosnstraint links path flows f_p^{rs} and link flows x_a with:

$$\delta^{rs}_{a,p} = \begin{cases} 1 \text{ if path p} \in P_{rs} \text{ uses link a,} \\ 0 \; otherwise \end{cases}$$

The leader and the follower problems are connected through the use of common variables, namely tolls T_a $(a \in \overline{A})$ and flows x_a $(a \in A)$. The profit of the leader cannot be computed until flows are known. These flows are not in the direct control of the manager, but the solution of a follower problem parameterized by the toll vector T. This gives the following BOP:

$$\begin{array}{lll} \underset{T,f,x}{Max} & \sum_{a \in \overline{A}} T_a x_a & \\ \text{subject to } l_a \leq T_a \leq u_a, \forall a \in \overline{A} & & \\ & \underset{f,x}{Min} & \sum_{a \in A} c_a x_a + \sum_{a \in \overline{A}} T_a x_a \\ & \text{subject to} & \sum_{p \in P_{rs}} f^{rs}_p = d_{rs}, \forall (r,s) \in \Theta \\ & & x_a = \sum_{(r,s) \in \Theta} \sum_{p \in P_{rs}} \delta^{rs}_{a,p} f^{rs}_p, \forall a \in A \\ & & f^{rs}_p \geq 0, \forall p \in P_{rs}, \forall (r,s) \in \Theta \end{array}$$

Hence, most of metaheuristics for solving BOPs are designed to deal with continuous BOPs. One of the reasons of this development is the availability of "standard" benchmarks for continuous BOPs. We expect in the near future a growing interest in solving combinatorial BOPs. Indeed many real-life and well-known academic problems (e.g. vehicle routing, knapsack, scheduling) can be modeled as combinatorial BOPs.

1.2.2 Complexity and Optimality Conditions

The difficulty of BOPs and their complexity is assessed by the simplest problems. The simplest family of BOPS, in which all the functions are continuous and linear, are strongly NP-hard [28] [4]. Even if all the functions defining the BOP are continuous and linear, the induced region is a non-convex set [21]. In the presence of upper-level constraints of the form $G(x) \leq 0$, the induced region is a connected set. If we consider upper-level constraints involving the lower-level variables, of the form $G(x,y) \leq 0$, then the induced region could become a disconnected set [21]. Moreover, checking the local optimality in a continuous linear BOP is a NP-hard problem [59]. It is very easy to construct a linear BOP problem where the number of local optima grows exponentially function of the number of variables [8].

Theorem 1.1. *For any $\varepsilon > 0$ it is NP-hard to find a feasible solution to the linear bi-level programming problem with no more than ε times the optimal value [19].*

Many research studies in the literature concern the determination of optimality conditions for a BOP. This is a central topic since the presence of the lower-level optimization problem as a constraint to the upper-level problem. In some conditions, a BOP can be transformed to a single-level optimization problem.

If the lower-level problem is convex and continuously differentiable in the lower-level variables, then the BOP admits a necessary and sufficient representation in terms of its first-order necessary conditions. The resulting problem is a single level optimization problem.

Several published approaches replace the lower-level optimization problem with its Karush-Kuhn-Tucker (KKT) conditions. This popular approach transforms the BOP into a single-level optimization problem with complementary constraints. KKT conditions are used to identify whether a solution is an optimum for a constrained optimization problem [88]. If a regularity condition is satisfied for the lower-level problem, then the KKT conditions are necessary optimality conditions. Those conditions are also sufficient when the BOP is a convex optimization problem in the $y-$variables for fixed parameters x. The problem can be transformed as:

$$\begin{cases} \underset{x,y,\lambda}{Min} & F(x,y) \\ \text{subject to} & \\ & G(x,y) \leq 0 \\ & \nabla_y f(x,y) + \lambda^t \nabla_y g(x,y) = 0 \\ & g(x,y) \leq 0 \\ & \lambda \geq 0, \lambda^t g(x,y) = 0 \end{cases}$$

where ∇_y denotes the gradient with respect to the variable y.

In a constrained optimization problem, two types of optimal solutions are possible. The first type of optimal solution lies inside the feasible region and the second type lies on the boundary of the feasible region. In the case the optimum is inside the feasible region and the problem does not contain equality constraints, the gradient of the objective function and the Lagrange multipliers μ are equal to zero. When equality constraints are present in the model, the gradient of the objective function and the Lagrange multipliers of the equality constraints λ can be different to zero. Moreover, if the optimal solution lies at the boundary, these terms take non zero values.

1.3 Relationships with Other Problems

In this section, the relationship of BOPs with other related problems, such as Stackelberg games and multi-objective optimization, is analyzed.

1.3.1 Bi-level versus Stackelberg Games

Stackelberg game is a leader-follower strategy and an N-people nonzero-sum game [43]. In two-person nonzero-sum games, the objectives of the players are neither exactly opposite nor do they coincide with each other, and the loss of one of them is not equal to the other. A bi-level optimization problem can be viewed as a static version of the non-cooperative two-person game introduced by Von Stackelberg in the context of unbalanced economic markets [55].

Indeed, BOPs are more or less similar in principle to Stackelberg games in game theory [23]. In Stackelberg games, the lower-level problem is an equilibrium problem while in bi-level optimization, an optimization problem arises in the lower level. A Stackelberg game may differ from a BOP when the reaction set of the lower-level decision maker is not a singleton for some decisions of the leader, then a solution of the static Stackelberg game may not be a solution for the BOP [54].

1.3.2 Bi-level versus Multi-objective Problems

The relationashio between BOPs and MOPs has been naturally investigated in the literature. A BOP may not be equivalent to a corresponding bi-objective problem composed with the upper-level and the lower-level objectives:

$$\begin{cases} \underset{x,y}{Min} F(x,y) \\ \underset{x,y}{Min} f(x,y) \\ \text{subject to} \quad G(x,y) \leq 0 \\ \qquad\qquad\quad\; g(x,y) \leq 0 \end{cases}$$

Indeed, the optimal solution of the BOP is not necessarily a Pareto optimal solution of the MOP and vice versa. At least one feasible solution of the BOP is Pareto optimal for the bi-objective optimization problem. Hence, solving the bi-level programming problem via the bi-objective optimization problem using the Pareto dominance will not work.

Many researchers attempted to establish a link between BOPs and multi-objective optimization problems (MOPs) [5] [58]. No conditions have been found which guarantee that the optimal solution of a BOP is a Pareto optimal solution for the upper-level and the lower-level optimization problems [54]. Using a counter example, many authors show that optimality in BOPs and Pareto optimality in MOP are two different concepts [65] [9] [12].

Example 1.5. **Bi-level versus bi-objective optimization:** this example illustrates the existence of optimal solution for a BOP not being on the Pareto frontier of the bi-objective problem. Hence, solutions on the Pareto frontier are not necessarily good quality solutions for a BOP. Let us consider the following continuous linear BOP:

$$\begin{cases} \underset{x\geq 0}{\text{Min}} \quad \text{F}(x,y) = x - 4y \\ \text{subject to} \\ \qquad \begin{cases} \underset{y\geq 0}{\text{Min}}\ \text{f}(x,y) = y \\ \text{subject to} \\ -x - y \leq -3 \\ -2x + y \leq 0 \\ 2x + y \leq 12 \\ -3x + 2y \leq -4 \end{cases} \end{cases}$$

Figure 1.1 shows the inducible region and the optimal solution of the BOP. In figure 1.2, all solutions in the triangle *ABC* dominate the optimal solutions of the bi-level optimization problem.

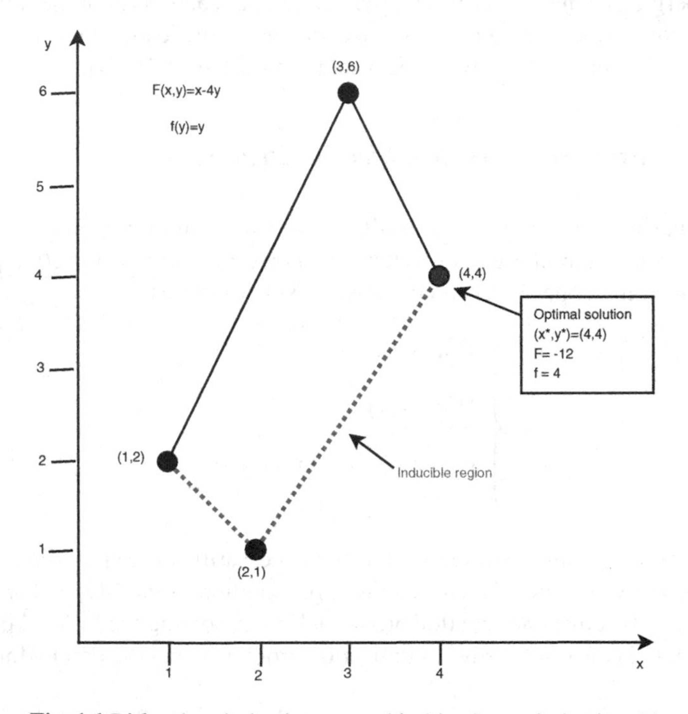

Fig. 1.1 Bi-level optimization versus bi-objective optimization (1)

1.3.3 Bi-level versus Set-Valued Optimization Problems

BOPs are also closely related to set-valued optimization problems (SVOP). SVOP can be defined as:

$$\underset{x}{Min}\{F(x) : x \in X\}$$

where $F : X \to 2^{R^p}$ is a point to set mapping transforming $x \in X \subseteq R^n$ to a subset of R^p. Assume that the function G does not depend on y, the solution set of the system $\{x : G(x) \leq 0\}$ is identified with the set X, and $F(x)$ corresponds to the set of all possible upper level objective function values

$$F(x) := \bigcup_{y \in M(x)} F(x,y)$$

Thus, a SVOP problem can be transformed into a BOP. In general, the considerations are restricted to optimization problems with set-valued objective functions and will not consider problems with constraints in the form of set inclusions.

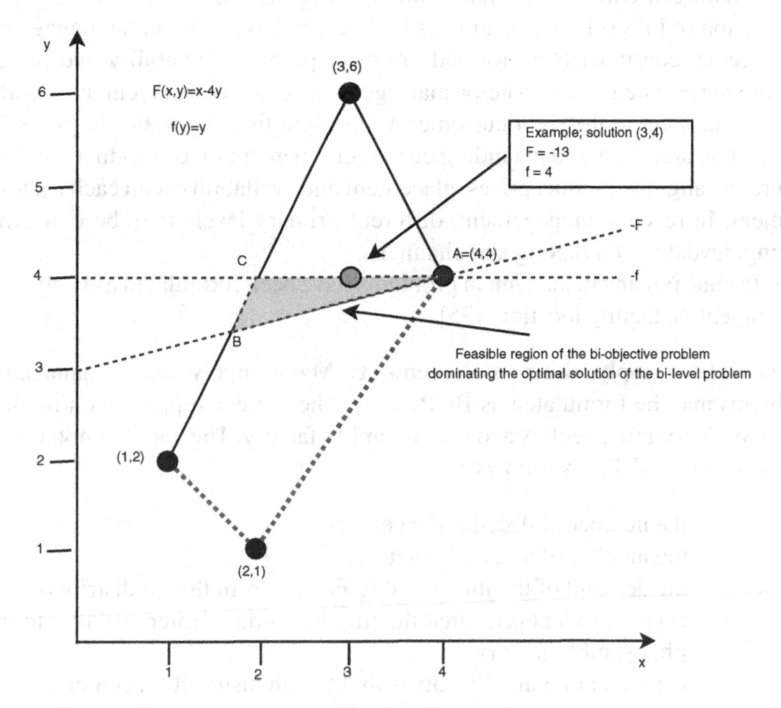

Fig. 1.2 Bi-level optimization versus bi-objective optimization (2)

1.4 Applications of Bi-level Optimization

Hierarchical problem structures appear in many real-life applications where low-level decisions depend on upper-level actions. Indeed, various applications of bi-level optimization problems arise in practice [13] [47]:

- **Transportation:** Many transportation models take the form of a BOP. In the upper-level model, the traffic planner takes decisions regarding management, control, design, and improvement investments to improve the performance of the system [47]. In the lower-level model, the network users make choices with regard to route, travel mode, origin and destination of their travel in response to the upper-level decision. Typical examples include road network design [37], logistic distribution network and supply-chain design [32], multi-depot vehicle routing problem [38], optimal congestion pricing [67]. In road network design, the upper-level optimization problem models the decision making of the network manager. For example, in the case one has to add a new link, the upper-level problem is how much toll to charge on a road or how to set traffic signals. In the lower-level optimization problem, the model characterizes the optimal decision of the user responding to these controls.

- **Yield management:** Yield management constitutes one of the most popular application of bi-level optimization [14]. The objective of revenue management is to predict consumer behavior and optimize product availability and price to maximize revenue growth. The primary goal of revenue management is selling the right product to the right customer at the right time for the right price. The core of this area is in understanding customers' perception of product value and accurately aligning product prices, placement and availability with each customer segment. In revenue management, different primary levels may be concerned: pricing, inventory, marketing and channels.
- **Supply chain:** many management problems have been formulated as BOPs, such as competitive facility location [35].

Example 1.6. **Supply chain management:** Many supply chain management problems may be formulated as BOPs [36]. The current supply chain model is with two distribution centers and one assembly factory. The variable notation for supply chain model is as follows:

$$\begin{cases} i: & \text{the number of distribution centers,} \\ j: & \text{the number of assembly factories,} \\ D_{ij}: & \text{the demand of the jth assembly factory from the ith distribution} \\ & \text{center, product price that the ith distribution center provides to the} \\ & \text{jth assembly factory,} \\ X_i: & \text{the total amount of products that the ith distribution center has,} \\ C_i(X_i): & \text{the unit cost for the ith distribution center to purchase product } X_i, \\ W_i: & \text{the capacity constraint for the ith distribution center,} \\ A_j: & \text{the total amount of products that the jth assembly factory needs,} \\ S_{ij}(D_{ij}): & \text{the unit cost of product that the jth assembly factory order from the} \\ & \text{ith distribution center,} \\ T_{ij}(D_{ij}): & \text{the unit transportation cost of product being delivered from the} \\ & \text{ith distribution center to the jth assembly factory,} \\ h_j: & \text{the unit holding cost for the jth assembly factory} \end{cases}$$

We assume that the distribution centers belong to the upper level, while assembly factories are the lower-level. The objective of the upper-level is to maximize the total profits for distribution centers, while the objective of the lower level is to minimize the total costs for assembly factories. The related parameters are set as: $i = 2$, $j = 1$, $P_{11} = 100$, $P_{21} = 150$, $C_1(X_1) = 40$, $C_2(X_2) = 50$, $W_1 = 30$, $W_2 = 20$, $S_{11}(D_{11}) = 10$, $S_{21}(D - 21) = 15$, $T_{11}(D_{11}) = 20$, $T_{21}(D_{21}) = 25$, and $h_1 = 5$. Thus, the corresponding model can be represented as a BOP:

$$\begin{cases} MAX\ f_1 = 100D_{11} + 150D_{21} - 40X_{11} - 50X_{22} \\ \quad Min \qquad\qquad\qquad\qquad f_2 = 135D_{11} + 195D_{21} \\ \quad \text{subject to:} \\ \qquad\qquad\qquad\qquad\qquad D_{11} \leq X_1 \\ \qquad\qquad\qquad\qquad\qquad D_{21} \leq X_2 \\ \qquad\qquad\qquad\qquad\qquad X_1 \leq 30 \\ \qquad\qquad\qquad\qquad\qquad X_2 \leq 20 \\ \qquad\qquad\qquad\qquad\qquad X_1 + X_2 \geq 50 \\ \qquad\qquad\qquad\qquad\qquad X_i \geq 0, i = 1,1 \\ \qquad\qquad\qquad\qquad\qquad Dij \geq 0, i = 1,2; j = 1 \end{cases}$$

- **Engineering design:** in engineering design (e.g. chemistry, mechanics), many practical problems involve an upper-level optimization problem which requires that a feasible solution must satisfy some given physical conditions (e.g. stability and equilibrium conditions) [31]. Such conditions are ensured by solving a given optimization problem which can be considered as the lower-level optimization problem of a BOP.
- **Security:** indeed, many security applications are naturally modeled by bi-level models such as the vulnerability analysis of a system (e.g. electric grid) under terrorist threat [2].

Example 1.7. **Terrorist threat problem:** the terrorist threat problem can be modeled as a BOP problem [3]. The vulnerability analysis involves two agents who try to make optimal decisions according to their respective objectives functions: the terrorists (i.e. destructive agents, leader) attack the system with the goal of maximizing the damage, whereas the network operator (i.e. lower-level agent, follower) reacts to minimize such damage (Fig.1.3). So the upper-level optimization problem is associated with the disruptive agent which will select the components of the system to be attacked in order to maximize the damage caused to the system (e.g. lines of an electrical grid). The damage is computed in terms of the level of system load shed. In the lower-level model, the system operator reacts to those attacks. The follower will determine the optimal power system operation (e.g. load shedding, generation redispatch, line switching) which will minimizes the damage caused by the disruptive agents.

1.5 Metaheuristics

The word *heuristic* has its origin from the Greek word *heuriskein*. This old Greek word means the art of discovering new strategies (rules) to solve problems. The suffix *meta* which is also a Greek word, means "upper-level methodology". The term *metaheuristic* has been introduced by F. Glover in the paper [81]. Metaheuristic search methods can be defined as upper-level general methodologies (templates) that can be used as guiding strategies in designing underlying heuristics to solve specific optimization problems.

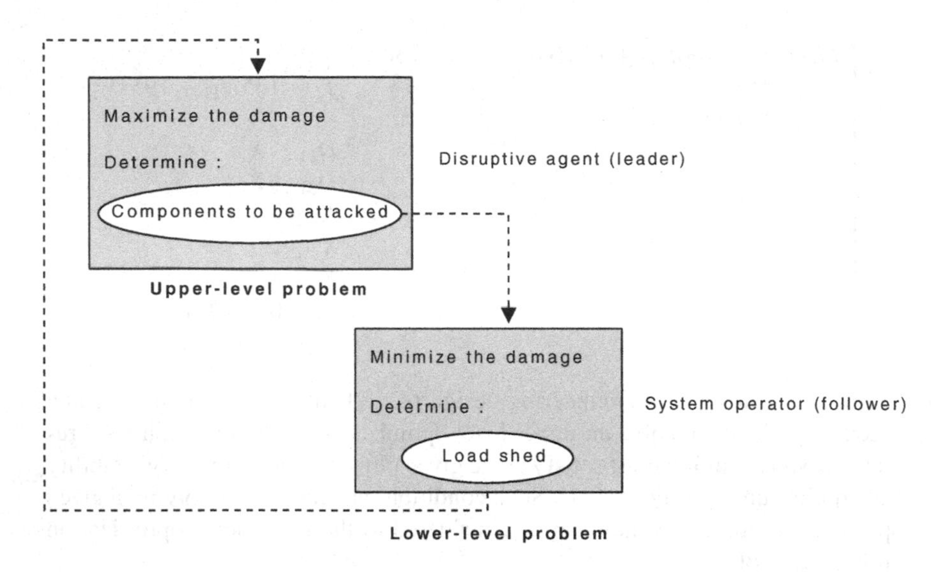

Fig. 1.3 Vulnerability analysis of the electric grid under terrorist threat

Unlike exact methods, metaheuristics allow to tackle large-size problem instances by delivering satisfactory solutions in a reasonable time. There is no guarantee to find global optimal solutions or even bounded solutions. Metaheuristics have received more and more popularity in the last 20 years. Their use in many applications show their efficiency and effectiveness to solve large and complex problems.

Figure 1.4 shows the genealogy of the numerous metaheuristics. The heuristic concept in solving optimization problems has been introduced by Polya in 1945 [71]. The simplex algorithm, created by G. Dantzig in 1947, can be seen as a local search algorithm for linear programming problems. J. Edmonds presents first the greedy heuristic in the combinatorial optimization literature in 1971 [72]. The original references of the following metaheuristics are based on their application to optimization and/or machine learning problems: ACO (Ant Colonies Optimization) [89], AIS (Artificial Immune Systems) [90] [91], BC (Bee Colony) [73] [92], CA (Cultural Algorithms) [93], CEA (Co-Evolutionary Algorithms) [94] [95], CMA-ES (Covariance Matrix Adaptation Evolution Strategy) [75], DE (Differential Evolution) [74] [96], EDA (Estimation of Distribution Algorithms) [76], EP (Evolutionary Programming) [99], ES (Evolution Strategies) [97] [98], GA (Genetic Algorithms) [100] [82], GDA (Great Deluge) [101], GLS (Guided Local Search) [77] [102], GP (Genetic Programming) [85], GRASP (Greedy Adaptive Search Procedure) [103], ILS (Iterated Local Search) [104], NM (Noisy Method) [105], PSO (Particle Swarm Optimization) [106], SA (Simulated Annealing) [80] [107], SM (Smoothing Method) [108], SS (Scatter Search) [86], TA (Threshold Accepting) [109], TS (Tabu Search) [81] [110], VNS (Variable Neighborhood Search) [78].

There are three common design questions related to all iterative metaheuristics: the representation (i.e encoding) of solutions handled by the algorithms, the

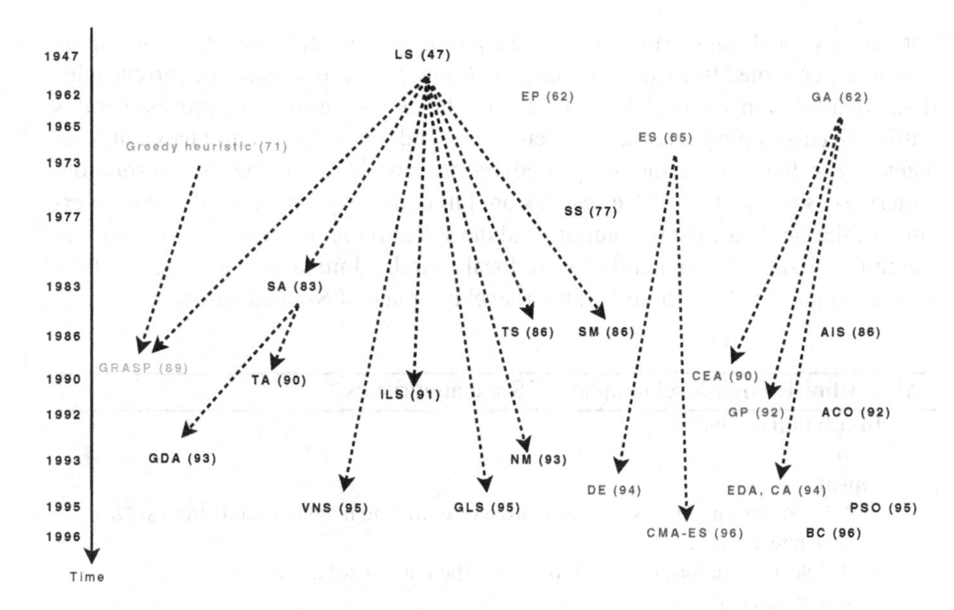

Fig. 1.4 Genealogy of metaheuristics. The application to optimization and/or machine learning is taken into account as the original date.

definition of the objective function that will guide the search and the strategies for handling constraints .

Population-based search vs. single-solution based search: single-solution based algorithms (e.g. local search, simulated annealing) manipulate and transform a single solution during the search while in population-based algorithms (e.g. particle swarm, evolutionary algorithms) a whole population of solutions is evolved. These two families have complementary characteristics: single-solution based metaheuristics are exploitation oriented; they have the power to intensify the search in local regions. Population-based metaheuristics are exploration oriented; they allow a better diversification in the whole search space.

1.5.1 S-metaheuristics

While solving optimization problems, single-solution based metaheuristics (S-metaheuristics) improve a single solution. They could be viewed as "walks" through neighborhoods or search trajectories through the search space of the problem at hand [112]. The walks (or trajectories) are performed by iterative procedures that move from the current solution to another one in the search space. S-metaheuristics show their efficiency in tackling various optimization problems in different domains.

S-metaheuristics iteratively apply the generation and replacement procedures from the current single solution (Fig. 1.5). In the generation phase, a set of candidate solutions are generated from the current solution s. This set $C(s)$ is generally obtained by local transformations of the solution. In the replacement phase[4], a selection is performed from the candidate solution set $C(s)$ to replace the current solution, i.e. a solution $s' \in C(s)$ is selected to be the new solution. This process iterates until a given stopping criteria. The generation and the replacement phases may be *memoryless*. In this case, the two procedures are based only on the current solution. Otherwise, some history of the search stored in a memory can be used in the generation of the candidate list of solutions and the selection of the new solution. Popular examples of such S-metaheuristics are local search, simulated annealing and tabu search. Algorithm 1 illustrates the high-level template of S-metaheuristics.

Algorithm 1: High-level template of S-metaheuristics.

Input: Initial solution s_0.
$t = 0$;
repeat
  /* Generate candidate solutions (partial or complete neighborhood) from s_t */
  Generate($C(s_t)$) ;
  /* Select a solution from $C(s)$ to replace the current solution s_t */
  s_{t+1} = Select($C(s_t)$) ;
  $t = t + 1$;
until Stopping criteria satisfied
Output: Best solution found.

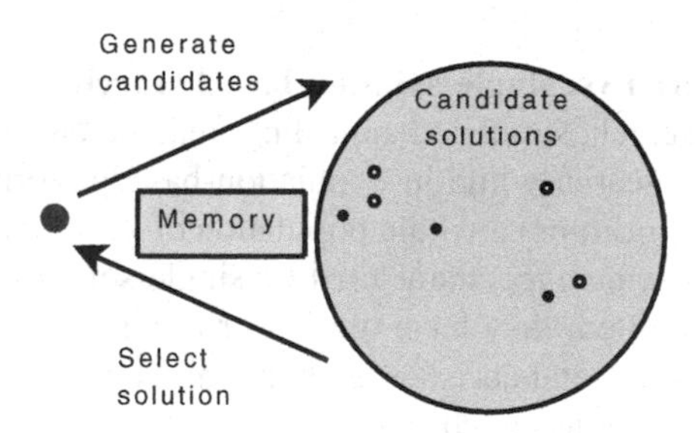

Fig. 1.5 Main principles of single-based metaheuristics

The common search concepts for *all* S-metaheuristics are the definition of the *neighborhood* structure and the determination of the *initial solution* .

1.5.2 P-metaheuristics

Population based metaheuristics (P-metaheuristics) start from an initial population of solutions[5]. Then, they iteratively apply the generation of a new population and

---

[4] Also named transition rule, pivoting rule and selection strategy.

[5] Some P-metaheuristics such as ant colony optimization start from partial or empty solutions.

the replacement of the current population (Fig. 1.6). In the generation phase, a new population of solutions is created. In the replacement phase, a selection is carried out from the current and the new populations. This process iterates until a given stopping criteria. The generation and the replacement phases may be *memoryless*. In this case, the two procedures are based only on the current population. Otherwise, some history of the search stored in a memory can be used in the generation of the new population and the replacement of the old population. Most of the P-metaheuristics are nature inspired algorithms. Popular examples of P-metaheuristics are evolutionary algorithms, ant colony optimization, scatter search, particle swarm optimization, bee colony and artificial immune systems. Algorithm 2 illustrates the high-level template of P-metaheuristics.

Algorithm 2: High-level template of P-metaheuristics.

$P = P_0$; /* Generation of the initial population */
$t = 0$;
repeat
  Generate(P'_t) ; /* Generation a new population */
  P_{t+1} = Select-Population($P_t \cup P'_t$) ; /* Select new population */
  $t = t+1$;
until Stopping criteria satisfied
Output: Best solution(s) found.

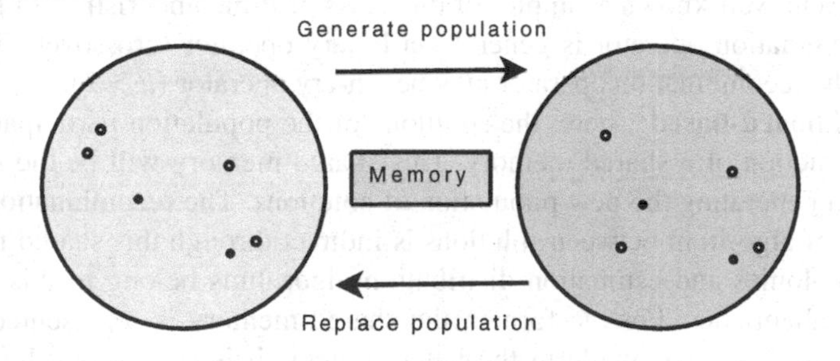

Fig. 1.6 Main principles of population based metaheuristics (P-metaheuristics)

P-metaheuristics differ from the way they perform the generation and the selection procedures and the search memory they are using during the search:

- **Search Memory:** the memory of a P-metaheuristic represents the set of information extracted and memorized during the search. The content of this memory varies from a P-metaheuristic to another one (Tab. 1.2). In most of the P-metaheuristics such as evolutionary algorithms and scatter search, the search memory is limited to the population of solutions. In ant colonies, the pheromone matrix is the main component of the search memory, whereas in estimation distribution algorithms, it is a probabilistic learning model which composes the search memory.

Table 1.2 Search memories of some P-metaheuristics

P-metaheuristic	Search memory
Evolutionary algorithms (EA)	Population of individuals
Scatter Search (SS)	Population of solutions, reference set
Ant colonies (AC)	Pheromone matrix
Estimation of Distribution Algorithms (EDA)	Probabilistic learning model
Particle Swarm Optimization (PSO)	Population of particles, best global and local solutions
Bee colonies (BC)	Population of bees
Artificial immune systems(AIS): Clonal selection	Population of antibodies

- **Generation:** in this step, a new population of solutions is generated. According to the generation strategy, P-metaheuristics may be classified into two main categories (Fig. 1.7):
  - **Evolutionary-based:** in this category of P-metaheuristics, the solutions composing the population are selected and reproduced using variation operators (e.g. mutation, recombination[6]) acting *directly* on their representations. A new solution is constructed from the different attributes of solutions belonging to the current population. Evolutionary algorithms (EAs) and scatter search (SS) represent well-known examples of this class of P-metaheuristics. In EAs, the recombination operator is generally a binary operator (crossover), while in SS, the recombination operator may be a n-ary operator ($n > 2$).
  - **Blackboard-based[7]:** here, the solutions of the population participate in the construction of a shared memory. This shared memory will be the main input in generating the new population of solutions. The recombination in this class of algorithm between solutions is indirect through this shared memory. Ant colonies and estimation distribution algorithms belong to this class of P-metaheuristics. For the former, the shared memory is represented by the pheromone matrix, while in the latter strategy, it is represented by a probabilistic learning model. For instance, in ant colonies, the generated solutions by past ants will affect the generation of solutions by future ants via the pheromone. Indeed, the previously generated solutions participate in updating the pheromone.
- **Selection:** the last step in P-metaheuristics consists in selecting the new solutions from the union of the current population and the generated population. The traditional strategy consists in selecting the generated population as the new population. Other strategies use some *elitism* in the selection phase where they provide

[6] Also called crossover and merge.

[7] A blackboard system is an artificial intelligence application based on the blackboard architectural model, where a shared knowledge base, the "blackboard", is iteratively updated by a diverse group of agents [111].

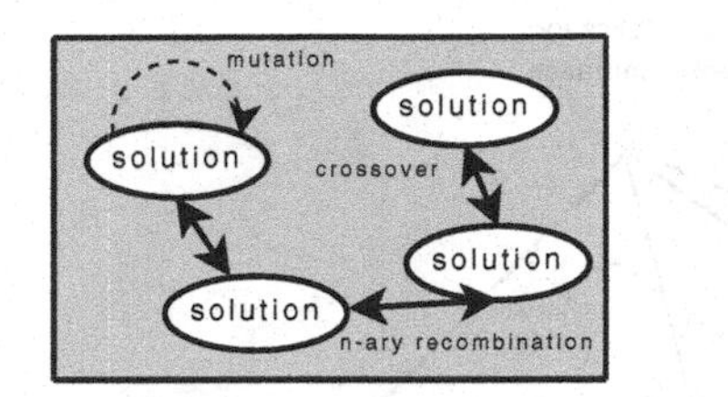

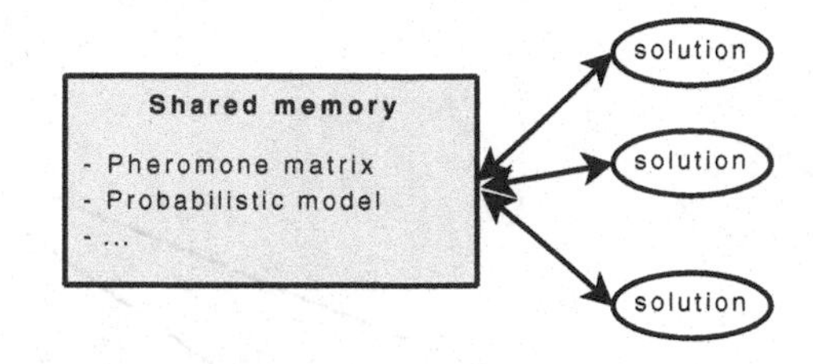

(a) Evolutionary-based P-metaheuristics: evolutionary algorithms, scatter search, ...

(b) Blackboard-based P-metaheuristics: ant colonies, estimation distribution algorithms, ...

Fig. 1.7 Evolutionary-based versus Blackboard-based strategies in P-metaheuristics

the best solutions from the two sets. In blackboard-based P-metaheuristics, there is no explicit selection. The new population of solutions will update the shared search memory (e.g. pheromone matrix for ant colonies, probabilistic learning model for estimation of distribution algorithms) which will affect the generation of the new population.

As for S-metaheuristics, the search components which allow to define and differentiate P-metaheuristics have been identified. The common search concepts for P-metaheuristics are the determination of the initial population and the definition of the stopping criteria.

1.6 Metaheuristics for Bi-level Optimization

Due to the intrinsic complexity of bi-level models, the problem has been recognized as one of the most difficult, yet challenging problems to solve. Hence, metaheuristic algorithms have been investigated to solve BOPs. They can be classified into the following type of strategies (Fig. 1.8):

- **Nested sequential approach:** in this class of metaheuristic strategies, the lower-level optimization problem is solved in a nested and sequential ways to evaluate the solutions generated at the upper-level of the BOP.
- **Single-level transformation approach:** the main characteristic of this class of metaheuristics is to reformulate the BOP into a single-level optimization problem. Then, any traditional metaheuristic can be used to solve the single-level problem.
- **Multi-objective approach:** in this class of metaheuristics strategies, the BOP is transformed to a multi-objective optimization problem. Then, any multi-objective metaheuristic can be used to solve the generated problem.
- **Co-evolutionary approach:** this is the most general methodology to solve BOPs in which many metaheuristics[8], solving the different levels of the problem, co-evolve in parallel and exchange information.

[8] The number of metaheuristics is generally equal to the number of levels.

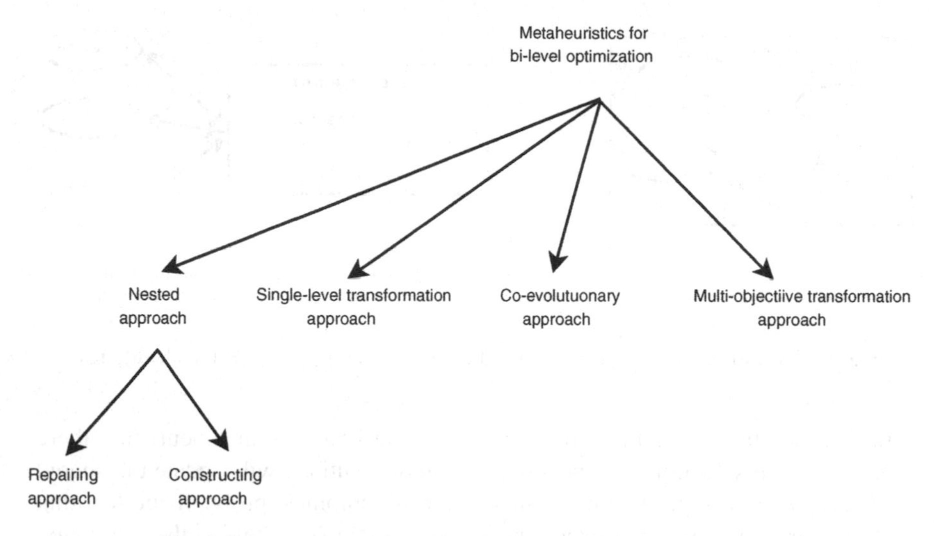

Fig. 1.8 A taxonomy of metaheuristics for bi-level optimization

1.6.1 Metaheuristics Based on Nested Approach

In the nested approach, hierarchical optimization algorithms try to solve the two levels sequentially, improving solutions on each level to get a good overall solution on both levels. Such algorithms include:

- **Repairing approach:** this approach considers the lower-level problem as a constraint and solve it during the evaluation step [34]. It is supposed that the lower-optimization problem has a given structure that can be taken into account in solving efficiently the problem.
- **Constructing approach:** this approach applies two improving algorithms on a population, one for each level, sequentially until meeting a stopping criterion (e.g given number of generations) [41].

As mentioned, the repairing approach considers the follower problem as a constraint. In this first phase, a solution is generated at the upper-level (i.e., generation of (x,y)). The solution (x,y) is sent to the lower-level problem and is considered as given initial solution for the lower-level problem (Fig.1.9). Then, an optimization algorithm (e.g. any metaheuristic) can be used to find a "good" solution $y*$ according to the lower-level optimization problem. At the lower-level, the variable x is used as a parameter and the is fixed. Afterwards, the solution of the lower-level $(x,y*)$ is transmitted to the upper-level. Then, the whole solution of the upper-level is replaced by $(x,y*)$ and the upper-level objective is evaluated. Those three phases proceed iteratively in a sequential way until a given stopping criteria.

In the constructive approach, the low-level problem is solved to improve a population of solutions (x,y) generated at the upper-level using the objective function F (Fig.1.10). This approach is generally used into a population based metaheuristic (i.e. P-metaheuristic). Then, this population is improved at the lower-level using the objective function f in which the decision variables x are fixed. Finally, after a given stopping criteria, the improved population of solutions $(x, y*)$ will constitute the initial population at the upper-level. This process iterates until a given stopping criteria.

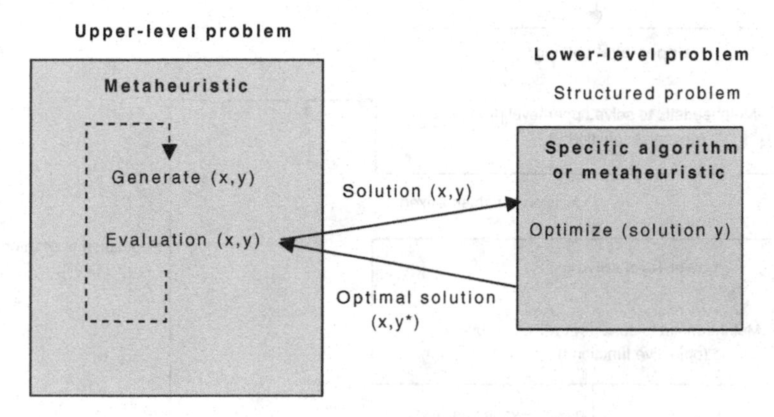

Fig. 1.9 The metaheuristics nested repairing approach for solving bi-level optimization problems

The main drawback of the nested approach is its computational complexity. Indeed, the nested procedure needs to solve an optimization problem (i.e. lower-level problem) for each solution of the problem generated at the upper-level. The efficiency of this class of strategies depends strongly on the difficulty in solving the lower-level problem. For complex lower-level problems, more efficient metaheuristic strategies must be designed in which more coordination is carried out between the two levels of optimization.

The nested approach is characterized by:

- **Upper-level solving approach:** the generation of solutions at the upper-level model needs to be considered carefully since it is the key for solving efficiently the whole BOP problem. At the upper-level model, two different approaches may be used:
  - **Exact method:** the nested exact approach is widely used when the set of feasible solutions can be explored in an exhaustive way. Indeed, in some BOPs, the set of feasible solutions can be enumerated completely [32].
  - **Metaheuristic:** any metaheuristic (or a specific heuristic) can be used to solve the upper-level problems (e.g. genetic algorithms [2] [20], particle swarm optimization (PSO) [27], Differentiel evolution [34]). Metaheuristics are widely used to solve the upper-level problem. For instance, genetic algorithms have been used to solve linear BOPs in which the follower's reaction is obtained from the solution of a linear programming problem [45] [62] [50]. Each individual represents a feasible solution but not necessarily an extreme point. In [68], a genetic algorithm has been used to solve the upper-level non-convex

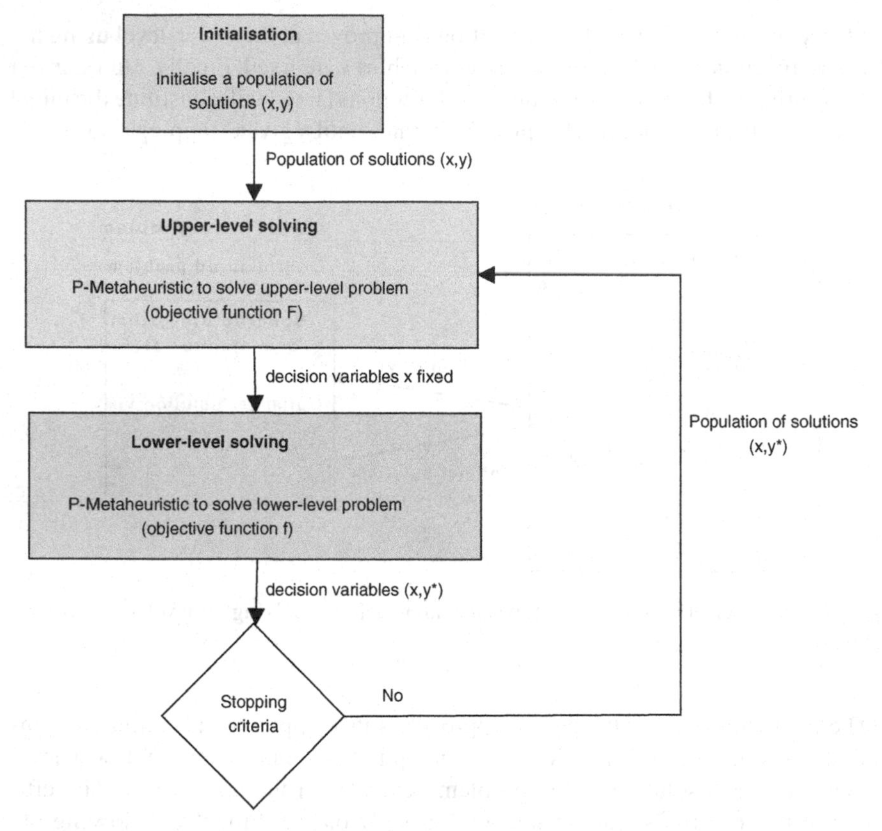

Fig. 1.10 Constructing nested approach to solve BOPs. The different optimization algorithms may be homogeneous or heterogeneous.

problem, while a Frank-Wolfe gradient based linearized optimization strategy has been used to solve the lower-level problem. A hybrid evolutionary algorithms (genetic algorithm with simplex algorithm) in which the upper-level problem is non-linear has been proposed in [39]. The basic idea is to generate an initial population satisfying the constraint of the upper-level problem. Then for each solution of the population, the optimal solutions according to the lower-level problem is generated using for instance a simplex method 1.11. Each solution will be evaluated according to the corresponding optimal solution. This lower-level optimal solution can be considered as a feasible solution at the upper-level.

- **Lower-level solving approach:** as for the upper-level, any traditional optimization strategy can be developped to solve the lower-level problem depending on the difficulty of the inner problem:

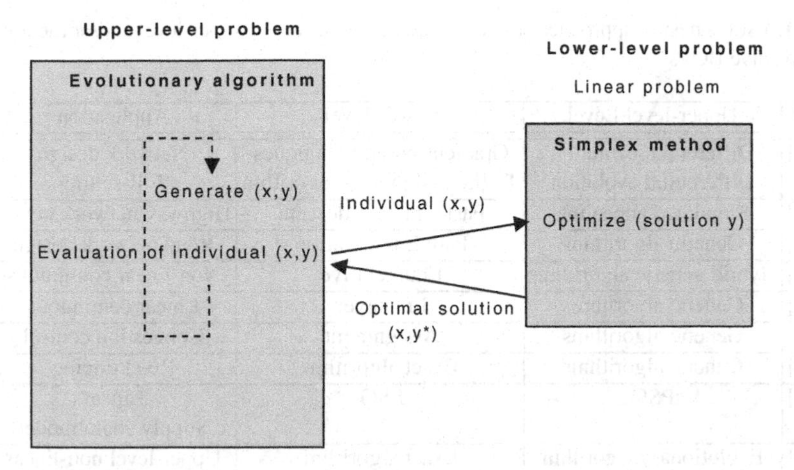

Fig. 1.11 The evolutionary algorithm nested approach for solving linear bi-level optimization problems

- **Exact:** according to the complexity of the problem, an exact optimization algorithm can be used to solve the lower-level problems (e.g. branch and cut solving a mixed-integer linear BOP [2], branch and bound solving a mixed-integer non linear BOP [35]).
  For some lower-level structured or easy problems, the nested approach may be practical. For instance, when the lower-level problem is a linear continuous problem, efficient linear programming techniques may be used to solve the lower-level problem and then the nested approach can be used.
- **Metaheuristic:** generally, when the lower-level problem is difficult, metaheuristics are used to solve the inner problem (e.g. evolutionary algorithm [32], gradient-based techniques [34]).

Table 1.3 shows some proposed nested approaches to solve BOPs. It is difficult to extend this approach to non-linear and large scale optimization problems because of it high computational complexity.

In homogeneous nested approach , the same metaheuristic is used at both levels (upper-level and lower-level). For instance, in [41], a particle swarm optimization based metaheuristic has been designed, while in [15], genetic algorithms have been developed to solve a continuous network design problem.

1.6.2 *Metaheuristics Based on Reformulation to Single-Level Optimization Problems*

Over the years, the most popular approaches for metaheuristics transform the BOP into a single-level optimization problem by using approximate or exact methodologies to replace the lower-level optimization problem. Several approaches, for

Table 1.3 Some nested approches based on metaheuristics at the upper-level and/or the lower-level to solve BOPs

Ref	Upper-level Level	Lower-level	Application
[34]	Differential evolution	Gradient-based techniques	Network design
-	Differential evolution	Traffic assignment algorithm	Toll-setting
[51]	Simulated annealing	Fast Gradient descent	Highway network layout
[20]	Genetic algorithms	Path Enumeration	Road network design
[39]	Hybrid genetic algorithms	Enumerative	Non-linear continuous
[63]	Genetic algorithms	LP solver	Linear continuous
[56]	Genetic algorithms	Assignment	Congestion control
[70]	Genetic algorithms	Exact algorithm	Road pricing
[36]	PSO	PSO	Linear Supply chain model
[40]	Evolutionary algorithm	Exact algorithm	Upper-level non-linear Lower-level linear
[35]	Tabu search	Branch and bound	Facility location
-	Gradient ascent		Mixed nonlinear
[24]	Particle swarm	Particle swarm	Pricing
[69]	Genetic algorithm	Assignment	Transit scheduling
[53]	Genetic algorithm	Assignment	Toll-setting

example, enumeration methods, penalty methods [1] [26], marginal function [46], method and trust-region methods, have been proposed for BOPs, under the assumptions that all functions are convex and twice differentiable.

In the case of differentiable objectives and constraints in the lower-level problem, a popular approach is to include the KKT conditions of the lower-level problem as constraints into the upper-level optimization problem [31]. Additional variables into the upper-level problem are represented by the Lagrange multipliers of the lower-level problem. Other conditions must be satisfied to ensure that the KKT solutions are optimal solutions.

Karush-Kuhn-Tucker (KKT) conditions: KKT conditions on the lower-level optimization problem are generally used as constraints in the formulation of the KKT conditions of the upper-level optimization problem. This will involve the second derivatives of the objectives and constraints of the lower-level problem as necessary conditions of the upper-level optimization problem. This methodology is difficult to apply in practical problems since the presence of many lower-level Lagrange multipliers and an abstract term containing coderivatives [18].

Once the BOP is transformed to a single-level optimization problem (e.g. using KKT conditions), any traditional metaheuristic can be used to solve the single-level problem: genetic algorithms [61] [30], differential evolution [34], simulated annealing [33] [22], evolutionary algorithms [68] [62] [29] [57] [64], local search algorithm [11] [52] [59], and hybrid metaheuristics (genetic algorithm with neural network) [66].

The basic idea of the penalty approach is the use of the concept of penalty function. In [26], a penalty function has been used to generate an initial solution and improving the current solution using a Tabu search algorithm. The current solution belongs always to the admissible region. Marcotte et al. transformed the network design problem into a single-level equivalent differentiable optimization problem, in which the required constraints involve all the extreme points of the closed convex polyhedron for the feasible acyclic multicommodity floow patterns [44].

Table 1.4 summarizes some metaheuristics applying the reformulation of the BOP to a single-level optimization problem.

Table 1.4 Some reformulation approches based on metaheuristics to solve the single-level optimization problem

Ref	Transformation applied	Used metaheuristic
[1]	Penalty function	
[46]	Marginal function	
[26]	Penalty function	Tabu search
[61]	KKT conditions	Genetic algorithms
[30]	KKT conditions	Genetic algorithms
[34]	KKT conditions	Differential evolution
[33]	KKT conditions	Simulated annealing
[22]	KKT conditions	Simulated annealing
[68]	KKT conditions	Evolutionary algorithms
[62]	KKT conditions	Evolutionary algorithms
[29]	KKT conditions	Evolutionary algorithms
[57]	KKT conditions	Evolutionary algorithms
[64]	KKT conditions	Evolutionary algorithms
[11]	KKT conditions	Local search
[52]	KKT conditions	Local search
[59]	KKT conditions	Local search
[66]	KKT conditions	Hybrid metaheuristics GA+neural network

1.6.3 Metaheuristics Based on Transformation to Multi-objective Optimization Problems

Being a problem with two different objective functions, a natural approach to tackle bi-level optimization problems would be to use a Pareto-based multi-objective approach . However bi-level optimization problems have a different structure. A good solution considering a similar problem approximating the Pareto frontier could be of bad quality in the bi-level way.

In [21], the authors propose a methodology in which a BOP is transformed to an equivalent multi-objective optimization problem. A specific cone dominance concept is used.

Hence, one can use any metaheuristic for multi-objective optimization to solve the problem. However, this approach is limited to differentiable problems since the derivatives of the objectives of the original BOP problem are used in the mathematical formulation of the MOP problem. These assumptions are clearly very restrictive and can be seldom satisfied.

1.6.4 Co-evolutionary Metaheuristics

In many cases, methodologies based on the nested, multi-objective or reformulation approaches may not be used or practically inefficient. Indeed, most of those traditional approaches are designed for specific versions of BOPs or based on specific assumptions (e.g. upper-level or lower-level problem differentiable, convex feasible region, low-level structured problems, upper-level reduced search space). Because of such deficiencies, those approaches cannot be used to solve real-life complex applications (e.g. BOPs with non-differentiable objective functions, complex combinatorial BOPs). Therefore, some co-evolutionary based metaheuristics approaches have been developed to solve general BOPs as bi-level programming problems without any transformation.

In co-evolutionary metaheuristics, the two levels proceed in parallel. At each level, an optimization strategy is applied. In general, the optimization strategy is a population-based metaheuristic. Each level try to maintain and improve his own population separately (Fig. 1.12). The two populations are evolving in parallel. Different populations evolve a part of the decision variables, and complete solutions are built by means of a cooperative exchange of individuals from populations. Hence, The two levels exchange information to keep the global view of the BOP.

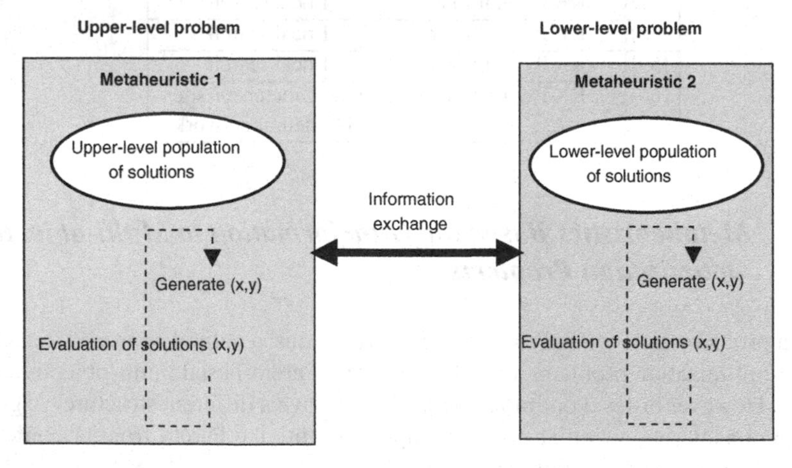

Fig. 1.12 Co-evolutionary approach for solving BOPs. The two metaheuristics evolve in parallel and cooperate via information exchange.

In designing a co-evolutionary model for any metaheuristic, the same design questions need to be answered:

- **The exchange decision criterion (When?):** the exchange of information between the metaheuristics can be decided either in a *blind* (periodic or probabilistic) way or according to an *"intelligent"* adaptive criterion. Periodic exchange occurs in each algorithm after a fixed number of iterations; this type of communication is synchronous. Probabilistic exchange consists in performing a communication operation after each iteration with a given probability. Conversely, adaptive exchanges are guided by some run-time characteristics of the search. For instance, it may depend on the evolution of the quality of the solutions or the search memory. A classical criterion is related to the improvement of the best found local solutions.
- **The information exchanged (What?):** this parameter specifies the information to be exchanged between the metaheuristics. In general, it may be composed of:
  - **Solutions:** this information deals with a selection of the generated and stored solutions during the search. In general, it contains elite solutions that have been found such as the best solution at the current iteration, local best solutions, global best solution, best diversified solutions. The number of solutions to exchange may be an absolute value or a given percentage of the population. Any selection mechanism can be used to select the solutions. The most used selection strategy consists in selecting the best solutions for a given criteria (e.g. objective function of the problem, diversity, age) or random ones.
  - **Search memory:** this information deals with any element of the search memory which is associated to the involved metaheuristic (Tab. 1.5). For ant colonies (resp. estimation distribution algorithms), the information may be related to the pheromone trails (resp. the probability model).

Table 1.5 Exchanged information while partitioning the population of some P-metaheuristics

Metaheuristic	Search memory
Evolutionary algorithms	Population of individuals
Ant colonies	Pheromone matrix
Particle swarm optimization	Population of particles
Scatter search	Reference set, population of solutions
Estimation of distribution algorithms	Probabilistic model

- **The integration policy (How?):** symmetrically to the information exchange policy, the integration policy deals with the usage of the received information. In general, there is a local copy of the received information. The local variables are updated using the received ones. For instance, the best found solution for a given solution (x,y) is recombined with the received solution $(x,y*)$.

Table 1.6 summarizes the characteristics of some proposed co-evolutionary metaheuristics to solve BOPs. A co-evolutionary approach has been used in [49]. Two different populations are maintained in a parallel and independent way. The first (resp. second) handles the decision variables x (resp. y). A limited asymmetric cooperation between the two players is carried out. An external elite population is maintained to identify the elite members of both populations after the co-evolutionary operator for every generation. Asymmetric cooperation is implied by a mono-directional cooperation. The follower cooperates with the leader but not otherwise. Cooperation is only allowed amongst the best solutions (evaluated using the objective function space) who are able to satisfy their own local objectives. This serves as the incentive to guide the search towards the optimal region. Co-evolution is carried out by copying the lower variables to the upper population via a crossover operator.

In [38], the authors proposed a more general bi-level co-evolutionary algorithm, which is an elitist optimisation algorithm developed to encourage cooperation between the two levels, to solve different classes of bi-level problems within a flexible framework. CoBRA is a coevolutionary algorithm using for each level a different population, and a different archive. The cooperation between the two players is symetric.

Table 1.6 Characteristics of some co-evolutionary metaheuristics

Reference	When?	What?	How ?
BiGA [49]	Periodic, asymmetric One generation	Elite solutions	Crossover
CoBRA [38]	Periodic, symetric Numerous generations	Elite solutions	Population's recombination

1.7 Performance Assessment of Bi-level Metaheuristics

Definition 1.9 (Relaxed BOP). The relaxed version of a BOP can be defined as:

$$\begin{cases} \underset{x,y}{\text{Min}} & F(x,y) \\ \text{subject to} & G(x,y) \leq 0 \\ & g(x,y) \leq 0 \end{cases}$$

The optimal solution for the relaxed BOP may not be feasible for the origianl BOP. Indeed, the optimal value of the relaxed problem is a lower bound for the BOP. The relaxed feasible region is defined as:

$$\Omega = \{(x,y) : G(x,y) \leq 0, g(x,y) \leq 0\}$$

When the low-level problem is solved to optimality, the performance assessment of bi-level metaheuristics is carried out using the upper-level objective. This can be done using traditional performance measures (e.g. efficiency, effectiveness, robustness) of the single-level metaheuristic community [84].

Performance assessment of bi-level metaheuristics in which the low-level problem is not solved to optimality is not an easy task. Indeed, bi-level optimization aims at identifying solutions in the form (x,y) which give good upper objective vectors, while being near the optimum regarding the lower objective in which the values of x are fixed. Figure 1.13 illustrates this difficulty. The solution X^* is considered to be the optimal solution of the bi-level problem, whereis A, B and C are considered to be approximate solutions found by metaheuristics. The two solutions A and B are non dominated in the multi-objective space composed of F and f. However, the solution C is dominated by both solution A and B.

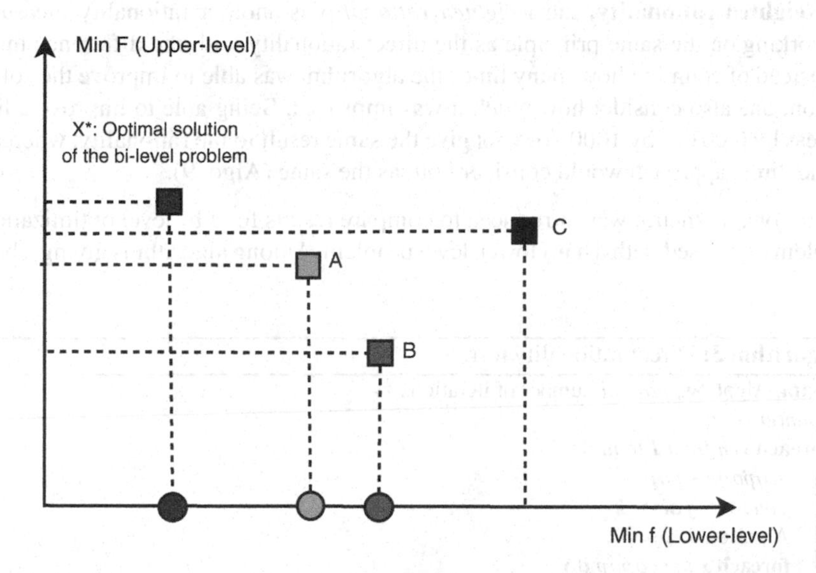

Fig. 1.13 Performance assessment of bi-level metaheuristics when the lower-level problem is not solved to optimality

1.7.1 Performance Indicators

If the optimal solution $(x*,y*)$ of the BOP is available, some performance metrics such as the error rates fot the upper-level objective F: $\frac{|F(x*,y*)-F(x,y)|}{F(x*,y*)}$ and the lower-level objective f: $\frac{|f(x*,y*)-f(x,y)|}{f(x*,y*)}$ may be used.

When the optimal solution of the BOP is not available, the *rationality* metric has been proposed in [38] for assessing the performance of metaheuristics in solving bi-level optimization problems. Rationality is based on the proximity from the optimum of the lower-level variables with the corresponding upper-level variables fixed. The notion of rationality corresponds to the difficulty to improve a solution (x,y), with the subset of upper-level variables x fixed, according to the lower-level objective function. A rational solution is a solution where the follower reaction is rational, seeking for the optimality of its own objective function(s).

Given the subset x fixed, if the optimal solution for the lower problem is known, the rationality can be measured by the error rate between the optimal solution and the approximated solution found by the metaheuristic. Otherwise, when the optimal solution is not known, two different rationality metrics can be used:

- **Direct rationality:** the direct rationality measure corresponds to the difficulty of improving a solution without regarding the actual improvement. One simply consider the "improvability". To evaluate it for a population, one can apply a parametrized number of time a "good" lower-level algorithm, and count how many times the algorithm did improve the solution (Algo. 8).
- **Weighted rationality:** the *weighted rationality* is another rationality measure working on the same principle as the direct rationality with the difference that, instead of counting how many times the algorithm was able to improve the solution, one also consider how much it was improved. Being able to improve a fitness by 0.001 or by 1000 does not give the same result to the rationality, whereas the direct approach would consider both as the same (Algo. 9).

The rationality metric was introduced to compare results for a bi-level optimization problem composed with a hard lower-level problem. Among algorithms giving a bad

Algorithm 3: Direct rationality test.

```
Data: AlgoLow, pop, ni number of iterations
counter ← 0;
foreach gen from 1 to ni do
    neopop ← pop;
    found ← false;
    AlgoLow(neopop);
    foreach x in neopop do
        if (not found) and (x dominates an element of pop) then
            counter++;
            found ← false;
        end
    end
end
return counter/ni
```

Algorithm 4: Weighted rationality test

```
Data: AlgoLow, pop, ni number of iterations
ratio ← 0;
foreach gen from 1 to ni do
    neopop ← pop;
    AlgoLow(neopop);
    ratio=ratio+ε_ind(pop,neopop)/ni;
end
return ratio
```

direct rationality, some algorithms can do better and were far nearer to the optimal on the lower-level than others. The weighted rationality is able to differentiate such algorithms.

Let us notice that rational metrics are not absolute, in the sense that we have to compare the metaheuristics using other optimization algorithms (e.g. metaheuristics), thus introducing a bias. Those performance indicators compare the capacity of a metaheuristic to use improvement optimization algorithms, but do not actually compare the overall capacity to tackle the whole BOP. To this end, one has to ensure that none of the tested metaheuristics is biased toward the improvement algorithm used by the rationality evaluation.

1.8 Conclusions and Perspectives

This chapter provides a unified view in designing metaheuristics for bi-level optimization problems. Moreover, a taxonomy of metaheuristics in solving BOPs is presented in an attempt to provide a common terminology and classification mechanisms. The class of bi-level optimization problems has an immense practical importance.

It is not surprising that most metaheuristic algorithms to date has focused on the simplest cases of bi-level optimization problems characterized by nice properties such as linear, quadratic or convex objective and constraint functions. There is still a need for extensive additional research in this area. First, different variants of metaheuristics should be compared in solving realistic large-scale problems. The comparisons should be made from the perspectives of both computational efficiency and effectiveness. Performance assessment of metaheuristics for bi-level optimization constitute an important scientific challenge for researchers.

Second, designing parallel an co-evolutionary metaheuristics to solve BOPs is a challenging issue. Indeed, solving hierarchical problems needs to revisit the design and implementation of parallel metaheuristics on different parallel architectures (e.g. GPUs, multi-core, clusters, Grids, heterogeneous computers).

It is also promising to extent metaheuristics to solve multi-level optimization problems with more than two levels. This is a challenging problem in terms of design and implementation of metaheuristics and also modeling real-life applications as multi-level optimization problems.

Finally, it will be also important to tackle BOP with uncertainty. Indeed, uncertainty is present in many real-life optimization problems (e.g. transportation, yield management, supply chain).

One hope that this chapter will inspire further applications and research in bi-level optimization using metaheuristics, to see in the future an increasing interest in designing efficient metaheuristics for large scale BOPs.

References

1. Aiyoshi, E., Shimizu, K.: A solution method for the static constrained stackelberg problem via penalty method. IEEE Transactions on Automatic Control 29, 1111–1114 (1984)
2. Arroyo, J.M., Fernandez, F.J.: A genetic algorithm approach for the analysis of electric grid interdiction with line switching. In: 15th International Conference on Intelligent System Applications to Power Systems (ISAP 2009), pp. 1–6 (November 2009)
3. Arroyo, J.M., Galiana, F.D.: On the solution of the bilevel programming formulation of the terrorist threat problem. IEEE Transactions on Power Systems 20(2), 789–797 (2005)
4. Bard, J.: Some properties of the bilevel programming problem. Optim. Theory Appl.
5. Bard, J.: Optimality conditions for the bilevel programming problems. Naval Research Logistics 31, 13–26 (1984)
6. Bard, J.F.: Practical bilevel optimization. Algorithms and applications. Kluwer Academic Publishers, Boston (1998)
7. Bracken, J., McGill, J.: Mathematical programs with optimization problems in the constraints. Operations Research 21, 37–44 (1973)
8. Calamai, P., Vicente, L.: Generating bilevel and linear-quadratic programming problems. SIAM Journal on Scientific and Statistical Computing 14, 770–782 (1993)
9. Candler, W.: A linear bilevel programming algorithm: a comment. Computers and Operations Research 15(3), 297–298 (1988)
10. Candler, W., Norton, R.: Multilevel programming. Technical Report 20, World Bank Development Research, Washington D. C (1977)
11. Chiou, S.: Bilevel programming formulation for the continuous network design problem. Transportation Research Part B 39(4), 361–383 (2005)
12. Clark, P.A., Westerberg, A.W.: A note on the optimality conditions for the bilevel programming problem. Naval Research Logistic Quarterly 35, 413–421 (1988)
13. Colson, B., Marcotte, P., Savard, G.: An overview of bilevel optimization. Annals of Operations Research (153), 235–256 (2007)
14. Côté, J.-P., Marcotte, P., Savard, G.: A bilevel modeling approach to pricing and fare optimization in the airline industry. Journal of Revenue and Pricing Management 2, 23–36 (2003)
15. Cree, N.D., Maher, M., Paechter, B.: The continuous equilibrium optimal network design problem: a genetic approach. In: Transportation Networks: Recent Methodological Advances, pp. 163–174. Elsevier (1998)
16. Dempe, S.: Foundations of bilevel programming. Kluwer Academic Publishers, Boston (2002)
17. Dempe, S.: Annotated bibliography on bilevel programming and mathematical programs with equilibrium constraints. Optimization 52(3), 333–359 (2003)
18. Dempe, S., Dutta, J., Lohse, S.: Optimality conditions for bilevel programming problems. Optimization 55(5), 505–524 (2006)
19. Deng, X.: Complexity issues in bilevel linear programming. In: Multilevel Optimization: Algorithms and Applications, pp. 149–164. Kluwer Academic Publishers, Dordrecht (1998)

20. Dimitriou, L., Tsekeris, T., Stathopoulos, A.: Genetic Computation of Road Network Design and Pricing Stackelberg Games with Multi-class Users. In: Giacobini, M., Brabazon, A., Cagnoni, S., Di Caro, G.A., Drechsler, R., Ekárt, A., Esparcia-Alcázar, A.I., Farooq, M., Fink, A., McCormack, J., O'Neill, M., Romero, J., Rothlauf, F., Squillero, G., Uyar, A.Ş., Yang, S. (eds.) EvoWorkshops 2008. LNCS, vol. 4974, pp. 669–678. Springer, Heidelberg (2008)
21. Fliege, J., Vicente, L.N.: Multicriteria approach to bilevel optimization. J. Optimization Theory Appl. 131(2), 209–225 (2006)
22. Friesz, T.L., Anandalingam, G., Mehta, N.J., Nam, K., Shah, S.J., Tobin, R.L.: The multiobjective equilibrium network design problem revisited: A simulated annealing approach. European Journal of Operational Research 65(1), 44–57 (1993)
23. Fudenberg, D., Tirole, J.: Game theory. MIT Press (1993)
24. Gao, J., Zhang, G., Lu, J., Wee, H.-M.: Particle swarm optimization for bi-level pricing problems in supply chains. Journal of Global Optimization 51, 245–254 (2010)
25. Gao, Y.: bi-level decision making with fuzzy sets and particle swarm optimization. PhD thesis, UTS, Australia (2010)
26. Gendreau, M., Marcotte, P., Savard, G.: A hybrid tabu-ascent algorithm for the linear bilevel programming problem. Journal of Global Optimization 8, 217–233 (1996)
27. Halter, W., Mostaghim, S.: Bilevel optimization of multi-component chemical systems using particle swarm optimization. In: World Congress on Computational Intelligence (WCCI 2006), pp. 1240–1247 (2006)
28. Hansen, P., Jaumard, B., Savard, G.: New branch and bound rules for linear bilevel programming. SIAM Journal on Scientific and Statistical Computing 13, 1194–1217 (1992)
29. Hecheng, L., Wang, Y.: A genetic algorithm for solving a special class of nonlinear bilevel programming problems. In: 7th International Conference on Computational Science (ICCS 2007), pp. 1159–1162 (2007)
30. Hejazia, S.R., Memariania, A., Jahanshahloob, G., Sepehria, M.M.: Linear bilevel programming solution by genetic algorithm. Computers and Operations Research 29, 1913–1925 (2002)
31. Herskovits, J., Leontiev, A., Dias, G., Santos, G.: Contact shape optimization:a bilevel programming approach. Struct. Multidisc. Optimization 20, 214–221 (2000)
32. Huang, B., Liu, N.: Bilevel programming approach to optimizing a logistic distribution network with balancing requirements. Journal of the Transportation Research, 188–197 (1894, 2004)
33. Ciric, A.R., Sahin, K.H.: A dual temperature simulated annealing approach for solving bilevel programming problem. Computers and Chemical Engineering 23(1), 11–25 (1998)
34. Koh, A.: Solving transportation bi-level programs with differential evolution. In: IEEE Congress on Evolutionary Computation (CEC 2007), pp. 2243–2250 (2007)
35. Küçükaydın, H., Aras, N., Altınel, İ.K.: A Hybrid Tabu Search Heuristic for a Bilevel Competitive Facility Location Model. In: Blesa, M.J., Blum, C., Raidl, G., Roli, A., Sampels, M. (eds.) HM 2010. LNCS, vol. 6373, pp. 31–45. Springer, Heidelberg (2010)
36. Kuo, R.J., Huang, C.C.: Application of particle swarm optimization algorithm for solving bi-level linear programming problem. Computers and Mathematics with Applications 58, 678–685 (2009)
37. LeBlanc, L., Boyce, D.E.: A bilevel programming for exact solution of the network design problem with user-optimal flows. Transportation Research 20(3), 259–265 (1986)
38. Legillon, F., Liefooghe, A., Talbi, E.-G.: Cobra: A cooperative coevolutionary algorithm for bi-level optimization. In: IEEE CEC 2012 Congress on Evolutionary Computation, Brisbane, Australia (June 2012)

39. Li, H., Wang, Y.: A hybrid genetic algorithm for solving nonlinear bilevel programming problems based on the simplex method. In: IEEE Third International Conference on Natural Computation (ICNC 2007), pp. 91–95 (2007)
40. Li, H., Wang, Y.: An evolutionary algorithm based on a new decomposition scheme for nonlinear bilevel programming problems. Int. J. Communications, Network and System Sciences 2, 87–93 (2009)
41. Li, X., Tian, P., Min, X.: A Hierarchical Particle Swarm Optimization for Solving Bilevel Programming Problems. In: Rutkowski, L., Tadeusiewicz, R., Zadeh, L.A., Żurada, J.M. (eds.) ICAISC 2006. LNCS (LNAI), vol. 4029, pp. 1169–1178. Springer, Heidelberg (2006)
42. Loridan, P., Morgan, J.: Weak via strong stackelberg problem: New results. Journal of Global Optimization 8, 263–287 (1996)
43. Luce, R., Raiffa, H.: Game and decisions. Wiley (1957)
44. Marcotte, P.: Network optimization with continuous control parameters. Transportation Science 17, 181–197 (1983)
45. Mathieu, R., Pittard, L., Anandalingam, G.: Genetic algorithm based approach to bi-level linear programming. Operations Research 28(1), 1–21 (1994)
46. Meng, Q., Yang, H., Bell, M.G.H.: An equivalent continuously differentiable model and a locally convergent algorithm for the continuous network design problem. Transportation Research Part B 35(1), 83–105 (2000)
47. Migdalas, A.: Bilevel programming in traffic planning: Models, methods and challenge. Journal of Global Optimization 7(4), 381–405 (1995)
48. Muu, L.D., Quy, N.V.: A global optimization method for solving convex quadratic bilevel programming problems. Journal of Global Optimization 26(2), 199–219 (2003)
49. Oduguwa, V., Roy, R.: Bi-level optimisation using genetic algorithm. In: IEEE International Conference on Artificial Intelligence Systems (ICAIS 2002), pp. 322–327 (2002)
50. Osman, M.S., Abdel-Wahed, W.F., El-Shafei, M.K., Abdel-Wahab, B.H.: A solution methodology of bi-level linear programming based on genetic algorithm. Journal of Mathematics and Statistics 5(4), 352–359 (2009)
51. Pei, Y., Ci, Y.: Study on bi-level planning model and algorithm optimizing highway network layout. In: Eastern Asia Society for Transportation Studies, pp. 750–761 (2005)
52. Savard, G., Gauvin, J.: The steepest descent direction for the nonlinear bilinear programming problem. Operations Research Letters 15, 275–282 (1994)
53. Shepherd, S.P., Sumalee, A.: A genetic algorithm based approach to optimal toll level and location problems. Network and Spatial Economics 4(2), 161–179 (2004)
54. Shi, C.: Linear bilevel programming technology - models and algorithms. PhD thesis, University of Technology, Sydney, Australia (2005)
55. Stackelberg, H.: The theory of market economy. Oxford University Press, Oxford (1952)
56. Sun, D.: A bi-level programming formulation and heuristic solution approach for traffic control optimization in networks with dynamic demand and stochastic route choice. PhD thesis, University of Illinois, Urbana Champaign, USA (2005)
57. Sun, D., Benekohal, R.F., Waller, S.T.: Bi-level programming formulation and heuristic solution approach for dynamic traffic signal optimization. Computer-Aided Civil and Infrastructure Engineering 21(5), 321–333 (2006)
58. Unlu, G.: A linear bilevel programming algorithm based on bicriteria programming. Computers and Operations Research 14, 173–179 (1987)
59. Vicente, L., Savarg, G., Judice, J.: Descent approaches for quadratic bilevel programming. Journal of Optimization Theory and Applications 81(2), 379–399 (1994)

60. Vicente, L.N., Calamai, P.H.: Bilevel and multilevel programming: A bibliography review. Journal of Global Optimization 5(3), 291–306 (1994)
61. Wang, G., Wan, Z., Wang, X., Fang, D.: Genetic algorithm for solving quadratic bilevel programming problem. Wuhan University Journal of Natural Sciences 12(3), 421–425 (2007)
62. Wang, G., Wan, Z., Wang, X., Lu, Y.: Genetic algorithm based on simplex method for solving linear-quadratic bilevel programming problem. Computers and Mathematics with Applications 56(10), 2550–2555 (2008)
63. Wang, G.-M., Wang, X.-J., Wan, Z.-P., Jia, S.-H.: An adaptive genetic algorithm for solving bilevel linear programming problem. Applied Mathematics and Mechanics 28(12), 1605–1612 (2007)
64. Wang, Y., Jiao, Y.-C., Li, H.: An evolutionary algorithm for solving nonlinear bilevel programming based on a new constraint-handling scheme. IEEE Transactions on Systems, Man, and Cybernetics, Part C: Applications and Reviews 35(2), 221–232 (2005)
65. Wen, U.P., Hsu, S.T.: A note on a linear bilevel programming algorithm based on bicriteria programming. Computers and Operations Research 16(1), 79–83 (1989)
66. Wu, C.-P.: Hybrid technique for global optimization of hierarchical systems. In: IEEE International Conference on Systems, Man and Cybernetics, Beijing, China, pp. 1706–1711. IEEE Press (October 2006)
67. Yang, H., Lam, H.K.: Optimal road tolls under conditions of queuing and congestion. Transportation Research 30(5), 319–332 (1996)
68. Yin, Y.: Genetic algorithm based approach for bilevel programming models. Journal of Transportation Engineering 126(2), 115–120 (2000)
69. Zhang, Y., Lam, W.H.K., Sumalee, A.: Transit schedule design in dynamic transit network with demand and supply uncertainties. Journal of the Eastern Asia Society for Transportation Studies 7 (2009)
70. Zuo, Z., Kanamori, R., Miwa, T., Morikawa, T.: A study of both optimal locations and toll levels road pricing using genetic algorithm. Journal of the Eastern Asia Society for Transportation Studies 8, 145–156 (2010)
71. Polya, G.: How to solve it. Princeton University Press (1945)
72. Edmonds, J.: Matroids and the greedy algorithm. Mathematical Programming 1(1), 127–136 (1971)
73. Yonezawa, Y., Kikuchi, T.: Ecological algorithm for optimal ordering used by collective honey bee behavior. In: 7th Int. Symposium on Micro Machine and Human Science, pp. 249–256 (1996)
74. Price, K.: Genetic annealing. Dr. Dobb's Journal, 127–132 (October 1994)
75. Hansen, N., Ostermeier Adapting, A.: arbitrary normal mutation distributions in evolution strategies: The covariance matrix adaptation. In: IEEE Conference on Evolutionary Computation, pp. 312–317 (1996)
76. Baluja, S.: Population based incremental learning: A method for integrating genetic search based function optimization and competitive learning. Carnegie Mellon University CMU-CS-94-163, Pittsburgh, Pennsylvania, USA (1994)
77. Voudouris, C., Tsang, E.: Guided local search. University of Essex CSM-247, UK (1995)
78. Mladenovic, N.: A variable neighborhood algorithm - a new metaheuristic for combinatorial optimization. Abstracts of papers presented at Optimization Days, Montreal, Canada (1995)
79. Aarts, E.H.L., Lenstra, J.K.: Local search in combinatorial optimization. John Wiley (1997)

80. Kirkpatrick, S., Gelatt, C.D., Vecchi, M.P.: Optimization by simulated annealing. Science 220(4598), 671–680 (1983)
81. Glover, F.: Future paths for integer programming and links to artificial intelligence. Comput. Ops. Res. 13(5), 533–549 (1986)
82. Holland, J.H.: Adaptation in natural and artificial systems. Michigan Press University, Ann Arbor (1975)
83. Beyer, H.-G.: The theory of evolution strategies. Springer (2001)
84. Talbi, E.-G.: Metaheuristics: from design to implementation. Wiley (2009)
85. Koza, J.R.: Genetic programming. MIT Press, Cambridge (1992)
86. Glover, F.: Heuristics for integer programming using surrogate constraints. Decision Sciences 8, 156–166 (1977)
87. Dorigo, M., Blum, C.: Ant colony optimization theory: A survey. Theoretical Computer Science 344, 243–278 (2005)
88. Kuhn, H., Tucker, A.: Non linear programming. In: Neyman, J. (ed.) 2nd Berkeley Symposium on Mathematical Statistics and Probability, Berkeley, CA, USA, pp. 481–492. Univ. of California (1951)
89. Dorigo, M.: Optimization, learning and natural algorithms. PhD thesis, Politecnico di Milano, Italy (1992)
90. Farmer, J.D., Packard, N., Perelson, A.: The immune system, adaptation and machine learning. Physica D 2, 187–204 (1986)
91. Bersini, H., Varela, F.J.: Hints for adaptive problem solving gleaned from immune networks. In: Schwefel, H.-P., Männer, R. (eds.) PPSN 1990. LNCS, vol. 496, pp. 343–354. Springer, Heidelberg (1991)
92. Seeley, T.D.: The wisdom of the hive. Harvard University Press, Cambridge (1995)
93. Reynolds, R.G.: An introduction to cultural algorithms. In: Sebald, A.V., Fogel, L.J. (eds.) 3rd Annual Conf. on Evolutionary Programming, River Edge, NJ, pp. 131–139. World Scientific (1994)
94. Hillis, W.D.: Co-evolving parasites improve simulated evolution as an optimization procedure. Physica D 42(1), 228–234 (1990)
95. Husbands, P., Mill, F.: Simulated co-evolution as the mechanism for emergent planning and scheduling. In: Belew, R., Booker, K. (eds.) 4th International Conference on Genetic Algorithms, San Diego, CA, USA, pp. 264–270. Morgan Kaufmann
96. Storn, R., Price, K.: Differential evolution - a simple and efficient adaptive scheme for global optimization over continuous spaces. Technical Report TR-95-012, Int CS Institute, Univ. of California (Mar 1995)
97. Rechenberg, I.: Cybernetic solution path of an experimental problem. Technical report, Royal Aircraft Establishment Library Translation No.1112, Farnborough, UK (1965)
98. Schwefel, H.-P.: Kybernetische evolution als strategie der experimentellen forschung in der strömungstechnik. Technical report, Diplomarbeit Hermann Fottinger Institut für Strömungstechnik, Technische universität, Berlin, Germany (1965)
99. Fogel, L.J.: Toward inductive inference automata. In: Proc. of the International Federation for Information Processing Congress, Munich, pp. 395–399 (1962)
100. Holland, J.H.: Outline for a logical theory of adaptive systems. J. ACM 3, 297–314 (1962)
101. Dueck, G.: New optimization heuristics: The great deluge algorithm and the record-to-record travel. Journal of Computational Physics 104(1), 86–92 (1993)
102. Voudouris, C.: Guided local search - an illustrative example in function optimization. BT Technology Journal 16(3), 46–50 (1998)
103. Feo, T.A., Resende, M.G.C.: A probabilistic heuristic for a computationally difficult set covering problem. Operations Research Letters 8, 67–71 (1989)

104. Martin, O., Otto, S.W., Felten, E.W.: Large-step markov chains for the traveling salesman problem. Complex Systems 5(3), 299–326 (1991)
105. Charon, I., Hudry, O.: The noising method: A new method for combinatorial optimization. Operations Research Letters 14, 133–137 (1993)
106. Kennedy, J., Eberhart, R.C.: Particle swarm optimization. In: IEEE Int. Conf. on Neural Networks, Perth, Australia, pp. 1942–1948 (1995)
107. Cerny, V.: A thermodynamical approach to the traveling salesman problem: An efficient simulation algorithm. Journal of Optimization Theory and Applications 45, 41–51 (1985)
108. Glover, F., Mc Millan, C.: The general employee scheduling problem: An integration of MS and AI. Computers and Operations Research 13(5), 563–573 (1986)
109. Duech, G., Scheuer, T.: Threshold accepting: A general purpose optimization algorithm appearing superior to simulated annealing. Journal of Computational Physics 90, 161–175 (1990)
110. Hansen, P.: The steepest ascent mildest descent heuristic for combinatorial programming. Congress on Numerical Methods in Combinatorial Optimization, Capri, Italy (1986)
111. Engelmore, R.S., Morgan, A.: Blackboard Systems. Addison-Wesley (1988)
112. Crainic, T.G., Toulouse, M.: Parallel strategies for metaheuristics. In: Glover, F.W., Kochenberger, G.A. (eds.) Handbook of Metaheuristics, pp. 475–513. Springer (2003)

Chapter 2
A Genetic Algorithm for Power System Vulnerability Analysis under Multiple Contingencies

José M. Arroyo and Francisco J. Fernández

Abstract. This chapter examines the use of a genetic algorithm to analyze the vulnerability of power systems. Recent blackouts worldwide have revealed the vulnerability of power systems and the inability of current security standards to cope with multiple contingencies. The need for new approaches for power system vulnerability assessment has given rise to the development of attacker-defender models, which are particular instances of bilevel programming. The upper-level optimization identifies a set of simultaneous outages in the power system whereas the lower-level optimization models the reaction of the system operator against the outages obtained in the upper level. The system operator reacts by determining the optimal power system operation under contingency. In general, attacker-defender models are characterized as mixed-integer nonlinear bilevel programs for which efficient solution procedures are yet to be explored.

A genetic algorithm is described in this chapter to assess power system vulnerability through an attacker-defender model. The modeling flexibility provided by genetic algorithms makes them suitable for this kind of bilevel programming problems. Numerical results demonstrate the effectiveness of the proposed approach in the identification of critical power system components.

2.1 Introduction

Power systems play a key role in the development of national economies worldwide. As any other critical infrastructure, power systems are subject to disruptions, either unintentional or deliberate, that may have a significant impact on their performance. Consequently, reliability and security are major factors in power system operation

José M. Arroyo · Francisco J. Fernández
Universidad de Castilla – La Mancha, ETSI Industriales, Campus Universitario s/n, 13071, Ciudad Real, Spain
e-mail: JoseManuel.Arroyo@uclm.es, e-mail: FcoJ.Fdez11@alu.uclm.es

E.-G. Talbi (Ed.): *Metaheuristics for Bi-level Optimization*, SCI 482, pp. 41–68.
DOI: 10.1007/978-3-642-37838-6_2 © Springer-Verlag Berlin Heidelberg 2013

and planning. Current reliability policy and associated security standards in power systems worldwide are limited to analyzing a reduced set of events where the outage of multiple components is typically neglected. However, recent blackouts in industrialized countries have uncovered the vulnerability of power systems. Moreover, such catastrophic events have been caused by the coincidence in time of the loss of several independent system components. Therefore, these disruptions reveal the need for considering a larger number of simultaneous outages in the traditional security assessment tools.

This chapter describes the work carried out by the authors at the Universidad de Castilla – La Mancha which is focused on the proposal of a genetic algorithm to analyze the vulnerability of power systems against multiple contingencies. The level of vulnerability is defined as the system load shed and a deterministic worst-case analysis is implemented. This is fundamental to deal with outages with low probability of occurrence but catastrophic impact on the system that can either result from unusually devastating natural disasters or even be targeted by strategic disruptive agents.

A general attacker-defender framework for power system vulnerability analysis under multiple contingencies is presented in this chapter. This general framework is formulated as a bilevel programming problem involving two optimization levels in which their respective objective functions are optimized over a jointly dependent set. The upper-level optimization determines a set of simultaneous outages in the power system whereas the lower-level optimization models the reaction of the system operator against the outages identified in the upper level.

Bilevel programs are complex problems for which no exact general solution techniques are currently available. This chapter reports experience in power system vulnerability analysis with a genetic algorithm. The modeling flexibility of this metaheuristic provides a suitable framework to address the complexity of this type of problems. The proposed genetic algorithm defines an individual as the vector of statuses for all components in a power system. The optimal solution to the lower-level problem is used to evaluate each individual. In order to find increasingly fitter solutions in terms of the upper-level objective function, individuals are constantly modified by the application of genetic operators.

The remainder of the chapter is organized as follows. Section 7.3 describes the problem of power system vulnerability analysis and motivates the need for new models accounting for multiple outages. Section 7.4 discusses the salient features of the bilevel programming framework used to model power system vulnerability analysis under multiple contingencies. Section 7.5 presents the structure of genetic algorithms from the perspective of metaheuristic techniques. This section also includes the description of the proposed genetic algorithm. Section 7.6 is devoted to a particular instance of power system vulnerability analysis. After mathematically formulating the problem, the specific features of the proposed genetic algorithm are examined. Section 7.7 gives numerical results to illustrate the performance of the proposed approach. In Section 6.7, relevant conclusions are drawn. Finally, the Appendix provides the equivalent linear expressions of nonlinear constraints appearing in the original formulation of the proposed application.

2.2 Power System Vulnerability Analysis

Due to the reliance of current societies on electric energy, power systems have become a strategic and critical infrastructure [30]. Hence, a secure and reliable operation is required. Moreover, power system planning should address the reduction of the overall vulnerability so that the correct flow of electricity from generation facilities to consumers is guaranteed even under the most adverse conditions.

The analysis of power system vulnerability allows identifying the set of contingencies to which the system is most vulnerable, i.e., those critical assets whose outage would yield the maximum damage to the system. The solution to this problem is relevant for the system planner and the system operator so that effective protective and corrective actions can be devised. According to [10, 11], several strategies for vulnerability reduction can be implemented such as (i) adding new assets for purposes of redundancy [1, 5, 14], and (ii) hardening the infrastructure or improving its active defenses so that the hardened or defended assets become invulnerable [9, 11, 15, 19, 37, 65]. Hardening and defense actions may include appropriate surveillance measures, patrolling localized assets, and undergrounding specific transmission components.

The vulnerability of power systems has been traditionally assessed through the well-established methodology of power system security analysis [22]. This methodology is based on the well-known $n-1$ and $n-2$ security criteria, by which the system is capable to withstand the loss of a single component or a couple of components, respectively [64]. Traditional approaches used by system operators rely on simulating a pre-specified set of contingencies, one at a time, based on the aforementioned security standards [23, 28, 33, 39, 40].

Several new factors are coming into play in the operation and planning of power systems. Over the last years, the consumption of electricity has increased above forecasts in many countries. Furthermore, power industry is currently immersed in a restructuring process where the main driving force is competition [57]. However, the transmission network has not been expanded accordingly due to economic, environmental, and political reasons. In addition, security standards have not been updated accordingly. As a consequence, power systems are being operated close to their static and dynamic limits, yielding a vulnerable operation. This vulnerability has been confirmed by recent large-scale events involving the outage of multiple system components [2, 7, 24, 41, 61].

Within this new framework of increasingly vulnerable power systems and unexpected contingencies comprising multiple outages, system planners and operators are exposed to a new challenge: how to assess power system vulnerability under the occurrence of multiple contingencies beyond currently used $n-1$ and $n-2$ security criteria.

The extension of vulnerability analysis tools to include an $n-K$ criterion with up to K out-of-service components poses computational challenges. As a matter of fact, the tremendous number of contingencies that should be examined in realistic power systems may exceed the computational capability of today's computers. Moreover, in selecting the set of credible contingencies in the new context of multiple simultaneous outages, the required experience and engineering judgment may not be available.

This issue constitutes one of the main targets of the initiatives that have been launched worldwide as a consequence of the concern raised by governments [16, 17, 30]. As a result, researchers have recently begun to address this problem with the development of approaches based on attacker-defender models for both intentional [4, 8, 9, 18, 48, 55, 56] and unintentional outages [3, 21, 52].

2.3 Attacker-Defender Models

Recently proposed attacker-defender models for power system vulnerability assessment perform a worst-case analysis considering both natural-occurring events and malicious attacks. Worst-case analysis is crucial for vulnerability assessment and mitigation of critical infrastructure such as power systems [11]. Unlike traditional simulation-based security assessment tools, approaches based on attacker-defender models address the problem from the perspective of mathematical programming. This change of paradigm is fundamental since rather than assessing the behavior of the system on an individual contingency basis, attacker-defender models implicitly consider the whole contingency set at the same time.

Attacker-defender models can be cast as bilevel programming problems [6, 20]. Bilevel programming is an appropriate framework to model optimization problems in which one or several constraints are optimization problems themselves. Therefore, the objective function of the upper-level problem is optimized considering that the lower-level problem optimizes its own objective function. Bilevel programs are particular instances of a sequential Stackelberg game [59] in which the game comprises a single round.

The upper-level problem is associated with either nature or a disruptive agent, whereas the lower-level problem is identified with the system operator. As shown in Fig. 2.1, out-of-service power system components are selected in the upper level. This optimization problem takes into account that in the lower level the system operator optimally reacts against the set of simultaneous contingencies determined in the upper level. This reaction may include redispatching generation; involuntary decreasing consumption, also known as load shedding; starting-up fast-acting generating units; or modifying the topology of the transmission network through the so-called transmission line switching.

Two bilevel programming models are available in the technical literature [3], namely a minimum vulnerability model and a maximum vulnerability model. In both models, vulnerability is measured in terms of the system load shed, i.e., the total power the system is unable to supply to the consumers under contingency.

In the minimum vulnerability model, the analysis of vulnerability is defined as the identification of the lowest number of simultaneous out-of-service components that result in a level of vulnerability (system load shed) greater than or equal to a pre-specified threshold set by the system planner or the system operator. The solution to this problem provides the system planner with relevant information on the

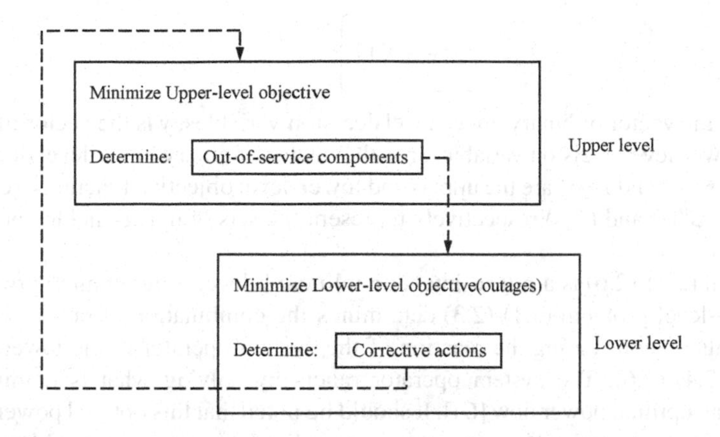

Fig. 2.1 Attacker-defender model

vulnerability of the system in a similar fashion as the currently used $n-1$ and $n-2$ security criteria. These traditional standards ensure that a minimum of 2 or 3 components should respectively be out of service to yield limit violations eventually requiring load shedding.

In contrast, the maximum vulnerability model determines the maximum level of vulnerability (system load shed) that can be attained with a number of simultaneous component outages less than or equal to a pre-specified limit. The solution to this problem is also useful for the system planner since it allows identifying the set of critical assets in a power system.

Both vulnerability analysis models can be formulated in a general compact way as follows:

(Upper-Level Problem)

$$\min_{v} F^u(v, y^*, z^*) \tag{2.1}$$

subject to:

$$G^u(v, y^*, z^*) \geq 0 \tag{2.2}$$

$$v \in \{0, 1\} \tag{2.3}$$

(Lower-Level Problem)

$$(y^*, z^*) \in \arg\left\{ \min_{y,z} \; F^l(v, y, z) \right. \tag{2.4}$$

subject to:

$$G^l(v, y, z) \geq 0 \tag{2.5}$$

$$z \in \{0,1\} \Bigg\}, \tag{2.6}$$

where v is the vector of binary upper-level decision variables; y is the vector of continuous lower-level decision variables; z is the vector of binary lower-level decision variables; $F^u(\cdot)$ and $F^l(\cdot)$ are the upper- and lower-level objective functions, respectively; and $G^u(\cdot)$ and $G^l(\cdot)$ respectively represent the sets of upper- and lower-level constraints.

Problem (2.1)-(2.6) is a mixed-integer nonlinear bilevel programming problem. The upper-level problem (2.1)-(2.3) determines the combination of out-of-service components, v, considering the reaction of the system operator in the lower-level problem (2.4)-(2.6). The system operator reacts by solving what is commonly known as an optimal power flow [64]. It should be noted that this optimal power flow (2.4)-(2.6) is parameterized in terms of the upper-level decision vector v. Moreover, it is worth emphasizing that the lower-level problem (2.4)-(2.6) is generally nonconvex due to the presence of discrete variables and nonlinearities characterizing the operation of power systems. Problem (2.1)-(2.6) is mathematically well posed based on the assumption that the corrective actions available to the system operator allow a feasible reaction against any feasible upper-level decision vector v. In other words, for every upper-level decision vector v satisfying upper-level constraints (2.2), there exist feasible lower-level decision vectors y and z.

The inherent structure of bilevel programs leads in general to nonconvex problems. Even in the simplest case with linear expressions and continuous variables, bilevel programs are strongly NP-hard [34]. Therefore, devising exact methodologies to solve this kind of problems is a complex task. Several solution techniques have been proposed in the technical literature to address the mixed-integer nonlinear bilevel problem (2.1)-(2.6). The most relevant works include decomposition-based approaches [8, 18, 55, 56], equivalent transformations to mixed-integer programs [3, 4, 21, 48], and approximate methods [9, 52].

However, all previous works share limitations from either the modeling or methodological perspectives. Thus, these models typically rely on a simplified representation of the system behavior leading to bilevel programs with convex lower-level problems. Furthermore, solution techniques (i) do not guarantee optimality, (ii) are specifically tailored to such simplified models, and (iii) are mainly applicable to systems of moderate size. Therefore, we still lack efficient tools addressing more realistic models that are also suitable for large-scale systems.

2.4 Genetic Algorithm: An Evolutionary Metaheuristic

Introduced by Holland [36], a genetic algorithm is a metaheuristic belonging to the class of evolutionary approaches. Metaheuristics [31, 60] are devised to address complex optimization problems for which conventional optimization methods are unable to be either effective or efficient. Among the various metaheuristics available, genetic algorithms are recognized as practical solution approaches for many real-world problems. Several factors have boosted the appeal of genetic algorithms as global optimization approaches:

- The need for more sophisticated models makes it difficult or even impossible the use of traditional optimization techniques. These models typically result in complex optimization problems with many local optima and little inherent structure to guide the search.
- Genetic algorithms are generally quite effective for rapid global search of large solution spaces. As a result, near-optimal solutions are likely to be attained in reasonable computation times.
- Genetic algorithms operate on a pool of individuals, thus multiple solutions are suggested.
- The search mechanism is intrinsically parallel, thus lending itself to a parallel implementation with the potential reduction in the computational requirement.

Particularly relevant applications of genetic algorithms in the field of power system operation and planning can be found in [27, 47, 58]. In addition, genetic algorithms have also been successfully applied to several instances of bilevel programming [12, 13, 35, 42–44, 49–51, 62, 63]. These previous works pave the way for the application of genetic algorithms to bilevel programs with nonconvex lower-level problems such as the attacker-defender model (2.1)-(2.6) characterizing power system vulnerability analysis.

2.4.1 Structure of a Genetic Algorithm

Within the general framework of metaheuristics, the genetic algorithm approach comprises two steps: (i) obtaining an initial set of solutions, and (ii) implementing an improving search driven by specific rules in order to yield new solution sets.

A genetic algorithm is a set-based method, i.e., in each step k, the current state of the algorithm is represented by $S_k \subseteq S$, where S_k denotes the set of solutions in step k and S is the solution space. For each solution set S_k, a neighborhood $N(S_k)$ is defined as all solutions resulting from the application of certain operators to the current solutions. A candidate solution set $C \subset N(S_k)$ is selected from the neighborhood $N(S_k)$ of the current solution set. The selected candidate solution set is subsequently evaluated through the calculation or estimation of the performance of the candidate solutions. Based on this evaluation, candidate solutions yield the solution set of the next step S_{k+1}. This process is repeated until a convergence criterion is met. Fig. 2.2 presents the general structure of a genetic algorithm under a metaheuristic framework.

As other evolutionary approaches, a genetic algorithm is an adaptive search technique inspired by the evolution of natural systems [36]. Based on the idea of natural selection, a genetic algorithm works by evolving or improving a constant-sized population of individuals over successive iterations called generations. Individuals represent samples of the search space and are typically referred to as chromosomes, which are encoded as strings of symbols. The position of a symbol and its value

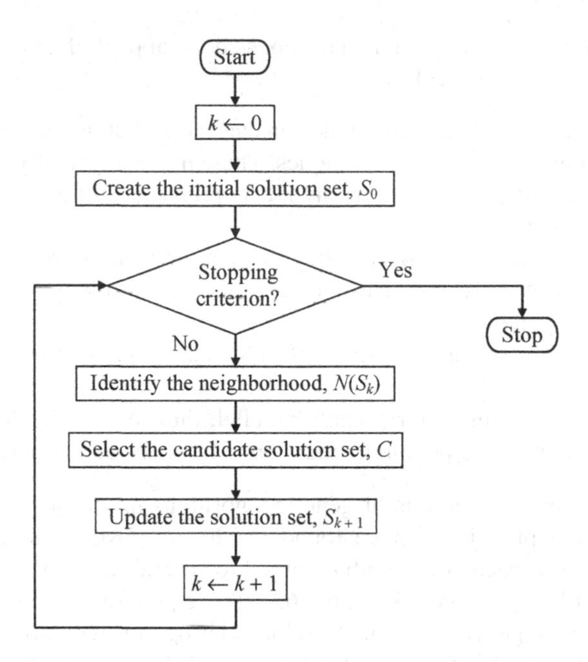

Fig. 2.2 General metaheuristic structure of a genetic algorithm

are respectively denoted as gene and allele. Each individual is evaluated in terms of its overall fitness with respect to the given application domain. High-performing individuals are selected to produce offspring that retain many of the features of their parents.

The evolution process is carried out through a series of genetic operators [26, 32, 36, 46]. Main genetic operators include (i) selection, which implements survival of the fittest or best solutions and determines the parents of the new generation; (ii) crossover, which randomly exchanges gene structures from two selected parents to produce new offspring; and (iii) mutation, which randomly changes one or more components of a selected individual, thus acting as a population perturbation operator.

From the perspective of the above general framework, the distinctive feature of a genetic algorithm with respect to other metaheuristics is the application of genetic operators to determine the neighborhood of a solution set. Thus, for a current solution set $S_k \subseteq S$, the neighborhood is defined as:

$$N(S_k) = N^{cro}(S_k) \cup N^{mut}(S_k) \cup S_k, \quad (2.7)$$

where $N^{cro}(S_k)$ and $N^{mut}(S_k)$ respectively denote the solution sets provided by crossover and mutation, which can be expressed as:

$$N^{cro}(S_k) = \{c \in S \mid c \text{ is the crossover of two solutions belonging to } S_k\} \quad (2.8)$$

$$N^{mut}(S_k) = \{m \in S \mid m \text{ is the mutation of some solution belonging to } S_k\}. \quad (2.9)$$

The selection of solutions is guided by their performance so that high-performing solutions, referred to as fit individuals, are more likely to survive and yield new solutions through crossover. As a consequence, fit solutions will be present in subsequent generations, either unchanged or as part of their offspring. In contrast, low-performing solutions will not survive. Selection is usually implemented using random strategies such as the roulette wheel mechanism [26, 32, 36, 46], by which the probability of being selected is proportional to the individual fitness.

Unlike selection, crossover and mutation are exclusively driven by the representation of the solutions. Hence, solutions yielded by these genetic operators may be worse than current solutions, which constitutes another salient aspect with respect to other metaheuristic approaches.

Genetic operators are intended to explore large portions of the solution space without getting trapped at local optima. An appropriate design of the genetic operators, which is specific for each application, is thus crucial to prevent stagnation of the evolution process.

2.4.2 Proposed Genetic Algorithm for Attacker-Defender Models

In addition to the general difficulties associated with standard combinatorial optimization, the bilevel programming framework for attacker-defender models (2.1)-(2.6) is challenging due to the existence of two optimization levels and due to the need for an accurate representation of the system operation.

Based on previous applications of genetic algorithms to several instances of bilevel programming [12, 13, 35, 42–44, 49–51, 62, 63], this evolution-inspired methodology is proposed to address the general mixed-integer nonlinear bilevel programming formulation for attacker-defender models (2.1)-(2.6). Unlike previously reported approaches [12, 13, 35, 43, 44, 49, 51, 62, 63], the proposed method considers the presence of binary variables in the lower-level problem.

The proposed genetic algorithm also differs from available genetic algorithms for bilevel programming [12, 13, 35, 42–44, 50, 51, 62, 63] by the use of specific repair procedures to enforce feasibility of solutions. Thus, the genetic algorithm deals with feasible solutions only, thereby avoiding the use of penalty terms in the objective functions, which are typically difficult to choose.

Similar to the approach presented in [44, 49, 62], the proposed genetic algorithm is based on the following nested procedure. The evolution process operates on a set of individuals, which are instances of the upper-level decision vector v. For each individual, the optimal solution to the lower-level problem yields vectors y and z, and eventually the values of the upper- and lower-level objective functions F^u and F^l. Fig. 2.3 presents the block diagram of the proposed genetic-algorithm-based approach for the general bilevel programming formulation (2.1)-(2.6) for power system vulnerability analysis. This figure also shows the relationship between the steps of the genetic algorithm and the associated optimization problem.

This nested implementation essentially decouples both optimization levels. The upper-level problem, i.e., the determination of decision vector v, is associated with the evolution process (initialization of the population and alteration of the individuals by the genetic operators). In contrast, the lower-level problem yielding decision variables y and z is solved to obtain the fitness of individuals. This decoupling allows preserving the original formulation of both optimization levels. In addition, this nested approach is advantageous since no assumption on convexity or differentiability of the lower-level problem is required, as done in [12, 35, 42, 43, 51, 63].

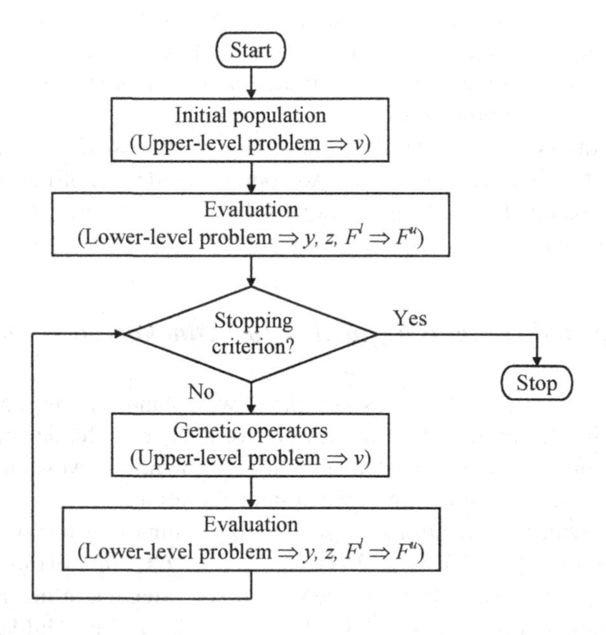

Fig. 2.3 Block diagram of the proposed genetic algorithm

The proposed genetic algorithm overcomes the major shortcomings characterizing previous works on power system vulnerability analysis [3, 4, 8, 9, 18, 21, 48, 52, 55, 56] as follows:

- The decoupling between the upper and lower levels provides a flexible modeling framework, thus allowing the consideration of sophisticated models including nonconvexities and nonlinearities to precisely represent the operation of power systems under contingency.
- Given the inherent parallel structure of the genetic algorithm, a parallel implementation of the approach is straightforward, thus enabling the application to large-scale systems with tractable computational effort.

When applying genetic algorithms, selecting an appropriate solution coding scheme and choosing a suitable evaluation procedure are both crucial. The information provided by the evaluation procedure on the fitness of each individual in the population is used to guide the search process. Moreover, the design of an adequate evolution procedure is essential to avoid getting trapped at local optima. These aspects are described next for the general formulation of the power system vulnerability analysis problem (2.1)-(2.6).

2.4.2.1 Coding

Each individual in the population represents a candidate solution to the vulnerability analysis problem, i.e., a vector v. Since decision vector v is binary, it is convenient to use binary coding. Thus, individuals are encoded as binary-valued strings of length n_v, where n_v is the size of vector v. Fig. 2.4 shows an example of binary coding for the proposed genetic algorithm.

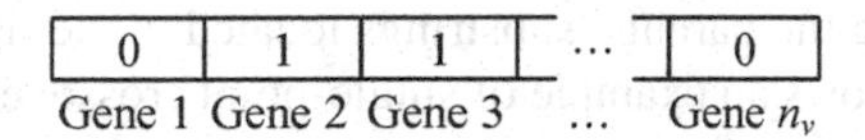

Fig. 2.4 Example of binary coding

2.4.2.2 Fitness Evaluation

The fitness function measures the quality of individuals. Let i and f_i respectively denote an individual and its associated fitness. For the vulnerability analysis problem (2.1)-(2.6), the fitness of i is the value of the objective function of the upper-level problem (2.1), i.e.:

$$f_i = F^u(v_i, y_i^*, z_i^*), \tag{2.10}$$

where v_i is the upper-level decision vector associated with individual i, whereas y_i^* and z_i^* denote the optimal lower-level decision vectors characterizing individual i. Note that y_i^* and z_i^* are obtained from the solution to the optimal power flow (2.4)-(2.6). This step may require the solution of a mixed-integer program.

2.4.2.3 Evolution Procedure

The initial population is created with random feasible individuals. For each generation, the population is updated by replacing all parent individuals with new potential solutions. This updating procedure is based on the individuals' fitness and comprises the application of several genetic operators, such as selection, crossover, mutation, and elitism. It should be noted that infeasible solutions violating upper-level constraints (2.2) go through a repair procedure in order to remove infeasibilities. Genetic operators are implemented as follows:

- **Selection.** Given a generation of feasible solutions, parents of the next generation are obtained by randomly selecting solutions of the current generation with probabilities proportional to their corresponding fitness. This selection scheme is known as the roulette wheel mechanism [46], where selection is implemented by using a roulette wheel with slots sized in proportion to individuals' fitness. For an individual i, the probability of being selected is expressed as:

$$p_i^{sel} = \frac{f_i}{\sum_{i \in I} f_i}, \tag{2.11}$$

where I is the set of individuals in the current generation.
The number of parents is the same as the size of the population. Once parent solutions are selected, they are arranged in couples in order to undergo crossover.

- **Crossover.** Based on a pre-specified crossover rate, crossover is applied to couples of parent solutions to produce two solutions of the next generation. For the sake of simplicity, a single-point crossover is implemented [46]. The single-point crossover randomly selects a location in the parent strings. Two offspring are created by swapping the parents' substrings located to the right of the crossover location. Fig. 2.5 shows an example of single-point crossover.

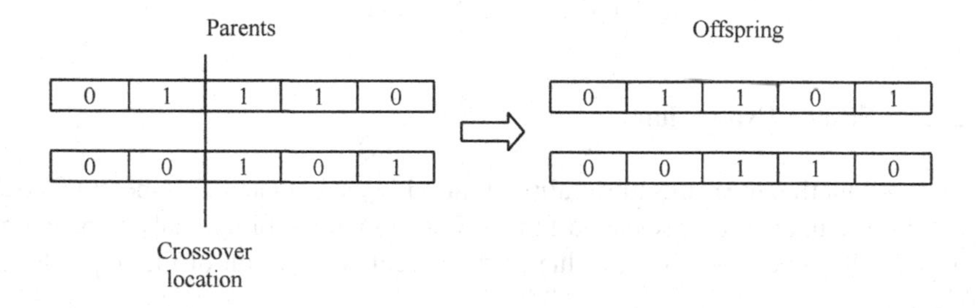

Fig. 2.5 Example of single-point crossover

- **Mutation.** To avoid the loss of potentially useful genetic material, individuals are randomly mutated according to a pre-specified mutation rate. The mutation operator is implemented by flipping a random element from the 0/1 vector from 0 to 1 or vice versa. The application of this operator is illustrated in Fig. 2.6.

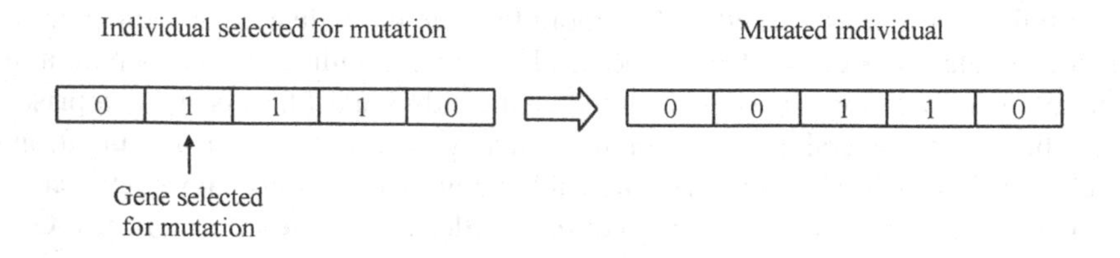

Fig. 2.6 Example of the mutation operator

- **Elitism.** The elitist operator preserves the best solutions found by maintaining a group of them in the next generation. As is shown in [26], this operator is necessary to prove the convergence to the optimum through a Markov chain analysis.
- **Feasibility.** The solutions resulting from crossover and mutation operators may be infeasible and specifically tailored procedures are used to regain feasibility. In order to enforce feasibility of the upper-level constraints (2.2), infeasible individuals are modified by flipping a randomly selected gene from 1 to 0 or vice versa according to the infeasibility level. This process is repeated until feasibility is attained. If the infeasibility was caused by the mutation operator, feasibility procedures are run without undoing what the mutation procedure did. As mentioned in Section 7.4, it should be noted that once a feasible upper-level vector v is obtained, its associated lower-level problem is always feasible.
- **Stopping criterion.** The genetic algorithm is stopped after a pre-specified number of generations.

2.5 Application: Maximum Vulnerability Model with Line Switching

This section presents the application of the proposed genetic algorithm approach to a particular instance of the general attacker-defender model described in Section 7.4. The problem addressed is a maximum vulnerability model considering transmission line switching as one of the corrective measures available to the system operator. For the sake of simplicity, the proposed model is based on the following modeling assumptions:

- A static planning model comprising a single period is considered. During this target period generation sites are known and a single load scenario is modeled, typically corresponding to the highest load demand forecast for the considered planning horizon.
- The model only considers failures of transmission lines and transformers, which are both characterized by their series reactance.
- A linearized model of the transmission network, referred to as dc network flow model [64], is used.
- The effects of unit commitment and decommitment are neglected, and thus the lower bound on the power output of each generator is zero.

Notwithstanding, the extension to a multiperiod setting including the restoration of damaged system components is straightforward. Similarly, the failure of other power system components, a nonlinear transmission network model, and generation scheduling could also be addressed.

After providing the mathematical formulation of this problem, the particular features of the genetic algorithm implementation are described.

2.5.1 Problem Formulation

The maximum vulnerability model with line switching can be formulated as the following bilevel programming problem:

$$\max_{v_\ell} \sum_{n\in N} \Delta P_n^{d*} \tag{2.12}$$

subject to:

$$\sum_{\ell\in L}(1-v_\ell)=K \tag{2.13}$$

$$v_\ell \in \{0,1\}; \quad \forall \ell \in L \tag{2.14}$$

$$\Delta P_n^{d*} \in \arg\left\{ \min_{P_\ell^f, P_j^g, w_\ell, \delta_n, \Delta P_n^d} \sum_{n\in N} \Delta P_n^d \right. \tag{2.15}$$

subject to:

$$P_\ell^f = v_\ell w_\ell \frac{1}{x_\ell}\left[\delta_{O(\ell)} - \delta_{R(\ell)}\right]; \quad \forall \ell \in L \tag{2.16}$$

$$\sum_{j\in J_n} P_j^g - \sum_{\ell|O(\ell)=n} P_\ell^f + \sum_{\ell|R(\ell)=n} P_\ell^f + \Delta P_n^d = P_n^d; \quad \forall n \in N \tag{2.17}$$

$$0 \le P_j^g \le \overline{P}_j^g; \quad \forall j \in J \tag{2.18}$$

$$-\overline{P}_\ell^f \le P_\ell^f \le \overline{P}_\ell^f; \quad \forall \ell \in L \tag{2.19}$$

$$\underline{\delta} \le \delta_n \le \overline{\delta}; \quad \forall n \in N \tag{2.20}$$

$$0 \le \Delta P_n^d \le P_n^d; \quad \forall n \in N \tag{2.21}$$

$$\left. w_\ell \in \{0,1\}; \quad \forall \ell \in L \right\}, \tag{2.22}$$

where v_ℓ is a 0/1 variable which is equal to 0 if transmission asset ℓ is out of service and otherwise is equal to 1; ΔP_n^d is the load shed at bus n; N is the set of bus indices; L is the set of indices of transmission assets; K is the number of simultaneous out-of-service transmission assets; P_ℓ^f is the power flow of transmission asset ℓ; P_j^g is the power output of generator j; w_ℓ is a 0/1 variable which is equal to 0 if transmission asset ℓ is disconnected by the system operator and 1 otherwise; δ_n is the phase angle at bus n; x_ℓ is the reactance of transmission asset ℓ; $O(\ell)$ and $R(\ell)$ are the sending and receiving buses of transmission asset ℓ, respectively; J_n is the set of indices of generators connected to bus n; P_n^d is the demand at bus n; $\overline{P}_j^g$ is the capacity of generator j; J is the set of generator indices; $\overline{P}_\ell^f$ is the power flow capacity of transmission asset ℓ; and $\underline{\delta}$ and $\overline{\delta}$ are the lower and upper bounds for the nodal phase angles, respectively.

Problem (2.12)-(2.22) comprises an upper-level problem (2.12)-(2.14) and a lower-level problem (2.15)-(2.22). The upper level controls binary variables v_ℓ. The system operator is represented by the optimal power flow in the lower-level problem (2.15)-(2.22), which is parameterized in terms of the upper-level decision variables v_ℓ. The system operator controls continuous variables P_ℓ^f, P_j^g, δ_n, and ΔP_n^d, as well as binary variables w_ℓ modeling the capability to modify the network topology.

The upper-level objective (2.12) is to maximize the system load shed. Note that ΔP_n^d are lower-level variables whose optimal values ΔP_n^{d*} are used in the upper-level objective function. The pre-specified number of failures in the transmission network is set in (2.13). Constraints (2.14) model the binary nature of variables v_ℓ.

The system operator reacts against the combination of out-of-service transmission assets v_ℓ determined by the upper-level problem by solving a dc optimal power flow modeled by (2.15)-(2.22). The objective of the system operator is to minimize the system load shed (2.15). Constraints (2.16) express the power flows in terms of the nodal phase angles δ_n, the line switching variables w_ℓ, and the upper-level variables v_ℓ. Note that if transmission asset ℓ is either out of service ($v_\ell = 0$) or disconnected ($w_\ell = 0$), the corresponding power flow is set to 0 by (2.16). Constraints (2.17) represent the power balance in each bus of the system. Upper and lower bounds on lower-level decision variables are imposed in constraints (2.18)-(2.21). Finally, constraints (2.22) model the integrality of variables w_ℓ. It should be noted that weights could be assigned to nodal loads shed to reflect the relative importance of each load.

Problem (2.12)-(2.22) is a mixed-integer nonlinear bilevel program. It is worth pointing out that the objective functions of both optimization levels are identical. Thus, the difficulties associated with the indifference of the lower level with respect to the upper-level decisions [6, 20] are not present, thereby guaranteeing the existence of an optimal solution.

Constraints (2.16) constitute the main difference with respect to the bilevel models presented in [3, 4, 9, 48, 55, 56]. These constraints make the lower-level problem nonconvex due to the presence of lower-level binary variables w_ℓ, and nonlinear due to the products of lower-level decision variables w_ℓ and δ_n. As a consequence, available methods are either inexact or even inapplicable. Furthermore, the number of possible contingency sets satisfying constraint (2.13) is equal to $\binom{|L|}{K}$, where $|L|$ denotes the total number of transmission assets that are susceptible to be out of service. Hence, an exhaustive search is not practical for large values of $|L|$ and K, and, consequently, new tools such as the proposed genetic algorithm are thus needed.

2.5.2 *Particular Features of the Genetic Algorithm Approach*

The application of the genetic algorithm approach described in Section 2.4.2 to problem (2.12)-(2.22) is characterized by several particular aspects related to coding, fitness evaluation, and constraint handling. These specific features are described in detail next.

2.5.2.1 Coding

As above mentioned, each individual in the population represents a candidate solution to the vulnerability analysis problem. Each contingency set is characterized by a binary-valued string of length $|L|$, associated with variables v_ℓ in problem (2.12)-(2.22). If the allele of a gene is equal to 0, it means that the corresponding asset is out of service. In contrast, a value of 1 indicates that the corresponding component is operative.

2.5.2.2 Fitness Evaluation

For a given individual i, the fitness is the value of the objective function of the upper-level problem (2.12), i.e., the system load shed:

$$f_i = \sum_{n \in N} \Delta P_n^d. \tag{2.23}$$

Since ΔP_n^d are lower-level variables, the fitness of individual i is obtained from solving the optimal power flow (2.15)-(2.22) associated with its vector of out-of-service transmission assets. For quick reference, the formulation of this optimization is:

$$\min_{P_\ell^f, P_j^g, w_\ell, \delta_n, \Delta P_n^d} \sum_{n \in N} \Delta P_n^d \tag{2.24}$$

subject to:

$$P_\ell^f = v_{i\ell} w_\ell \frac{1}{x_\ell} \left[\delta_{O(\ell)} - \delta_{R(\ell)}\right]; \quad \forall \ell \in L \tag{2.25}$$

$$\sum_{j \in J_n} P_j^g - \sum_{\ell | O(\ell) = n} P_\ell^f + \sum_{\ell | R(\ell) = n} P_\ell^f + \Delta P_n^d = P_n^d; \quad \forall n \in N \tag{2.26}$$

$$0 \leq P_j^g \leq \overline{P}_j^g; \quad \forall j \in J \tag{2.27}$$

$$-\overline{P}_\ell^f \leq P_\ell^f \leq \overline{P}_\ell^f; \quad \forall \ell \in L \tag{2.28}$$

$$\underline{\delta} \leq \delta_n \leq \overline{\delta}; \quad \forall n \in N \tag{2.29}$$

$$0 \leq \Delta P_n^d \leq P_n^d; \quad \forall n \in N \tag{2.30}$$

$$w_\ell \in \{0, 1\}; \quad \forall \ell \in L, \tag{2.31}$$

where $v_{i\ell}$ is a parameter denoting the on/off state of transmission asset ℓ for individual i.

Problem (2.24)-(2.31) is a mixed-integer nonlinear programming problem. Nonlinearities arise in (2.25) due to the products of binary variables w_ℓ and continuous variables $\delta_{O(\ell)}$ and $\delta_{R(\ell)}$. However, the product of a binary variable and a continuous variable can be equivalently transformed into linear expressions [25]. The equivalent linear formulation of constraints (2.25) is provided in the Appendix. By transforming nonlinear expressions (2.25) into linear equivalents, the resulting optimal power flow becomes a mixed-integer linear programming problem suitable for off-the-shelf branch-and-cut software [38].

2.5.2.3 Feasibility Repair Procedure

Crossover and mutation operators may produce infeasible individuals violating upper-level constraint (2.13). In order to restore feasibility, the following repair procedure is implemented on each infeasible individual:

1. If the number of out-of-service transmission assets is less than K, a randomly selected gene with allele equal to 1 is flipped from 1 to 0. This process is repeated until the number of out-of-service assets is equal to K.
2. If the number of out-of-service transmission assets is greater than K, a randomly selected gene with allele equal to 0 is flipped from 0 to 1. This process is repeated until the number of out-of-service assets is equal to K.

2.6 Numerical Results

This section presents a case study based on the IEEE One Area Reliability Test System-1996 (RTS-96) [54]. This test system is a widely adopted benchmark by the power system community since its size allows both reproducibility and a comprehensive analysis of the results. The one-line diagram of RTS-96 is depicted in Fig. 2.7. This system comprises 24 buses, 38 transmission assets, 32 generators, and 17 loads. The corresponding number of feasible contingency sets and associated resolutions of the lower-level problem is $\binom{38}{K}$. For large values of K, methods based on complete enumeration may require excessive computational effort even for this benchmark system.

For illustration purposes, the data of RTS-96 [54] are slightly modified as reported in Tables 2.1 and 2.2. In addition, circuits sharing the same towers are treated as independent lines; e.g., line 20-23 has two circuits: 20-23A and 20-23B. For the sake of simplicity, line switching is restricted to the disconnection of transmission assets.

Several instances of the maximum vulnerability model (2.12)-(2.22) have been solved to illustrate the effectiveness of the proposed genetic algorithm. The input parameter defining the instances is K, representing the number of simultaneous out-of-service transmission assets. The proposed genetic algorithm has been applied for values of K ranging between 6 and 12, which are well beyond current $n-1$ and $n-2$ security criteria.

Some parameters of the genetic algorithm affect the quality of the solution and the computation time. These parameters are selected empirically. Thus, the population size is 100 and the maximum number of generations is 150, which are large

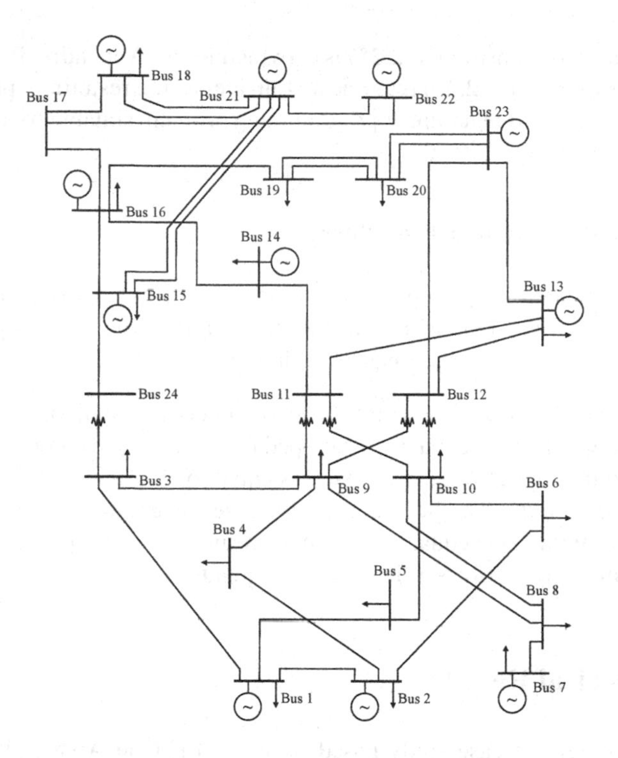

Fig. 2.7 IEEE One Area Reliability Test System

enough values guaranteeing a vast exploration of the search space. In order to promote a higher exchange of genetic information among the individuals, the crossover rate is 1.0, i.e., crossover is applied to all selected pairs of parent solutions. Furthermore, in each generation a member of the population is randomly selected to be subject to mutation. Finally, the elitist operator maintains the best solution into the next generation.

In order to test the robustness and accuracy of the method with the selected parameter configuration, the algorithm was run 100 times with different initial populations randomly created from scratch. The simulations have been run on a Sun Fire X4140 X64 at 2.3 GHz and 8 GB of RAM using MATLAB 7.7 [45] . The mixed-integer linear programming optimal power flows associated with each individual have been solved using CPLEX 11.2 [38] under GAMS 23.0 [29], which was called from MATLAB using a recently developed interface [53].

The performance of the proposed genetic algorithm in terms of solution accuracy is assessed through the comparison with the upper and lower bounds on the vulnerability level provided by two available methods. The upper bound is obtained from the solution to the maximum vulnerability problem without considering line switching in the lower-level problem. An exact solution approach for this relaxed problem was presented in [48]. The lower bound, i.e., the best known result, is

Table 2.1 Power flow capacity

Asset	$\overline{P}^f_\ell$ (MW)	Asset	$\overline{P}^f_\ell$ (MW)
1-2	87.5	12-13	250.0
1-3	87.5	12-23	250.0
1-5	87.5	13-23	50.0
2-4	87.5	14-16	50.0
2-6	87.5	15-16	50.0
3-9	87.5	15-21A	250.0
3-24	80.0	15-21B	250.0
4-9	100.0	15-24	80.0
5-10	100.0	16-17	250.0
6-10	87.5	16-19	50.0
7-8	87.5	17-18	250.0
8-9	50.0	17-22	250.0
8-10	87.5	18-21A	250.0
9-11	50.0	18-21B	250.0
9-12	200.0	19-20A	250.0
10-11	50.0	19-20B	250.0
10-12	200.0	20-23A	250.0
11-13	250.0	20-23B	250.0
11-14	50.0	21-22	250.0

Table 2.2 Nodal power demand

Bus	P^d_n (MW)	Bus	P^d_n (MW)
1	108	10	170
3	100	13	265
4	74	14	100
5	50	15	317
6	136	16	100
7	125	18	333
8	137	19	181
9	155	20	128

attained by a heuristic Benders-decomposition-based approach that was recently presented in [18]. Both bounds are relevant since they allow measuring the maximum distance to the optimal solution, referred to as optimality gap. In other words, the optimality gap gives insight on the maximum improvement attainable by the proposed genetic algorithm.

The results of this assessment are summarized in Table 2.3. Columns 2-4 respectively list the aforementioned lower and upper bounds as well as the corresponding optimality gap. Columns 5-7 present the results from the genetic algorithm, namely the best vulnerability level, the improvement over the best known result, and the new optimality gap, respectively. In addition, Table 2.4 lists the sets of critical transmission assets identified by the heuristic method of [18] and the proposed genetic algorithm.

As shown in Table 2.3, the proposed genetic algorithm always finds a solution better than or equal to the best known solution in terms of vulnerability level. For K equal to 7, 11, and 12, the lower and upper bounds for the vulnerability level are identical and thus equal to the optimal value, which is also achieved by the genetic algorithm (Tables 2.3 and 2.4). For K equal to 6 and 9, the genetic algorithm attains the same solutions found by the heuristic approach described in [18]. However, for K equal to 8 and 10, the genetic algorithm significantly outperforms the heuristic method. The vulnerability level is substantially increased by 1.05% and 3.34%, respectively, through the identification of different sets of critical assets, as listed in Table 2.4. Moreover, in these cases where the genetic algorithm improves upon the best known solution, the optimality gap is considerably reduced to tight values of 0.48% and 0.19%, respectively, thus allowing us to claim that the genetic algorithm is able to achieve high-quality near-optimal, probably optimal, solutions.

Table 2.3 Accuracy assessment of the proposed genetic algorithm

K	Available methods			Genetic algorithm		
	Lower bound (MW)	Upper bound (MW)	Gap (%)	Best solution (MW)	Improvement (%)	New gap (%)
6	775.0	794.9	2.50	775.0	0.00	2.50
7	855.0	855.0	0.00	855.0	0.00	0.00
8	905.0	918.9	1.51	914.5	1.05	0.48
9	1002.0	1003.0	0.10	1002.0	0.00	0.10
10	1017.0	1053.0	3.42	1051.0	3.34	0.19
11	1131.0	1131.0	0.00	1131.0	0.00	0.00
12	1194.0	1194.0	0.00	1194.0	0.00	0.00

As an illustrative example, Fig. 2.8 shows the sets of critical transmission assets found by the proposed genetic algorithm and the heuristic approach for $K = 10$ (Table 2.4). Note that the genetic algorithm includes in the critical contingency set assets 1-3, 1-5, 9-12, and 10-12 instead of assets 1-2, 12-13, 12-23, and 16-19, which are identified as critical by the heuristic method. This slight modification in the critical contingency set increases the level of vulnerability by 3.34%, thus providing a more effective vulnerability analysis.

The robustness of the proposed genetic algorithm is backed by the statistical results over all runs presented in Table 2.5. This table lists the minimum, maximum, and mean values as well as the standard deviation of the best vulnerability level attained at the last generation. The relatively small values of the standard deviation indicate that the genetic algorithm systematically finds the best solution in most runs.

Table 2.4 Critical transmission assets

K	Heuristic approach	Genetic algorithm
6	7-8, 11-13, 12-13, 12-23, 20-23A, 20-23B	7-8, 11-13, 12-13, 12-23, 20-23A, 20-23B
7	7-8, 11-13, 12-13, 12-23, 15-24, 20-23A, 20-23B	7-8, 11-13, 12-13, 12-23, 15-24, 20-23A, 20-23B
8	7-8, 11-13, 12-13, 12-23, 14-16, 15-24, 20-23A, 20-23B	11-13, 12-13, 12-23, 15-21A, 15-21B, 16-17, 20-23A, 20-23B
9	7-8, 11-13, 12-13, 12-23, 15-21A, 15-21B, 16-17, 20-23A, 20-23B	7-8, 11-13, 12-13, 12-23, 15-21A,15-21B, 16-17, 20-23A, 20-23B
10	1-2, 2-4, 2-6, 7-8, 11-13, 12-13, 12-23, 16-19, 20-23A, 20-23B	1-3, 1-5, 2-4, 2-6, 7-8, 9-12, 10-12, 11-13, 20-23A, 20-23B
11	1-3, 1-5, 2-4, 2-6, 7-8, 11-13, 12-13, 12-23, 15-24, 20-23A, 20-23B	1-3, 1-5, 2-4, 2-6, 7-8, 11-13, 12-13, 12-23, 15-24, 20-23A, 20-23B
12	1-2, 2-4, 2-6, 7-8, 11-13, 12-13, 12-23, 15-21A, 15-21B, 16-17, 20-23A, 20-23B	1-2, 2-4, 2-6, 7-8, 11-13, 12-13, 12-23, 15-21A, 15-21B, 16-17, 20-23A, 20-23B

Table 2.5 Best vulnerability level (MW) attained by the genetic algorithm

K	Minimum	Maximum	Mean	Standard deviation
6	775.0	775.0	775.0	0.00
7	855.0	855.0	855.0	0.00
8	879.5	914.5	910.7	5.31
9	963.5	1002.0	972.6	14.41
10	1019.0	1051.0	1046.8	9.96
11	1097.0	1131.0	1120.1	12.68
12	1149.0	1194.0	1188.6	9.39

As for the computational performance of the genetic algorithm, Table 2.6 provides statistical information on the computation times required over all trials. High-quality solutions are achieved in less than 39 minutes for all cases, which is moderate bearing in mind that a planning problem is being solved for which computational issues are not a primary concern. It should also be noted that a slight reduction in computation time is experienced as K increases due to the faster attainment of the optimal solutions to the lower-level problems.

Finally, Fig. 2.9 illustrates the convergence behavior of the genetic algorithm by showing the evolution of the mean best vulnerability level along generations for all cases. The mean best vulnerability level is defined as the average among all runs of the best vulnerability level achieved at each generation. As a general result, it should be noted that the quality of the best solutions is rapidly improved in the first 20 generations. From Fig. 2.9, it might also be inferred that convergence is mostly attained

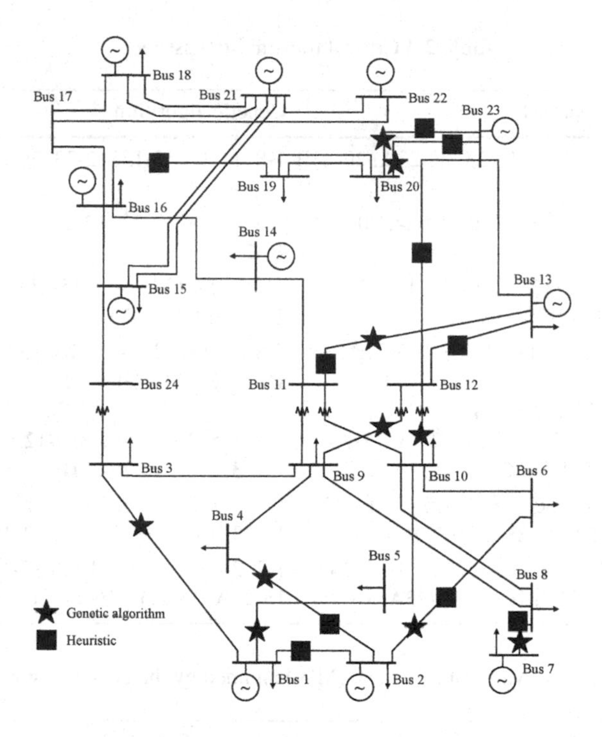

Fig. 2.8 Solutions for $K = 10$

Table 2.6 Computation time (min) required by the genetic algorithm

K	Minimum	Maximum	Mean	Standard deviation
6	33.8	38.9	36.5	1.09
7	34.4	37.5	35.9	0.63
8	34.0	36.8	35.6	0.52
9	34.2	36.0	35.0	0.37
10	33.0	35.0	34.1	0.30
11	33.1	34.4	33.7	0.25
12	32.8	33.8	33.3	0.20

after 80 generations. Notwithstanding, it should be noted that finding the best solution required at least 100 generations in 19.4% of the simulations. In other words, improvements between generations took place along the whole evolution process, thereby supporting both the effectiveness of the genetic operators in avoiding stagnation as well as the choice of 150 generations as the stopping criterion.

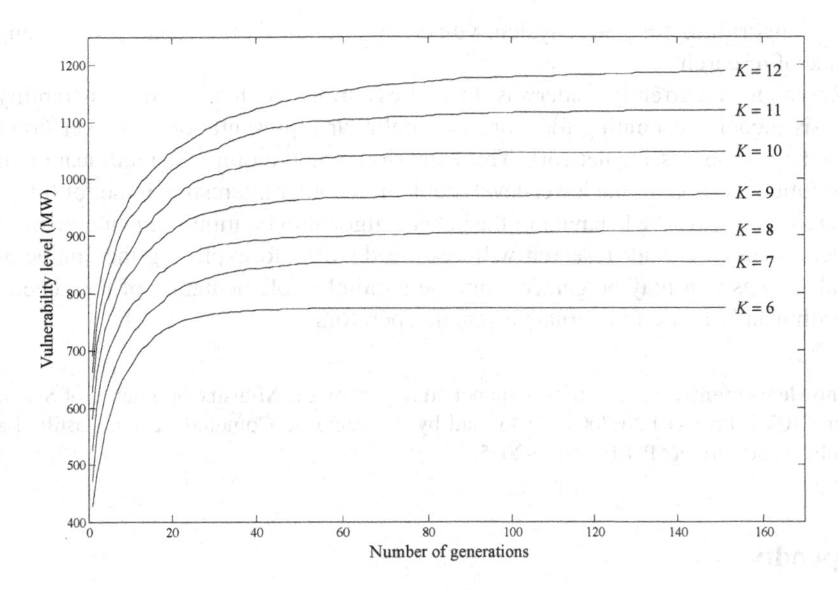

Fig. 2.9 Convergence patterns

2.7 Conclusions

This chapter has presented a genetic algorithm to solve the bilevel programming models used for the analysis of vulnerability of power systems under multiple contingencies. The proposed approach provides a flexible modeling framework that allows considering the nonlinearities and nonconvexities associated with the operation of power systems.

The proposed genetic algorithm has been applied to a maximum vulnerability model in which the upper-level problem identifies the set of critical components, whereas the lower-level problem selects the corrective actions implemented by the system operator in order to minimize the system damage. As a complicating factor, the set of corrective actions includes the modification of the network topology through the connection and disconnection of transmission assets by the system operator. The resulting problem is characterized as a mixed-integer nonlinear bilevel programming problem for which no exact solution techniques are available.

Numerical simulations conducted on the IEEE One Area Reliability Test System revealed an effective performance by the genetic algorithm in terms of solution quality and computational burden. Solution quality is measured by a lower and an upper bound for the optimal value of the vulnerability level, which are both attained by previously reported methods. It is worth mentioning that the genetic algorithm always provided solutions within those bounds in moderate computation time. Moreover, in several cases the genetic algorithm outperformed the best known solutions, which is of utmost relevance for the system planner. Based on these results, the application of genetic algorithms for power system vulnerability analysis represents a promising avenue of research.

Research is currently underway to address more sophisticated vulnerability analysis models accounting for a precise nonlinear representation of power flows through the transmission network. This extension would require the modification of the solution method for the lower-level problem. Another interesting issue for future research is to study the behavior of the genetic algorithm on minimum vulnerability models. Finally, further research will also be devoted to exploring the computational savings that may be gained from the parallel implementation of the genetic algorithm and the use of alternative genetic operators.

Acknowledgements. This work was supported in part by the Ministry of Science of Spain, under CICYT Project ENE2009-07836; and by the Junta de Comunidades de Castilla-La Mancha, under Project POII11-0130-2055.

Appendix

In (2.25) there are two products of binary and continuous variables per transmission asset: (i) w_ℓ and the phase angle of the sending node of asset ℓ, denoted as $\delta_{O(\ell)}$; and (ii) w_ℓ and the phase angle of the receiving node of asset ℓ, denoted as $\delta_{R(\ell)}$. As explained in [25], by introducing four new sets of continuous variables $\delta^{\mathrm{Q}}_{O(\ell)}$, $\delta^{\mathrm{Q}}_{R(\ell)}$ (representing the products $w_\ell \delta_{O(\ell)}$ and $w_\ell \delta_{R(\ell)}$, respectively), $\delta^{\mathrm{A}}_{O(\ell)}$, and $\delta^{\mathrm{A}}_{R(\ell)}$, nonlinear constraints (2.25) are equivalently replaced by:

$$P^f_\ell = v_{i\ell}\frac{1}{x_\ell}\left[\delta^{\mathrm{Q}}_{O(\ell)} - \delta^{\mathrm{Q}}_{R(\ell)}\right]; \quad \forall \ell \in L \tag{2.32}$$

$$\delta^{\mathrm{Q}}_{O(\ell)} = \delta_{O(\ell)} - \delta^{\mathrm{A}}_{O(\ell)}; \quad \forall \ell \in L \tag{2.33}$$

$$\delta^{\mathrm{Q}}_{R(\ell)} = \delta_{R(\ell)} - \delta^{\mathrm{A}}_{R(\ell)}; \quad \forall \ell \in L \tag{2.34}$$

$$\underline{\delta} w_\ell \le \delta^{\mathrm{Q}}_{O(\ell)} \le \overline{\delta} w_\ell; \quad \forall \ell \in L \tag{2.35}$$

$$\underline{\delta} w_\ell \le \delta^{\mathrm{Q}}_{R(\ell)} \le \overline{\delta} w_\ell; \quad \forall \ell \in L \tag{2.36}$$

$$\underline{\delta}(1 - w_\ell) \le \delta^{\mathrm{A}}_{O(\ell)} \le \overline{\delta}(1 - w_\ell); \quad \forall \ell \in L \tag{2.37}$$

$$\underline{\delta}(1 - w_\ell) \le \delta^{\mathrm{A}}_{R(\ell)} \le \overline{\delta}(1 - w_\ell); \quad \forall \ell \in L. \tag{2.38}$$

Constraints (2.32) are the new linear expressions of the power flows. Expressions (2.33) and (2.34) relate the nodal phase angles with the new variables $\delta^{\mathrm{Q}}_{O(\ell)}$, $\delta^{\mathrm{A}}_{O(\ell)}$, and $\delta^{\mathrm{Q}}_{R(\ell)}$, $\delta^{\mathrm{A}}_{R(\ell)}$, respectively. Finally, lower and upper bounds on variables $\delta^{\mathrm{Q}}_{O(\ell)}$, $\delta^{\mathrm{Q}}_{R(\ell)}$, $\delta^{\mathrm{A}}_{O(\ell)}$, and $\delta^{\mathrm{A}}_{R(\ell)}$ are imposed in (2.35)-(2.38), respectively.

If transmission asset ℓ is disconnected by the system operator ($w_\ell = 0$), variables $\delta^{\mathrm{Q}}_{O(\ell)}$ and $\delta^{\mathrm{Q}}_{R(\ell)}$ are set to 0 by (2.35)-(2.36), and, consequently, the power flow through asset ℓ is equal to 0 by (2.32). In addition, variables $\delta^{\mathrm{A}}_{O(\ell)}$ and $\delta^{\mathrm{A}}_{R(\ell)}$ are respectively equal to the phase angles at the sending and receiving nodes by (2.33)-(2.34).

Similarly, if transmission asset ℓ is not disconnected ($w_\ell = 1$), variables $\delta^{\mathrm{A}}_{O(\ell)}$ and $\delta^{\mathrm{A}}_{R(\ell)}$ are both equal to 0 by (2.37)-(2.38), and variables $\delta^{\mathrm{Q}}_{O(\ell)}$ and $\delta^{\mathrm{Q}}_{R(\ell)}$ are respectively equal to $\delta_{O(\ell)}$ and $\delta_{R(\ell)}$ by (2.33)-(2.34). Hence, the power flow is determined by the difference of phase angles at the sending and receiving nodes (2.32).

References

1. Alguacil, N., Carrión, M., Arroyo, J.M.: Transmission network expansion planning under deliberate outages. Int. J. Electr. Power Energy Syst. 31, 553–561 (2009)
2. Andersson, G., Donalek, P., Farmer, R., Hatziargyriou, N., Kamwa, I., Kundur, P., Martins, N., Paserba, J., Pourbeik, P., Sanchez-Gasca, J., Schulz, R., Stankovic, A., Taylor, C., Vittal, V.: Causes of the 2003 major grid blackouts in North America and Europe, and recommended means to improve system dynamic performance. IEEE Trans. Power Syst. 20, 1922–1928 (2005)
3. Arroyo, J.M.: Bilevel programming applied to power system vulnerability analysis under multiple contingencies. IET Gener. Transm. Distrib. 4, 178–190 (2010)
4. Arroyo, J.M., Galiana, F.D.: On the solution of the bilevel programming formulation of the terrorist threat problem. IEEE Trans. Power Syst. 20, 789–797 (2005)
5. Arroyo, J.M., Alguacil, N., Carrión, M.: A risk-based approach for transmission network expansion planning under deliberate outages. IEEE Trans. Power Syst. 25, 1759–1766 (2010)
6. Bard, J.F.: Practical bilevel optimization. Algorithms and applications. Kluwer Academic Publishers, Dordrecht (1998)
7. Berizzi, A.: The Italian 2003 blackout. In: Proc. 2004 IEEE PES Gen Meeting, Denver, pp. 1673–1679 (2004)
8. Bienstock, D., Verma, A.: The N-k problem in power grids: New models, formulations, and numerical experiments. SIAM J. Optim. 20, 2352–2380 (2010)
9. Bier, V.M., Gratz, E.R., Haphuriwat, N.J., Magua, W., Wierzbicki, K.R.: Methodology for identifying near-optimal interdiction strategies for a power transmission system. Reliab Eng. Syst. Saf. 92, 1155–1161 (2007)
10. Brown, G.G., Carlyle, W.M., Salmerón, J., Wood, K.: Analyzing the vulnerability of critical infrastructure to attack and planning defenses. In: Greenberg, H.J., Smith, J.C. (eds.) Tutorials in Operations Research. Emerging Theory, Methods, and Applications, INFORMS, Hanover, pp. 102–123 (2005)
11. Brown, G., Carlyle, M., Salmerón, J., Wood, K.: Defending critical infrastructure. Interfaces 36, 530–544 (2006)
12. Calvete, H.I., Galé, C., Mateo, P.M.: A new approach for solving linear bilevel problems using genetic algorithms. Eur. J. Oper. Res. 188, 14–28 (2008)
13. Calvete, H.I., Galé, C., Mateo, P.M.: A genetic algorithm for solving linear fractional bilevel problems. Ann. Oper. Res. 166, 39–56 (2009)

14. Carrión, M., Arroyo, J.M., Alguacil, N.: Vulnerability-constrained transmission expansion planning: A stochastic programming approach. IEEE Trans. Power Syst. 22, 1436–1445 (2007)
15. Chen, G., Dong, Z.Y., Hill, D.J., Xue, Y.S.: Exploring reliable strategies for defending power systems against targeted attacks. IEEE Trans. Power Syst. 26, 1000–1009 (2011)
16. Commission of the European Communities, Green paper on a European programme for critical infrastructure protection (2005), http://eur-lex.europa.eu/LexUriServ/site/en/com/2005/com2005_0576en01.pdf
17. Critical Infrastructure Protection Committee, CIPC (2011) North American Electric Reliability Council, NERC. http://www.nerc.com/page.php?cid=6$vert$69
18. Delgadillo, A., Arroyo, J.M., Alguacil, N.: Analysis of electric grid interdiction with line switching. IEEE Trans. Power Syst. 25, 633–641 (2010)
19. Delgadillo, A., Arroyo, J.M., Alguacil, N.: Power system defense planning against multiple contingencies. In: Proc. 17th Power Syst Comput Conf, PSCC 2011, Paper no. 269, Stockholm (2011)
20. Dempe, S.: Foundations of bilevel programming. Kluwer Academic Publishers, Dordrecht (2002)
21. Donde, V., López, V., Lesieutre, B., Pinar, A., Yang, C., Meza, J.: Severe multiple contingency screening in electric power systems. IEEE Trans. Power Syst. 23, 406–417 (2008)
22. Dy Liacco, T.E.: The adaptive reliability control system. IEEE Trans. Power App. Syst. PAS 86, 517–531 (1967)
23. Ejebe, G.C., Wollenberg, B.F.: Automatic contingency selection. IEEE Trans. Power App. Syst. PAS 98, 97–109 (1979)
24. E.ON Netz GmbH, Report on the status of the investigations of the sequence of events and causes of the failure in the continental European electricity grid on Saturday, after 22:10 hours (November 4, 2006)
25. Floudas, C.A.: Nonlinear and mixed-integer optimization: Fundamentals and applications. Oxford University Press, New York (1995)
26. Fogel, D.B.: Evolutionary computation. Toward a new philosophy of machine intelligence, 3rd edn. John Wiley & Sons, Inc., Hoboken (2005)
27. Fozdar, M., Arora, C.M., Gottipati, V.R.: Recent trends in intelligent techniques to power systems. In: Proc. 42nd International Universities Power Engineering Conf, UPEC 2007, Brighton, pp. 580–591 (2007)
28. Galiana, F.D.: Bound estimates of the severity of line outages in power system contingency analysis and ranking. IEEE Trans. Power App. Syst. PAS 103, 2612–2624 (1984)
29. GAMS Development Corporation (2011), http://www.gams.com
30. Gheorghe, A.V., Masera, M., Weijnen, M., de Vries, L.: Critical infrastructures at risk. Securing the European electric power system. Springer, Dordrecht (2006)
31. Glover, F., Kochenberger, G.A.: Handbook of metaheuristics. Kluwer Academic Publishers, Dordrecht (2003)
32. Goldberg, D.E.: Genetic algorithms in search, optimization, and machine learning. Addison-Wesley, Reading (1989)
33. Halpin, T.F., Fischl, R., Fink, R.: Analysis of automatic contingency selection algorithms. IEEE Trans. Power App. Syst. PAS 103, 938–945 (1984)
34. Hansen, P., Jaumard, B., Savard, G.: New branch-and-bound rules for linear bilevel programming. SIAM J. Sci. Stat. Comput. 13, 1194–1217 (1992)
35. Hejazi, S.R., Memariani, A., Jahanshahloo, G., Sepehri, M.M.: Linear bilevel programming solution by genetic algorithm. Comput. Oper. Res. 29, 1913–1925 (2002)

36. Holland, J.H.: Adaptation in natural and artificial systems: An introductory analysis with applications to biology, control, and artificial intelligence. University of Michigan Press, Ann Arbor (1975)
37. Holmgren, Å.J., Jenelius, E., Westin, J.: Evaluating strategies for defending electric power networks against antagonistic attacks. IEEE Trans. Power Syst. 22, 76–84 (2007)
38. IBM ILOG CPLEX (2011), `http://www-01.ibm.com/software/integration/optimization/cplex-optimizer/`
39. Irisarri, G., Sasson, A.M., Levner, D.: Automatic contingency selection for on-line security analysis–Real-time tests. IEEE Trans. Power App. Syst. PAS 98, 1552–1559 (1979)
40. Kaye, R.J., Wu, F.F.: Analysis of linearized decoupled power flow approximations for steady-state security assessment. IEEE Trans. Circuits Syst. CAS 31, 623–636 (1984)
41. Larsson, S., Ek, E.: The black-out in southern Sweden and eastern Denmark. In: Proc. 2004 IEEE PES Gen Meeting, Denver, September 23, 2003, pp. 1668–1672 (2004)
42. Li, H., Wang, Y.: Exponential distribution-based genetic algorithm for solving mixed-integer bilevel programming problems. J. Syst. Eng. Electron 19, 1157–1164 (2008)
43. Li, H., Jiao, Y., Zhang, L.: Orthogonal genetic algorithm for solving quadratic bilevel programming problems. J. Syst. Eng. Electron 21, 763–770 (2010)
44. Mathieu, R., Pittard, L., Anandalingam, G.: Genetic algorithm based approach to bi-level linear programming. Recherche Opérationnelle/Operations Research 28, 1–21 (1994)
45. The MathWorks, Inc. (2011), `http://www.mathworks.com`
46. Michalewicz, Z.: Genetic algorithms + Data structures = Evolution programs, 3rd edn. Springer, Berlin (1996)
47. Miranda, V., Srinivasan, D., Proença, L.M.: Evolutionary computation in power systems. Int. J. Electr. Power Energy Syst. 20, 89–98 (1998)
48. Motto, A.L., Arroyo, J.M., Galiana, F.D.: A mixed-integer LP procedure for the analysis of electric grid security under disruptive threat. IEEE Trans. Power Syst. 20, 1357–1365 (2005)
49. Nishizaki, I., Sakawa, M.: Computational methods through genetic algorithms for obtaining Stackelberg solutions to two-level mixed zero-one programming problems. Cybern Syst. 31, 203–221 (2000)
50. Nishizaki, I., Sakawa, M.: Computational methods through genetic algorithms for obtaining Stackelberg solutions to two-level integer programming problems. Cybern Syst. 36, 565–579 (2005)
51. Nishizaki, I., Sakawa, M., Niwa, K., Kitaguchi, Y.: A computational method using genetic algorithms for obtaining Stackelberg solutions to two-level linear programming problems. Electr. Commun Jpn 85, 55–62 (2002)
52. Pinar, A., Meza, J., Donde, V., Lesieutre, B.: Optimization strategies for the vulnerability analysis of the electric power grid. SIAM J. Optim. 20, 1786–1810 (2010)
53. Power System Analysis Toolbox (2011), `http://www.uclm.es/area/gsee/Web/Federico/psat.htm`
54. Reliability Test System Task Force, The IEEE Reliability Test System - 1996. IEEE Trans. Power Syst. 14, 1010–1020 (1999)
55. Salmeron, J., Wood, K., Baldick, R.: Analysis of electric grid security under terrorist threat. IEEE Trans. Power Syst. 19, 905–912 (2004)
56. Salmeron, J., Wood, K., Baldick, R.: Worst-case interdiction analysis of large-scale electric power grids. IEEE Trans. Power Syst. 24, 96–104 (2009)
57. Shahidehpour, M., Yamin, H., Li, Z.: Market operations in electric power systems: Forecasting, scheduling, and risk management. John Wiley & Sons, Inc., New York (2002)
58. da Silva, A.P.A., Abrão, P.J.: Applications of evolutionary computation in electric power systems. In: Proc. 2002 Congress on Evolutionary Computation, CEC 2002, Honolulu, pp. 1057–1062 (2002)

59. von Stackelberg, H.: The theory of the market economy. William Hodge, London (1952)
60. Talbi, E.-G.: Metaheuristics: From design to implementation. John Wiley & Sons, Inc., Hoboken (2009)
61. U.S.-Canada Power System Outage Task Force, Final report on the August 14, 2003 blackout in the United States and Canada: Causes and recommendations (2004), http://www.nerc.com/filez/blackout.html
62. Wang, G., Wang, X., Wan, Z., Jia, S.: An adaptive genetic algorithm for solving bilevel linear programming problem. Appl. Math. Mech-Engl Ed 28, 1605–1612 (2007)
63. Wang, G., Wan, Z., Wang, X., Lv, Y.: Genetic algorithm based on simplex method for solving linear-quadratic bilevel programming problem. Comput. Math Appl. 56, 2550–2555 (2008)
64. Wood, A.J., Wollenberg, B.F.: Power generation, operation, and control, 2nd edn. John Wiley & Sons, Inc., New York (1996)
65. Yao, Y., Edmunds, T., Papageorgiou, D., Alvarez, R.: Trilevel optimization in power network defense. IEEE Trans. Syst. Man Cybern Part C-Appl. Rev. 37, 712–718 (2007)

Chapter 3
A Bilevel Particle Swarm Optimization Algorithm for Supply Chain Management Problems

Yannis Marinakis and Magdalene Marinaki

Abstract. Nature inspired methods are approaches that are used in various fields and for the solution for a number of problems. In this study, a new bilevel particle swarm optimization algorithm is proposed for solving two well known supply chain management problems, the Vehicle Routing Problem and the Location Routing Problem. The results of the algorithms are compared with the results of algorithms that solve these problems with a single objective function and with a bilevel genetic algorithm. As most of the decisions in Supply Chain Management are taken in different levels, the study presented in this paper has two main goals. The first one is to give to the decision maker the possibility to formulate the supply chain management problems as bilevel or multilevel problems and the second one is to propose an efficient nature inspired algorithm that solves this kind of problems.

3.1 Introduction

Significant attention concerning companies' organization related to the entire supply chain is paid by all companies that aim to be competitive on the market. In particular, companies have to analyze the supply chain in order to improve the customer service level without an uncontrolled growth of costs [41].

The decisions for supply chain management are classified into three broad categories - strategic, tactical and operational [1]. *Strategic planning* represents the

Yannis Marinakis
Decision Support Systems Laboratory, Department of Production Engineering and Management, Technical University of Crete, 73100 Chania, Crete, Greece
e-mail: marinakis@ergasya.tuc.gr

Magdalene Marinaki
Industrial Systems Control Laboratory, Department of Production Engineering and Management, Technical University of Crete, 73100 Chania, Crete, Greece
e-mail: magda@dssl.tuc.gr

E.-G. Talbi (Ed.): *Metaheuristics for Bi-level Optimization*, SCI 482, pp. 69–93.
DOI: 10.1007/978-3-642-37838-6_3 © Springer-Verlag Berlin Heidelberg 2013

highest level of the hierarchy of decision making activities which occur within a firm or an organization. These decisions are concerned with the definition of the long term objectives of a firm, the charting of the long term course which will allow a firm to meet its defined objectives, and the assurance that a firm has the proper resources and assets necessary to support its long term objectives [41]. Thus, facility locations, missions and relationships (i.e. network infrastructure and design), new facility locations and sizes, facility closings and facility capacity levels are some typical strategic logistics issues and problems which firms must address. *Tactical planning* represents the second or intermediate level of decision making activities that occur in a firm. The decision making process primarily focuses on resource allocation and resource utilization at this level while quite often annual planning is viewed as a subset of tactical planning. *Operational planning and scheduling* represents the third and lowest level of the hierarchical planning process. At this level, the firm must carry out the resource allocation and utilization decisions made at the tactical level in the daily and weekly activities which occur at the operational level [41]. At this level, an extraordinary number of individual manufacturing and distribution decisions occur regularly, a small sample of which are the customer order processing and scheduling, the facility operations scheduling, and the vehicle scheduling and routing.

In this paper, two classic NP-hard Supply Chain Management problems, the Vehicle Routing Problem and the Location Routing Problem are formulated as a bilevel programming problems. For the Vehicle Routing Problem the decisions are made in the operational level. When the problem is formulated as a bilevel programming problem, in the first level, the decision maker assigns customers to the vehicles checking the feasibility of the constructed routes (vehicle capacity constraints) and without taking into account the sequence by which the vehicles will visit the customers. In the second level, the decision maker finds the optimal routes of these assignments. The decision maker of the first level, once the cost of each routing has been calculated in the second level, estimates which assignment is the better one to choose. For the Location Routing Problem, the decisions are made in the strategic level and in the operational level. Thus, we formulate the problem in such a way that in the first level, the decisions of the strategic level are made, namely, the top manager finds the optimal location of the facilities, while in the second level, the operational level decisions are made, namely, the operational manager finds the optimal routing of vehicles.

For the solution of these bilevel problems a hybrid Particle Swarm Optimization algorithm is used. **Particle Swarm Optimization (PSO)** is a population-based swarm intelligence algorithm that was originally proposed by Kennedy and Eberhart [23] and simulates the social behavior of social organisms by using the physical movements of the individuals in the swarm. Its mechanism enhances and adapts to the global and local exploration. Most applications of PSO have concentrated on the optimization in continuous space while some work has been done to the discrete optimization [24, 56]. Recent complete surveys for the Particle Swarm Optimization can be found in [2, 3, 48]. The Particle Swarm Optimization (PSO) is a very popular optimization method and its wide use, mainly during the last years, is due to the number of advantages that this method has, compared to other optimization methods. Some of the key advantages are that this method does not need the calculation of derivatives, that the knowledge of good solutions is retained by all particles and that particles in the swarm share information between them. PSO is less sensitive to the nature of the objective function, can be used for stochastic objective functions and can easily escape from local minima. Concerning its implementation, PSO can easily be programmed, has few parameters to regulate and the assessment of the optimum is independent of the initial solution.

As there are not any nature inspired methods based on Particle Swarm Optimization, at least to our knowledge, for the solution of bilevel Supply Chain Management problems focusing on routing decisions, we would like to develop such an algorithm and to test its efficiency compared to other evolutionary algorithms for bilevel formulated problems. Also, the results will be compared with the results of a Particle Swarm Optimization algorithm when this algorithm is used for the solution of these supply chain management problems when they are not formulated as bilevel problems. Thus, in this paper, we demonstrate how a nature inspired intelligent technique, the Particle Swarm Optimization (PSO) [23] and the Expanding Neighborhood Search strategy [29] can be incorporated in a hybrid scheme, in order to give very good results for the Bilevel Vehicle Routing Problem and the Bilevel Location Routing Problem.

The rest of the paper is organized as follows. In section 3.2 a brief introduction of Bilevel Optimization is given. In section 3.3, the formulation of the Supply Chain Management Problems is given while in section 3.4, an analytical description of the proposed algorithm is given. In section 3.5, the computational results of the algorithm are presented and analyzed. Finally, in the last section some general conclusions and the future research are given.

3.2 Bilevel Programming

Bilevel optimization problems arise in hierarchical decision making, where players of different ranks are involved. The situation is described by the so-called Stackelberg game . Because of the inherent nonconvexity of these problems, it is not easy to find globally optimal solutions [39, 40].

The bilevel programming problem describes a hierarchical system which is composed of two levels of decision makers. The higher level decision maker, known as *leader*, controls the decision variables y, while the lower level decision maker, known as *follower*, controls the decision variables x. The interaction between the two levels is modeled in their respective *loss functions* $\varphi(x,y)$ and $f(x,y)$ and often in the feasible regions. The leader and the follower play a *Stackelberg duopoly game* . The idea of the game is as follows: The first player, the leader, chooses y to minimize the loss function $\varphi(x,y)$, while the second player, the follower, reacts to leader's decision by selecting an admissible strategy x that minimizes his loss function $f(x,y)$. Thus, the follower's decision depends upon the leader's decision, i.e. $x = x(y)$, and the leader is in full knowledge of this.

The general bilevel programming problem is stated as follows:

$$\textbf{(BP)} \min_{y \in Y} \quad \varphi(x(y), y) \tag{3.1}$$

$$\text{subject to } \psi(x(y), y) \leq 0, \tag{3.2}$$

$$\text{where } x(y) = \arg\min_{x \in X} f(x, y) \tag{3.3}$$

$$\text{subject to } g(x, y) \leq 0, \tag{3.4}$$

where $X \subset R^n$ and $Y \subset R^m$ are closed sets, $\psi : X \times Y \to R^p$ and $g : X \times Y \to R^q$ are multifunctions, φ and f are real-valued functions.

The *upper level* (3.1)-(3.2) corresponds to the leader, while the *lower level* (3.3)-(3.4) corresponds to the follower. The set $\mathscr{S} = \{(x,y) : x \in X, y \in Y, \psi(x,y) \leq 0, g(x,y) \leq 0\}$ is the *constraint set* of **BP**. For fixed $y \in Y$, the set $X(y) = \{x \in X : g(x,y) \leq 0\}$ is the *feasible set* of the follower. The set $\mathscr{R}(y) = \{x \in X : x \in \arg\min_{w \in X(y)} f(w,y)\}$ is called the *rational reaction set* of **BP**. The *feasible set* of **BP** is $\mathscr{F} = \{(x,y) \in \mathscr{S} : x \in \mathscr{R}(y)\}$. A feasible point $(x^\star, y^\star) \in \mathscr{F}$ is a *Stackelberg equilibrium* (with the first player as the leader) if $\varphi(x^\star, y^\star) \leq \varphi(x,y)$ for all $(x,y) \in \mathscr{F}$.

3.3 Supply Chain Management Problems

3.3.1 Vehicle Routing Problem

The **Vehicle Routing Problem (VRP)** or the **Capacitated Vehicle Routing problem (CVRP)** is often described as the problem in which vehicles based on a central depot are required to visit geographically dispersed customers in order to fulfill known customer demands. The vehicle routing problem was first introduced by Dantzig and Ramser (1959) [9]. Since then, a number of variants of the classic Vehicle Routing Problem has been proposed in order to incorporate more constraints like time windows, multi-depot, stochastic or dynamic demands. The reader can find more detailed descriptions of the algorithms proposed for the CVRP and its variants in the survey papers [5, 6, 14, 17, 18, 26–28, 57] and in the books [19, 21, 45, 59].

In this section, the Vehicle Routing Problem is formulated as a problem of two decision levels [31]. In the first level, the decision maker assigns customers to the vehicles checking the feasibility of the constructed routes (vehicle capacity constraints) and without taking into account the sequence by which the vehicles will visit the customers. In the second level, the decision maker finds the optimal routes of these assignments. The decision maker of the first level, once the cost of each routing has been calculated in the second level, estimates which assignment is the better one to choose.

One of the most important formulations that has ever been proposed for the solution of the problem is the formulation of Fisher and Jaikumar [13]. This formulation belongs to the category of vehicle flow models. Fisher and Jaikumar proved that the constraints of the problem can be separated in two sets. The first set of constraints are the constraints of a generalized assignment problem and they ensure that each route begins and ends at the depot, that every customer is served by some vehicle, and that the load assigned to a vehicle is within capacity. The second set of constraints corresponds to the constraints of a Traveling Salesman Problem for all customers of each vehicle. So, they solved a generalized assignment problem approximation of the Vehicle Routing Problem in order to obtain an assignment of customers. Subsequently, the customers assigned to each vehicle can be sequenced using any Traveling Salesman algorithm.

The formulation of Fisher and Jaikumar gave us the idea that the Vehicle Routing Problem can be formulated as a problem of two decision phases or as a problem of two decision levels. By saying two levels it is meant that in each level a different problem is solved, but the solution of the one level depends on the solution of the other. In particular, it is assumed that the decisions in the second level are reacting to the decisions of the first level and that the decisions in the first level must be made by taking this fact into consideration. This kind of formulation is called bilevel formulation. Thus, the Vehicle Routing Problem can be viewed as a bilevel decision problem where in the first level decisions must be made concerning the assignment of the customers to the routes and in the second level decisions must be made concerning the routing of the customers.

Let $G = (N, E)$ be a graph where N is the vertex set and E is the arc set. The customers are indexed $i = 2, \cdots, n$, $j = 1, \cdots, n$ and $i = 1$ refers to the depot. The vehicles are indexed $k = 1, \cdots, K$. The capacity of vehicle k is Q_k. If the vehicles are homogeneous, the capacity for all vehicles is equal and denoted by Q. A demand q_j and a service time st_j are associated with each customer node j. The travel cost between customers i and j is c_{ij}. The problem is to construct a low cost, feasible set of routes - one for each vehicle (starting and finishing at the depot). A route is a sequence of locations that a vehicle must visit along with the indication of the serve it provides [6]. The vehicle must start and finish its tour at the depot. Customer orders cannot be split. The customers are assigned to a single route until the routes reach capacity or time limits, subsequently a new customer is selected as a seed customer for the new route and the process is continued. A seed customer is a customer not yet assigned in a route that is used in order to initialize a new route. The distance of the k seed customer from the depot is d_k.

In order to present the bilevel model for the problem, we define the following variables:

$$z_k = \begin{cases} 1 & \text{if } k \text{ is a seed customer} \\ 0 & \text{otherwise,} \end{cases}$$

$$x_{kj} = \begin{cases} 1 & \text{if customer } j \text{ belongs in the same route} \\ & \text{with the seed customer } k \\ 0 & \text{otherwise,} \end{cases}$$

and

$$y_{ij} = \begin{cases} 1 & \text{if edge } (i,j) \text{ is in the route} \\ 0 & \text{otherwise.} \end{cases}$$

The bilevel formulation for the Vehicle Routing Problem is then:

$$\text{(leader)} \min_{x,z} \sum_{k=1}^{K} d_k z_k + \sum_{k=1}^{K}\sum_{j=1}^{n} c_{kj} x_{kj} + \sum_{i=1}^{n}\sum_{j=1}^{n} c_{ij} y_{ij} \tag{3.5}$$

s.t.

$$\sum_{k=1}^{n} z_k = K \tag{3.6}$$

$$\sum_{j=1}^{n} q_j x_{jk} \leq (Q - q_k) z_k, \qquad \forall k \in K \tag{3.7}$$

$$\sum_{k=1}^{K} x_{kj} = 1, \qquad \forall j = 1, \cdots, n \tag{3.8}$$

where

$$\text{(follower)} \qquad \min_{y|x,z} \sum_{i=1}^{n}\sum_{j=1}^{n} c_{ij} y_{ij} \tag{3.9}$$

s.t.

$$y_{ij} \leq x_{ik}, \qquad i,j = 1, \cdots, n, \quad k = 1, \cdots, K \tag{3.10}$$

$$\sum_{i=1}^{n} y_{ij} = 1, \qquad j = 1, \cdots, n \tag{3.11}$$

$$\sum_{j=1}^{n} y_{ij} = 1, \qquad i = 1, \cdots, n \tag{3.12}$$

$$\sum_{j \in V}\sum_{i \in V} y_{ij} \leq |S| - 1, \qquad \forall S \subset V, S \neq \emptyset. \tag{3.13}$$

The objective function of the leader (Eq. 3.5) minimizes the sum of the seed customers' costs from depot, the sum of the costs of assigning customers to the routes, i.e. the assignment of the customers to the seed customers and the routing cost. Constraints (Eq. 3.6) requires z_k to be set equal to the number of vehicles. Constraints (Eq. 3.7) are the vehicle capacity constraints. Finally, constraints (Eq. 3.8) state that every customer j must be on exactly one route (vehicle). The objective function of the follower (Eq. 3.9) describes the routing cost of each vehicle based on the assignment by the leader. Constraints (Eq. 3.10) require that the Traveling Salesman Problem should be solved only for the group of customers that have been assigned to the specific vehicle. Constraints (Eq. 3.11), (Eq. 3.12) are degree constraints specifying that each node is entering exactly once and is leaving it exactly once. Constraints (Eq. 3.13) are subtour elimination constraints.

3.3.2 Location Routing Problem

In the most location models, it is assumed that the customers are served directly from the facilities being located. Each customer is served on his or her own route. In many cases, however, customers are not served individually from the facilities. Rather, customers are consolidated into routes which may contain many customers. One of the reasons for the added difficulty in solving these problems is that there are far more decisions that need to be made by the model. These decisions include:

- how many facilities to locate,
- where the facilities should be,
- which customers to assign to which depots,
- which customers to assign to which routes,
- in what order customers should be served on each route.

In the **Location Routing Problem (LRP)**, a number of facilities are located among candidate sites and delivery routes are established to a set of users in such a way that the total system cost is minimized. As Perl and Daskin [46] pointed out, location routing problems involve three inter-related, fundamental decisions: where to locate the facilities, how to allocate customers to facilities, and how to route the vehicles to serve customers. The difference of the Location Routing Problem from the classic vehicle routing problem is that not only routing must be designed but also, the optimal depot location must be simultaneously determined. The main difference between the location routing problem and the classical location-allocation problem is that, once the facility is located, the former requires a visitation of customers through tours while the later assumes that the customer will be visited from the vehicle directly and, then, it will return to the facility without serving any other customer [42]. In general terms, the combined location routing model solves the joint problem of determining the optimal number, capacity and location of facilities serving more than one customer and finding the optimal set of vehicle routes. In the location routing problem the distribution cost is decreased due to the assignment of the customers to vehicles while the main objective is the design of the appropriate routes of the vehicles. An extended recent literature review is included in the survey paper published by Nagy and Salhi [43].

In this section, a formulation for the Location Routing Problem based on the Bilevel Programming is given [32]. Location Routing Problem consists of decisions taken at two different levels, at the strategic level and at the operational level. The strategic level is considered as the first level of the proposed bilevel formulation and the operational level is considered as the second level. There are two decisions makers that solve a different problem at each level. The decision maker of the first level (leader), namely, the decision maker that will decide where the facilities will be located, calculates the alternative solutions for the location of the facilities and the assignment of the customers in each facility. For each proposed solution, the decision maker of the second level (the follower) reacts and calculates the routing cost and the order that the vehicles will visit the customers. The leader based on the solutions of the follower's problem decides the optimal location of the facilities. In the following, the proposed formulation is given and analyzed in detail.

Let $G = (N, E)$ be an undirected graph, where $N = \{1, ..., n\}$ is the set of nodes and E is the set of edges. Each node can be used either as facility node or customer node or both. Let $C = (c_{ij})$ be a matrix of costs, distances or travel times associated with the number of edges. If $c_{ij} = c_{ji}$ for all $i, j \in N$, the matrix and the problem is said to be symmetrical, otherwise it is asymmetrical. C satisfies the triangle inequality if and only if $c_{ij} + c_{jl} \geq c_{il}$ for all $i, j, l \in N$. There can be, at most, k identical vehicles of capacity Q_k based at facility l. It is assumed that c_{ij} are nonnegative. Every customer has a nonnegative demand q_j.

For the modeling of the problem, the following sets are used:

- N is the set of the demand nodes,
- U is the set of candidate facility sites, and
- K is the set of all vehicles that can be used.

The inputs to the model are the following:

- q_j is the demand at customer node j,
- F_l is the fixed cost of locating a facility at candidate site l,
- c_{ij} is the cost of traveling between node i and node j,
- Q_k is the capacity of vehicle k, and
- QF_l is the capacity of each facility l.

Finally, the decision variables of the problem are:

$$y_j = \begin{cases} 1, & \text{if the facility is located at candidate site } l \\ 0, & \text{otherwise,} \end{cases}$$

$$z_{lk} = \begin{cases} 1, & \text{if vehicle } k \text{ operates out of a facility at candidate} \\ & \text{site } l \\ 0, & \text{otherwise,} \end{cases}$$

and

$$x_{ijk} = \begin{cases} 1, & \text{if node } i \text{ immediately precedes node } j \text{ on a route} \\ & \text{using vehicle } k \\ 0, & \text{otherwise.} \end{cases}$$

Taking into account the previous definitions, inputs and variables, the proposed bilevel formulation of the Location Routing Problem is the following:

$$\text{(leader)}\ \min \sum_{l\in U} F_l y_l + \sum_{i\in N}\sum_{j\in N}\sum_{k\in K} c_{ij}x_{ijk} \tag{3.14}$$

s.t.

$$\sum_{k\in K} Q_k z_{lk} \le QF_l y_l, \qquad \forall l\in U \tag{3.15}$$

$$z_{lk} \le y_l, \qquad \forall l\in U, \forall k\in K \tag{3.16}$$

$$\sum_{l\in U} z_{lk} \le 1, \qquad \forall k\in K \tag{3.17}$$

where

$$\text{(follower)}\ c = \min \sum_{i\in N}\sum_{j\in N}\sum_{k\in K} c_{ij}x_{ijk} \tag{3.18}$$

s.t.

$$\sum_{k\in K}\sum_{i\in N} x_{ijk} = 1, \qquad \forall j\in N \tag{3.19}$$

$$\sum_{i\in N} q_i \sum_{j\in N} x_{ijk} \le Q_k \sum_{l\in U} z_{lk}, \quad \forall k\in K \tag{3.20}$$

$$\sum_{i\in N} x_{ijk} = z_{jk}, \qquad \forall k\in K, \forall j\in N \tag{3.21}$$

$$\sum_{i\in N} x_{jik} = z_{jk}, \qquad \forall k\in K, \forall j\in N \tag{3.22}$$

$$\sum_{i\in S}\sum_{l\in S}\sum_{k\in K} x_{ilk} \le |S|-1, \quad S=\{2,3,\cdots,n\}. \tag{3.23}$$

The objective function of the leader problem (Eq. 3.14) minimizes the sum of the fixed facility location costs and the distance related routing costs. Constraints (Eq. 3.15) state that the total quantity of products that the vehicles can carry in each route must not exceed the capacity of the facility. Constraints (Eq. 3.16) state that a vehicle can be assigned to a route originating from site l only if a facility is located at site l. Constraints (Eq. 3.17) state that each vehicle can be assigned to at most one facility. The objective function of the follower (Eq. 3.18) minimizes the sum of the distance related routing costs. Constraints (Eq. 3.19) state that every customer i must be on exactly one route. Constraints (Eq. 3.20) are the vehicle capacity constraints. Constraints (Eq. 3.21) and (Eq. 3.22) state that if vehicle k is assigned to a route originating from a facility at site j, then at least one link goes into node j (Eq. 3.21) and one leaves node j (Eq. 3.22) and are used, also, as flow conservation constraints. They state that if a vehicle k enters at node j (from any node i), then it must depart from node j (to some other node i). To eliminate the possibility of subtours, constraints (Eq. 3.23) are used that are known as subtour elimination constraints.

3.4 Bilevel Particle Swarm Optimization Algorithm

3.4.1 General Description of the Bilevel Particle Swarm Optimization Algorithm (PSOBilevel)

For the solution of these problems, a Bilevel Particle Swarm Optimization Algorithm, the PSOBilevel, is proposed. In the first level of the algorithm, a Capacitated Facility Location Problem is solved for the Location Routing Problem and a Generalized Assignment Problem is solved for the Vehicle Routing Problem in order to create the initial population of individuals (solutions). The follower's problem requires the solution of a Vehicle Routing Problem (for the Location Routing Problem) and of a Traveling Salesman Problem (for the Vehicle Routing Problem) for each individual of the population. The outline of the proposed algorithm is presented in the following.

Initialization

First Level Problem

1. **Select** the number of swarms.
2. **Select** the number of particles in each swarm.
3. **Initialize** the solutions with a random way.
3. **Convert** particles' positions in continuous form.
4. **Initialize** the position and velocity of each particle.
5. **Calculate** the initial fitness function of each particle.
6. **Find** the best solution of each particle.
7. **Find** the best particle of the entire swarm.
8. Call for each particle the second level algorithm.

Second Level Problem

1. For each particle of the initial population solve a Vehicle Routing Problem (for the Location Routing Problem) and a Traveling Salesman Problem (for the Vehicle Routing Problem) using a nearest neighborhood algorithm.
2. Improve the solution of each particle using the Expanding Neighborhood Search Method.

Main Algorithm

1. Set the number of iterations equal to zero.
2. **Do while** stopping criteria are not satisfied (the maximum number of iterations has not been reached):

   First Level Problem

   2.1 **Calculate** the velocity of each particle.
   2.2 **Calculate** the new position of each particle.
   2.3 **Convert** particles' positions in integer form.

2.4 **Calculate** the new fitness function of each particle.
2.5 Call for each particle the second level algorithm.
Second Level Problem
2.5.1 For each particle of the initial population solve a Vehicle Routing Problem (for the Location Routing Problem) and a Traveling Salesman Problem (for the Vehicle Routing Problem) using a nearest neighborhood algorithm.
2.5.2 Improve the solution of each particle using the Expanding Neighborhood Search Method.
2.6 **Update** the best solution of each particle.
2.7 **Find** the best particle of the whole swarm.
2.8 **Convert** particles' positions in continuous form.

3. **Enddo**
4. Return the best particle.

In the following, a detailed description of each step of the proposed algorithm is given.

3.4.2 Particle Swarm Optimization

The Particle Swarm Optimization (PSO) algorithm was proposed by Kennedy and Eberhart [23–25] to simulate the social behavior of social organisms such as bird flocking and fish schooling. This method has been identified as very useful in many problems. The reason is that the implementation is easy and it gives good results, especially in problems with continuous variables.

Some of the advantages of this method is that:

- it has memory which is important because the information from past good solutions passes on to future generations,
- there is cooperation between particles (solutions) of the swarm because they work together to create solutions.

In this section, the proposed Particle Swarm Optimization (PSO) algorithm for the solution of the Bilevel Supply Chain Management Problems is given . In PSO algorithm, initially a set of particles is created randomly where each particle corresponds to a possible solution. Each particle has a position in the space of solutions and moves with a given velocity. One of the key issues in designing a successful PSO for the bilevel problem is to find a suitable mapping between the bilevel problems and the particles in PSO. In the first level of the problem, where the Particle Swarm Optimization method is applied, the Capacitated Facility Location Problem (for the Location Routing Problem) and the Generalized Assignment Problem (for the Vehicle Routing Problem) are solved, respectively. In these problems the representation of a solution is mapped into a binary particle where the bit 1 denotes that the corresponding location is opened for the first problem and the customer is assigned to the vehicle for the second problem and the bit 0 denotes otherwise.

The position of each individual (called particle) is represented by a d-dimensional vector in problem space $x_i = (x_{i1}, x_{i2}, ..., x_{id})$, $i = 1, 2, ..., N$ (N is the population size and n is the number of the vector's dimension), and its performance is evaluated on the predefined fitness function ($f(x_{ij})$). The velocity v_{ij} represents the changes that will be made to move the particle from one position to another. Where will move the particle depends on the dynamic interaction of its own experience and the experience of the whole swarm. There are three possible directions that a particle can follow: to follow its own path, to move towards the best position it had during the iterations ($pbest_{ij}$) or to move to the best particle's position ($gbest_j$).

In the literature, a number of different variants of the Particle Swarm Optimization have been proposed for the calculation of the velocities. In this paper it is used the one called Inertia Particle Swarm Optimization [56]:

$$v_{ij}(t+1) = wv_{ij}(t) + c_1 rand_1 (pbest_{ij} - x_{ij}(t)) + c_2 rand_2 (gbest_j - x_{ij}(t)) \quad (3.24)$$

where c_1 and c_2 are the acceleration coefficients, $rand_1$ and $rand_2$ are two random variables in the interval [0, 1]. The acceleration coefficients c_1 and c_2 control how far a particle will move in a single iteration. Low values allow particles to roam far from target regions before being tugged back, while high values result in abrupt movement towards, or past, target regions [23]. If $c_1 = c_2 = 0$ the particles are directed where their velocity indicates, if $c_1 > 0$ and $c_2 = 0$, then, each particle is influenced only by its previous moves and not from the other particles in the swarm and if $c_2 > 0$ and $c_1 = 0$ all the particles follow the best particle. Most of the time the researchers select $c_1 = c_2$. In this case, the particle is influenced equally by both factors. Also, sometimes the values of c_1 and c_2 are changed so that the influence of the two factors can vary during the iterations. In our algorithm, we use the following definition for c_1 and c_2. Let $c_{1,min}, c_{1,max}, c_{2,min}, c_{2,max}$ be the minimum and maximum values that c_1, c_2 can take, respectively, then:

$$c_1 = c_{1,min} + \frac{c_{1,max} - c_{1,min}}{iter_{max}} \times t \quad (3.25)$$

$$c_2 = c_{2,min} + \frac{c_{2,max} - c_{2,min}}{iter_{max}} \times t \quad (3.26)$$

where t is the number of current iteration and $iter_{max}$ the maximum number of the iterations. In the first iterations of the algorithm, the values of c_1 and c_2 are small and, then, they increase until they reach to their maximum values. The advantage of this definition for c_1 and c_2 is that in the first iterations there is a great freedom of movement in the particles' (solutions') space.

The inertia weight w is used to control the impact of previous histories of velocities on the current velocity. The particle adjusts its trajectory based on information about its previous best performance and the best performance of its neighbors. The inertia weight w is, also, used to control the convergence behavior of the PSO.

A number of different alternatives for the definition of w have been proposed. These alternatives vary from constant values to different ways of increasing or decreasing of w during the iterations. In this paper, in order to exploit more areas in the solution space, the inertia weight w is updated according to the following equation:

$$w = w_{max} - \frac{w_{max} - w_{min}}{iter_{max}} \times t \tag{3.27}$$

where w_{max}, w_{min} are the maximum and minimum values of inertia weight. Initially, the particles' velocities are initialized with zeros.

The basic PSO and its variants have successfully operated for continuous optimization functions. As both of the problems should have binary values we use the extension for discrete spaces proposed by Kennedy and Eberhart [24]. In this version, a particle moves in a state space restricted to zero and one on each dimension where each v_i represents the probability of bit x_i taking the value 1. Thus, the particles' trajectories are defined as the changes in the probability and v_i is a measure of individual's current probability of taking 1. If the velocity is higher, it is more likely to choose 1, and lower values favor choosing 0. A sigmoid function is applied to transform the velocity from real number space to probability space:

$$sig(v_{ij}) = \frac{1}{1 + exp(-v_{ij})} \tag{3.28}$$

The position of a particle changes using the following equation:

$$x_{id}(t+1) = \begin{cases} 1, & \text{if } rand3 < sig(v_{ij}) \\ 0, & \text{if } rand3 \geq sig(v_{ij}) \end{cases} \tag{3.29}$$

A particle's best position ($pbest_{ij}$) in a swarm is calculated from the equation:

$$pbest_{ij} = \begin{cases} x_{ij}(t+1), & \text{if } f(x_{ij}(t+1)) < f(x_{ij}(t)) \\ pbest_{ij}, & \text{otherwise} \end{cases} \tag{3.30}$$

The optimal position of the whole swarm in the Bilevel PSO at time t is calculated by the equation:

$$\begin{aligned} gbest_j \in \{pbest_{1j}, pbest_{2j}, \cdots, pbest_{Nj} | f(gbest_j)\} = \\ min\{f(pbest_{1j}), f(pbest_{2j}), \cdots, f(pbest_{Nj})\} \end{aligned} \tag{3.31}$$

In the second level of the problem the Expanding Neighborhood Search algorithm is solved for each particle (see section 3.4.3). In each iteration of the algorithm, the optimal solution of the whole swarm and the optimal solution of each particle are kept. All the solutions in the second level are represented with the path representation of the tour. The algorithm stops when a maximum number of iterations has been reached.

3.4.3 Expanding Neighborhood Search

The local search method that is used in this paper is the Expanding Neighborhood Search [29]. Expanding Neighborhood Search (ENS) is a metaheuristic algorithm [29–31, 34, 35] that can be used for the solution of a number of combinatorial optimization problems with remarkable results. The main features of this algorithm are:

- the use of the Circle Restricted Local Search Moves Strategy,
- the use of an expanding strategy, and,
- the ability of the algorithm to change between different local search strategies.

These features are explained in detail in the following.

In the Circle Restricted Local Search Moves - CRLSM strategy , the computational time is decreased significantly compared to other heuristic and metaheuristic algorithms because all the edges that are not going to improve the solution are excluded from the search procedure. This happens by restricting the search space into circles around the candidate for deletion edges. It has been observed [29, 30, 35], for example, in the 2-opt local search algorithm that there is only one possibility for a trial move to reduce the cost of a solution, i.e. when at least one new (candidate for inclusion) edge has cost less than the cost of one of the two old edges (candidate for deletion edges) and the other edge has cost less than the sum of the costs of the two old edges. Thus, in the Circle Restricted Local Search Moves strategy, for all selected local search strategies, circles are created around the end nodes of the candidate for deletion edges and only the nodes that are inside these circles are used in the process of finding a better solution.

In Expanding Neighborhood Search strategy, the size of the neighborhood is **expanded** in each iteration. In order to decrease even more the computational time and because it is more possible to find a better solution near to the end nodes of the candidate for deletion edge, the largest possible circle is not used from the beginning but the search for a better solution begins with a circle with a small radius. For example, in the 2-opt algorithm if the length of the candidate for deletion edge is equal to A, the initial circle has radius $A/2$, then, the local search strategies are applied and if the solution can not be improved inside this circle, the circle is expanding by a percentage θ (θ is determined empirically) and the procedure continues until the circle reaches the maximum possible radius which is set equal to $A+B$, where B is the length of one of the other candidate for deletion edges.

The ENS algorithm has the ability to change between different local search strategies . The idea of using a larger neighborhood to escape from a local minimum to a better one, had been proposed initially by Garfinkel and Nemhauser [15] and recently by Hansen and Mladenovic [22]. Garfinkel and Nemhauser proposed a very simple way to use a larger neighborhood . In general, if with the use of one neighborhood a local optimum was found, then a larger neighborhood is used in an attempt to escape from the local optimum. Hansen and Mladenovic proposed a more systematical method to change between different neighborhoods, called Variable Neighborhood Search .

In the Expanding Neighborhood Search, a number of local search strategies are applied inside the circle. The procedure works as follows: initially an edge of the current solution is selected (for example the edge with the worst length) and the first local search strategy is applied. If with this local search strategy a better solution is not achieved, another local search strategy is selected for the same edge. This procedure is continued until a better solution is found or all local search strategies have been used. In the first case the solution is updated, a new edge is selected and the new iteration of the Expanding Neighborhood Search strategy begins, while in the second case the circle is expanded and the local search strategies are applied in the new circle until a better solution is found or the circle reach the maximum possible radius. If the maximum possible radius has been reached, then a new candidate for deletion edge is selected.

The local search strategies for the Vehicle Routing Problem and the Location Routing Problem are distinguished between local search strategies for a single route and local search strategies for multiple routes. The local search strategies that are chosen and belong to the category of the single route interchange (strategies that try to improve the routing decisions) are the well known methods for the TSP, the 2-opt and the 3-opt [6]. In the single route interchange all the routes have been created in the initial phase of the algorithm. The local search strategies for multiple route interchange try to improve the assignment decisions. This, of course, increases the complexity of the algorithms but gives the possibility to improve even more the solution. The multiple route interchange local search strategies that are used are the 1-0 relocate, 2-0 relocate, 1-1 exchange, 2-2 exchange and crossing [59].

3.5 Results

The algorithm was implemented in Fortran 90 and was compiled using the Lahey f95 compiler on a Intel Core 2 DUO CPU T9550 at 2.66 GHz, running Suse Linux 9.1. The parameters of the proposed algorithm are selected after thorough testing. A number of different alternative values were tested and the ones selected are those that gave the best computational results concerning both the quality of the solution and the computational time needed to achieve this solution. Thus, the selected parameters are given in Table 3.1.

The algorithm for the Vehicle Routing Problem was tested on two sets of benchmark problems. The 14 benchmark problems proposed by Christofides ([8]) and the 20 large scale vehicle routing problems proposed by Golden ([20]). Each instance of the first set contains between 51 and 200 nodes including the depot. Each problem includes capacity constraints while the problems 6-10, 13 and 14 have, also, maximum route length restrictions and non zero service times. For the first ten problems, nodes are randomly located over a square, while for the remaining ones, nodes are distributed in clusters and the depot is not centered. The second set of instances contains between 200 and 483 nodes including the depot. Each problem instance includes capacity constraints while the first eight have, also, maximum

Table 3.1 Parameter Values

Parameter	Value
Number of swarms	1
Number of particles	20
Number of generations	100
c_1	2
c_2	2
w_{min}	0.01
w_{max}	0.9
Size of RCL	50

route length restrictions but with zero service times. The efficiency of the PSOBilevel algorithm is measured by the quality of the produced solutions. The quality $\omega_{PSOBilevel}$ is given in terms of the relative deviation from the best known solution ($\omega = \frac{100(c_{PSOBilevel} - c_{opt})}{c_{opt}}$, where $c_{PSOBilevel}$ denotes the cost of the solution found by PSOBilevel and c_{opt} is the cost of the best known solution).

In the first column of Tables 3.2 and 3.3 the number of nodes of each instance is presented, while in the second, third and fourth columns the most important characteristics of the instances, namely the maximum capacity of the vehicles (Cap. - column 2), the maximum tour length of each vehicle (m.t.l. - column 3) and the service time of each customer (s.t. - column 4) are presented. In the last four columns, the results of the proposed algorithm (column 5), the best known solution (BKS - column 6), the quality of the solution of the proposed algorithm ($\omega_{PSOBilevel}$ - column 7) and the CPU time need to find the solution by the proposed algorithm for each instance (column 8) are presented, respectively. It can be seen from Table 3.2 that for the first set, the algorithm has reached the best known solution in nine out of the fourteen instances. For the other five instances the quality of the solutions is between 0.01% and 0.32% and the average quality for the fourteen instances is 0.06%. For the 20 large scale vehicle routing problems (Table 3.3), the algorithm has found the best known solution in one of them, for the rest the quality is between 0.16% and 1.08% and the average quality of the solutions is 0.47%. Also, in these Tables the computational time needed (in minutes) for finding the best solution by the proposed algorithm is presented. The CPU time needed is significantly low for the first set of instances and only for two instances (instance 5 and 10) is somehow increased but still is very efficient. In the second set of instances, the problems are more complicated and, thus, the computational time is increased but is still less than 10 min in all instances. These results denote the efficiency of the proposed algorithm.

In Tables 3.4 and 3.5, a comparison of the proposed algorithm with other three algorithms from the literature is presented. In the first algorithm (VRPBilevel) [31], the Vehicle Routing Problem is formulated, also, as a bilevel programming model solving a Generalized Assignment Problem in the first level and a Traveling Salesman Problem in the second level. The algorithm that is used for the solution of the problem is a bilevel genetic algorithm following the same idea that is behind the proposed Particle Swarm Optimization algorithm that is presented in this paper. The

Table 3.2 Results of PSOBilevel in Christofides benchmark instances for the VRP

Nodes	Cap.	m.t.l.	s.t.	PSOBilevel	BKS	$\omega_{PSOBilevel}$ (%)	CPU (min)
51	160	∞	0	524.61	524.61 [55]	0.00	0.09
76	140	∞	0	835.26	835.26 [55]	0.00	0.27
101	200	∞	0	826.14	826.14 [55]	0.00	0.35
151	200	∞	0	1028.42	1028.42 [55]	0.00	1.02
200	200	∞	0	1295.38	1291.45 [55]	0.32	2.25
51	160	200	10	555.43	555.43 [55]	0.00	0.10
76	140	160	10	909.68	909.68 [55]	0.00	0.27
101	200	230	10	865.94	865.94 [55]	0.00	0.85
151	200	200	10	1165.58	1162.55 [55]	0.26	1.37
200	200	200	10	1396.05	1395.85 [55]	0.01	2.38
121	200	∞	0	1043.28	1042.11 [55]	0.11	1.20
101	200	∞	0	819.56	819.56 [55]	0.00	0.25
121	200	720	50	1544.07	1541.14 [55]	0.19	0.45
101	200	1040	90	866.37	866.37 [55]	0.00	0.35

Table 3.3 Results of PSOBilevel in the 20 benchmark Golden instances for the VRP

Nodes	Cap.	m.t.l.	s.t.	PSOBilevel	BKS	$\omega_{PSOBilevel}$ (%)	CPU (min)
240	550	650	0	5688.31	5627.54 [38]	1.08	2.01
320	700	900	0	8458.24	8444.50 [53]	0.16	2.25
400	900	1200	0	11095.21	11036.22 [54]	0.53	6.08
480	1000	1600	0	13682.48	13624.52 [49]	0.43	7.25
200	900	1800	0	6460.98	6460.98 [58]	0.00	1.18
280	900	1500	0	8457.35	8412.8 [49]	0.53	1.45
360	900	1300	0	10201.37	10181.75 [47]	0.19	2.38
440	900	1200	0	11728.27	11643.90 [53]	0.72	6.08
255	1000	∞	0	585.95	583.39 [38]	0.44	1.41
323	1000	∞	0	744.25	741.56 [38]	0.36	2.24
399	1000	∞	0	925.28	918.45 [38]	0.74	3.08
483	1000	∞	0	1115.38	1107.19 [38]	0.74	7.45
252	1000	∞	0	861.21	859.11 [38]	0.24	3.08
320	1000	∞	0	1083.25	1081.31 [38]	0.18	2.42
396	1000	∞	0	1355.28	1345.23 [38]	0.75	7.15
480	1000	∞	0	1635.18	1622.69 [38]	0.77	9.11
240	200	∞	0	710.98	707.79 [38]	0.45	2.24
300	200	∞	0	1002.32	997.52 [38]	0.48	2.15
360	200	∞	0	1370.25	1366.86 [38]	0.25	3.05
420	200	∞	0	1825.68	1820.09 [38]	0.31	5.37

second algorithm is a hybrid Particle Swarm Optimization (HybPSO) for the solution of the Vehicle Routing Problem using the classic formulation of the VRP and a hybridized version of the Particle Swarm Optimization algorithm [37]. The reason that we compare the proposed algorithm with HybPSO is that the two approaches have a number of common characteristics but also they have a lot of different characteristics. Both methods use a hybrid version of the Particle Swarm Optimization algorithm but in the proposed algorithm the problem is solved in two levels while in HybPSO the problem is solved in one level. This issue has led to a different mapping of the particles. In the proposed algorithm the PSO is used for the solution of the Generalized Assignment Problem and, thus, each solution is mapped into a binary particle where the bit 1 denotes that the corresponding customer is assigned to the vehicle and the bit 0 denotes otherwise. From the other hand in the HybPSO, the mapping of the particles corresponds to the path representation of the tour. In the two algorithms, a number of procedures are similar like the Expanding Neighborhood Search algorithm that is used as a local search phase in the HybPSO algorithm and for the solution of the second level in the proposed algorithm. The third algorithm used for the comparisons is a Hybridization version of Particle Swarm Optimization with a Genetic Algorithm (HybGENPSO) for the solution of the classic version of the Vehicle Routing Problem [36]. The representation of the solution is similar with the representation used in HybPSO. For more information about the algorithms, please see the papers [31], [37] and [36].

Table 3.4 Comparison of the proposed algorithm with other approaches in the 14 Christofides benchmark instances for the VRP

VRPBilevel		HybPSO		HybGENPSO		PSOBilevel	
cost	ω (%)	cost	ω (%)	cost	ω (%)	cost	ω (%)
524.61	0.00	524.61	0.00	524.61	0.00	524.61	0.00
835.26	0.00	835.26	0.00	835.26	0.00	835.26	0.00
826.14	0.00	826.14	0.00	826.14	0.00	826.14	0.00
1028.42	0.00	1029.54	0.11	1028.42	0.00	1028.42	0.00
1306.17	1.15	1294.13	0.22	1294.21	0.23	1295.38	0.32
555.43	0.00	555.43	0.00	555.43	0.00	555.43	0.00
909.68	0.00	909.68	0.00	909.68	0.00	909.68	0.00
865.94	0.00	868.45	0.29	865.94	0.00	865.94	0.00
1177.76	1.31	1164.35	0.15	1163.41	0.07	1165.58	0.26
1404.75	0.64	1396.18	0.02	1397.51	0.12	1396.05	0.01
1051.73	0.92	1044.03	0.18	1042.11	0.00	1043.28	0.11
825.57	0.73	819.56	0.00	819.56	0.00	819.56	0.00
1555.39	0.92	1544.18	0.20	1544.57	0.22	1544.07	0.19
875.35	1.04	866.37	0.00	866.37	0.00	866.37	0.00

The most important comparison is the comparison with the VRPBilevel algorithm as both algorithms solve the bilevel version of the problem and are the only, at least to our knowledge, algorithms that solve the bilevel version of the Vehicle Routing Problem. For the first set of instances, the VRPBilevel algorithm finds the best known results in seven out of fourteen instances while the proposed PSO-Bilevel algorithm performs better as it finds the best known results in nine out of fourteen instances. In the other instances the PSOBilevel performs better as there is no instance that the deviation from the best known solution is more than 1% but in the VRPBilevel there are three instances with the deviation from the optimum

Table 3.5 Comparison of the proposed algorithm with other approaches in the 20 Golden instances for the VRP

VRPBilevel		HybPSO		HybGENPSO		PSOBilevel	
cost	ω (%)	cost	ω (%)	cost	ω (%)	cost	ω (%)
5702.48	1.33	5695.14	1.20	5670.38	0.76	5688.31	1.08
8476.64	0.38	8461.32	0.20	8459.73	0.18	8458.24	0.16
11117.38	0.74	11098.35	0.56	11101.12	0.59	11095.21	0.53
13706.78	0.60	13695.51	0.52	13698.17	0.54	13682.48	0.43
6482.67	0.34	6462.35	0.02	6460.98	0.00	6460.98	0.00
8501.15	1.05	8461.18	0.58	8470.64	0.69	8457.35	0.53
10254.35	0.71	10202.41	0.20	10215.14	0.33	10201.37	0.19
11957.15	2.69	11715.35	0.61	11750.38	0.91	11728.27	0.72
589.12	0.98	586.29	0.50	586.87	0.60	585.95	0.44
749.15	1.02	743.57	0.27	746.56	0.67	744.25	0.36
934.24	1.72	928.49	1.09	925.52	0.77	925.28	0.74
1138.92	2.87	1118.57	1.03	1114.31	0.64	1115.38	0.74
868.80	1.13	862.35	0.38	865.19	0.71	861.21	0.24
1096.18	1.38	1088.37	0.65	1089.21	0.73	1083.25	0.18
1367.25	1.64	1352.21	0.52	1355.28	0.75	1355.28	0.75
1645.24	1.39	1632.28	0.59	1632.21	0.59	1635.18	0.77
711.07	0.46	710.87	0.44	712.18	0.62	710.98	0.45
1015.12	1.76	1002.59	0.51	1006.31	0.88	1002.32	0.48
1389.15	1.63	1368.57	0.13	1373.24	0.47	1370.25	0.25
1842.17	1.21	1826.74	0.37	1831.17	0.61	1825.68	0.31

larger than 1% and in two others the deviation is near to 1%. In the second set of instances, the proposed algorithm finds the best known solution in one instance while the VRPBilevel did not find the solution in any instance. For the other instances, the PSOBilevel performs better as the improvement in the deviation from the best known results in the solutions is between 0.01% and 2.13% with average improvement compared to the VRPBilevel in all instances equal to 0.78%. HybPSO performs equally well compared to the proposed algorithm as in the first set the proposed algorithm finds the best known solution in nine instances and the HybPSO in seven. For the other instances, in two instances the HybPSO finds better solutions from the PSOBilevel while for all the other instances the PSOBilevel finds better solutions. The average improvement of PSOBilevel compared to the HybPSO in all instances is 0.02%. For the second set of instances, the PSOBilevel performs better in fourteen instances while the HybPSO performs better in the other six instances. The average improvement of the PSOBilevel compared to the HybPSO in all instances is 0.04%. The results with the HybGENPSO is almost similar to the results of the PSOBilevel as in the first set the proposed algorithm finds the best known solution in nine instances and the HybGENPSO in ten. For the other instances, in three instances the HybGENPSO finds better solutions from the PSOBilevel while for all the others the PSOBilevel finds better solutions. The average improvement of the HybGENPSO compared to PSOBilevel is 0.01%. For the second set of instances, the PSOBilevel performs better in fourteen instances, in one the two algorithms found the same solutions while the HybGENPSO performs better in the other five instances. The average improvement of the PSOBilevel compared to HybGENPSO is 0.13%. Thus, in general, the PSOBilevel algorithm is a very competitive algorithm compared to other algorithms for the solution of the Vehicle Routing Problem.

The algorithm for the Bilevel Location Routing problem was tested on one set of benchmark problems. It should be noted that there are not many papers in the literature that analyze and test the efficiency of the algorithms proposed in the past for the solution of location routing problems. Thus, a set of instances is used based on instances that most researchers have used (see [4]). In Table 3.6, the first column shows the researcher that proposed each instance and the paper that the instance was, firstly, described. The second column shows the number of customers, the third column shows the number of facilities, the fourth column shows the vehicle capacity, the fifth column shows the solution given by the proposed PSOBilevel algorithm and the last three columns show the Best Known Solution (BKS), the quality of the solution of the proposed algorithm and the computational time needed (in minutes) for finding the best solution by PSOBilevel. The quality is given in terms of the relative deviation from the best known solution, that is $\omega = \frac{100(c_{PSOBilevel} - c_{opt})}{c_{opt}}$, where $c_{PSOBilevel}$ denotes the cost of the solution found by PSOBilevel and c_{opt} is the cost of the best known solution. From Table 3.6, it can be seen that the PSOBilevel algorithm, in three out of nineteen instances has found the best known solution. For the

Table 3.6 Results of PSOBilevel in benchmark instances for the LRP

Name of Researchers	Customers	Facilities	Vehicle Capacity	PSO Bilevel	BKS	ω (%)	CPU (min)
Christofides and Eilon [7]	50	5	160	575.6	565.6 [51]	1.77	0.06
Christofides and Eilon [7]	75	10	140	855.8	844.4 [60]	1.35	0.35
Christofides and Eilon [7]	100	10	200	862.3	833.4 [12]	3.47	0.52
Daskin [10]	88	8	9000000	368.7	355.8 [51]	3.63	1.15
Daskin [10]	150	10	8000000	44415.8	43919.9 [60]	1.13	2.28
Gaskell [16]	21	5	6000	428.7	424.9 [52]	0.89	0.02
Gaskell [16]	22	5	4500	586.8	585.1 [50]	0.29	0.09
Gaskell [16]	29	5	4500	512.1	512.1 [4]	0.00	0.15
Gaskell [16]	32	5	8000	568.5	562.2 [11]	1.12	0.19
Gaskell [16]	32	5	11000	507.3	504.3 [50]	0.59	0.21
Gaskell [16]	36	5	250	468.5	460.4 [50]	1.76	0.19
Min et al. [42]	27	5	2500	3062	3062 [4]	0.00	0.52
Min et al. [42]	134	8	850	5978.1	5709 [60]	4.71	1.18
Perl and Daskin [46]	12	2	140	204	204 [4]	0.00	0.02
Perl and Daskin [46]	55	5	120	1129.8	1112.1 [60]	1.59	0.26
Perl and Daskin [46]	85	7	160	1656.9	1622.5 [60]	2.12	0.59
Perl and Daskin [46]	318	4	25000	570489.8	557275.2 [60]	2.37	3.15
Perl and Daskin [46]	318	4	8000	706539.8	673297.7 [60]	4.94	3.21
Or [44]	117	14	150	12474.2	12290.3 [60]	1.50	1.35

rest of the instances, the quality of the solution is between 0.29% and 4.94% with average deviation from the best known solution equal to 1.75%.

In Table 3.7, a comparison of the proposed algorithm with other two algorithms of the literature is presented. In the first algorithm (LRPBilevel) [32], the Location Routing Problem is formulated, also, as a bilevel programming model solving a Capacitated Facility Location Problem in the first level and a Vehicle Routing Problem in the second level. The algorithm that is used for the solution of the problem is the bilevel genetic algorithm used for the solution of the Bilevel Vehicle Routing Problem presented in [31], modified properly for the solution of the bilevel Location Routing Problem. The second algorithm used for the comparisons is a hybrid Particle Swarm Optimization (HybPSO) algorithm for the solution of the Location Routing Problem. In this algorithm the classic formulation of the LRP is used and a hybridized version of the Particle Swarm Optimization algorithm is applied [33]. The algorithm that was used for the solution of the Location Routing Problem is the algorithm that was described previously for the solution of the Vehicle Routing Problem (presented in [37]), modified properly for the solution of the Location Routing Problem. For more information about the algorithms and how they applied in the Location Routing Problem, please see the papers [32] and [33]. As in the case of the bilevel Vehicle Routing Problem, the interesting comparison is the one with the LRPBilevel as both algorithms solve the bilevel version of the problem and are the only, at least to our knowledge, algorithms that solve the bilevel version of the Location Routing Problem. Both algorithms find the best known solution in three

Table 3.7 Comparison of the proposed algorithm with other approaches for the LRP

LRPBilevel		HybPSO		PSOBilevel	
cost	ω (%)	cost	ω (%)	cost	ω (%)
582.7	3.02	582.7	3.02	575.6	1.77
886.3	4.96	886.3	4.96	855.8	1.35
889.4	6.72	889.4	6.72	862.3	3.47
384.9	8.18	384.9	8.18	368.7	3.63
46642.7	6.20	46642.7	6.20	44415.8	1.13
432.7	1.84	432.9	1.88	428.7	0.89
587.9	0.48	588.5	0.58	586.8	0.29
512.1	0.00	512.1	0.00	512.1	0.00
570.5	1.48	570.8	1.53	568.5	1.12
510.9	1.31	511.1	1.35	507.3	0.59
470.7	2.24	470.7	2.24	468.5	1.76
3062	0.00	3062	0.00	3062	0.00
6229	9.11	6230	9.13	5978.1	4.71
204	0.00	204	0.00	204	0.00
1135.8	2.13	1135.9	2.14	1129.8	1.59
1656.9	2.12	1656.9	2.12	1656.9	2.12
580680.2	4.20	580680.2	4.20	570489.8	2.37
747619	11.04	747619	11.04	706539.8	4.94
12474.2	1.50	12474.2	1.50	12474.2	1,50

out of nineteen instances while for the other instances the PSOBilevel algorithm performs better as the improvement in the deviation from the best known results in the solutions is between 0.18% and 6.10% with average improvement of the PSOBilevel compared to LRPBilevel is equal to 1.75%. The performance of the proposed algorithm compared to HybPSO is almost the same as previously as the two algorithms find the best known solution in three instances and for the rest instances, the improvement in the deviation from the best known results in the solutions is between 0.29% and 6.10% with average improvement of PSOBilevel compared to HybPSO is equal to 1.75%.

3.6 Conclusions and Future Research

In this paper a new bilevel version of the Particle Swarm Optimization algorithm was presented and was used for the solution of two well known Supply Chain Management problems, the Vehicle Routing Problem and the Location Routing Problem. These two problems are formulated as bilevel programming problems. The results of the algorithms were very satisfactory as they found in a number of instances the best known solution and in the other instances the deviation from the optimum was not larger than 1.08% for the Vehicle Routing Problem and 4.94% for the Location Routing Problem. The algorithms were thoroughly compared with the VRPBilevel and LRPBilevel algorithms which are two genetic based algorithms that have been applied in the past for the bilevel version of the VRP and LRP problems, respectively. The results of the proposed algorithm show that the Particle Swarm Optimization method can be applied effectively in this kind of problems. Our future research will be focused on the analysis of the Location Routing Problem in more than 2 levels, on the formulation as a multilevel programming problem and on the development of solution methods for multilevel formulations of other more complicated problems in Supply Chain Management.

References

1. Ballou, R.H.: Business Logistics Management, Planning, Organizing and Controlling the Supply Chain, 4th edn. Prentice-Hall International, Inc. (1999)
2. Banks, A., Vincent, J., Anyakoha, C.: A review of particle swarm optimization. Part I: background and development. Natural Computing 6(4), 467–484 (2007)
3. Banks, A., Vincent, J., Anyakoha, C.: A review of particle swarm optimization. Part II: hybridisation, combinatorial, multicriteria and constrained optimization, and indicative applications. Natural Computing 7, 109–124 (2008)
4. Barreto, S., Ferreira, C., Paixao, J., Santos, B.S.: Using Clustering Analysis in a Capacitated Location-Routing Problem. European Journal of Operational Research 179 (3), 968–977 (2007)
5. Bodin, L., Golden, B.: Classification in vehicle routing and scheduling. Networks 11, 97–108 (1981)

6. Bodin, L., Golden, B., Assad, A., Ball, M.: The state of the art in the routing and scheduling of vehicles and crews. Computers and Operations Research 10, 63–212 (1983)
7. Christofides, N., Eilon, S.: An Algorithm for the Vehicle Dispatching Problem. Operational Research Quarterly 20, 309–318 (1969)
8. Christofides, N., Mingozzi, A., Toth, P.: The vehicle routing problem. In: Christofides, N., Mingozzi, A., Toth, P., Sandi, C. (eds.) Combinatorial Optimization. Wiley, Chichester (1979)
9. Dantzig, G.B., Ramser, J.H.: The truck dispatching problem. Management Science 6(1), 80–91 (1959)
10. Daskin, M.: Network and Discrete Location. Models, Algorithms and Applications. John Wiley and Sons, New York (1995)
11. Duhamel, C., Lacomme, P., Prins, C., Prodhon, C.: A Memetic Approach for the Capacitated Location Routing Problem. In: EU/MEeting 2008 - Troyes, France, October 23-24 (2008)
12. Duhamel, C., Lacomme, P., Prins, C., Prodhon, C.: A GRASP × ELS approach for the capacitated location-routing problem. Computers and Operations Research 37, 1912–1923 (2010)
13. Fisher, M.L., Jaikumar, R.: A generalized assignment heuristic for vehicle routing. In: Golden, B., Bodin, L. (eds.) Proceedings of the International Workshop on Current and Future Directions in the Routing and Scheduling of Vehicles and Crews, pp. 109–124. Wiley and Sons (1979)
14. Fisher, M.L.: Vehicle routing. In: Ball, M.O., Magnanti, T.L., Momma, C.L., Nemhauser, G.L. (eds.) Network Routing, Handbooks in Operations Research and Management Science, vol. 8, pp. 1–33. North Holland, Amsterdam (1995)
15. Garfinkel, R., Nemhauser, G.: Integer Programming. John Wiley and Sons, New York (1972)
16. Gaskell, T.J.: Bases for Vehicle Fleet Scheduling. Operational Research Quarterly 18, 281–295 (1967)
17. Gendreau, M., Laporte, G., Potvin, J.Y.: Vehicle routing: modern heuristics. In: Aarts, E.H.L., Lenstra, J.K. (eds.) Local search in Combinatorial Optimization, pp. 311–336. Wiley, Chichester (1997)
18. Gendreau, M., Laporte, G., Potvin, J.Y.: Metaheuristics for the Capacitated VRP. In: Toth, P., Vigo, D. (eds.) The Vehicle Routing Problem, Monographs on Discrete Mathematics and Applications, pp. 129–154. SIAM (2002)
19. Golden, B.L., Assad, A.A.: Vehicle Routing: Methods and Studies. North Holland, Amsterdam (1988)
20. Golden, B.L., Wasil, E.A., Kelly, J.P., Chao, I.M.: The impact of metaheuristics on solving the vehicle routing problem: algorithms, problem sets, and computational results. In: Crainic, T.G., Laporte, G. (eds.) Fleet Management and Logistics, pp. 33–56. Kluwer Academic Publishers, Boston (1998)
21. Golden, B.L., Raghavan, S., Wasil, E.: The Vehicle Routing Problem: Latest Advances and New Challenges. Springer LLC (2008)
22. Hansen, P., Mladenovic, N.: Variable neighborhood search: Principles and applications. European Journal of Operational Research 130, 449–467 (2001)
23. Kennedy, J., Eberhart, R.: Particle swarm optimization. In: Proceedings of 1995 IEEE International Conference on Neural Networks, vol. 4, pp. 1942–1948 (1995)
24. Kennedy, J., Eberhart, R.: A discrete binary version of the particle swarm algorithm. In: Proceedings of 1997 IEEE International Conference on Systems Man and Cybernetics, vol. 5, pp. 4104–4108 (1997)

25. Kennedy, J., Eberhart, R., Shi, Y.: Swarm Intelligence. Morgan Kaufmann Publisher, San Francisco (2001)
26. Laporte, G., Semet, F.: Classical heuristics for the capacitated VRP. In: Toth, P., Vigo, D. (eds.) The Vehicle Routing Problem, Monographs on Discrete Mathematics and Applications, pp. 109–128. SIAM (2002)
27. Laporte, G., Gendreau, M., Potvin, J.Y., Semet, F.: Classical and modern heuristics for the vehicle routing problem. International Transactions on Operations Research 7, 285–300 (2000)
28. Marinakis, Y., Migdalas, A.: Heuristic solutions of vehicle routing problems in supply chain management. In: Pardalos, P.M., Migdalas, A., Burkard, R. (eds.) Combinatorial and Global Optimization, pp. 205–236. World Scientific Publishing Co. (2002)
29. Marinakis, Y., Migdalas, A., Pardalos, P.M.: Expanding neighborhood GRASP for the traveling salesman problem. Computational Optimization Applications 32, 231–257 (2005)
30. Marinakis, Y., Migdalas, A., Pardalos, P.M.: A hybrid Genetic-GRASP algorithm using langrangean relaxation for the traveling salesman problem. Journal of Combinatorial Optimization 10, 311–326 (2005)
31. Marinakis, Y., Migdalas, A., Pardalos, P.M.: A New Bilevel Formulation for the Vehicle Routing Problem and a Solution Method Using a Genetic Algorithm. Journal of Global Optimization 38, 555–580 (2007)
32. Marinakis, Y., Marinaki, M.: A Bilevel Genetic Algorithm for a Real Life Location Routing Problem. International Journal of Logistics: Research and Applications 11(1), 49–65 (2008)
33. Marinakis, Y., Marinaki, M.: A Particle Swarm Optimization Algorithm with Path Relinking for the Location Routing Problem. Journal of Mathematical Modelling and Algorithms 7(1), 59–78 (2008)
34. Marinakis, Y., Marinaki, M., Dounias, G.: Honey bees mating optimization algorithm for the vehicle routing problem. In: Krasnogor, N., Nicosia, G., Pavone, M., Pelta, D. (eds.) Nature Inspired Cooperative Strategies for Optimization, NICSO 2007. SCI, vol. 129, pp. 139–148. Springer, Heidelberg (2008)
35. Marinakis, Y., Marinaki, M., Dounias, G.: Honey bees mating optimization algorithm for large scale vehicle routing problems. Natural Computing 9, 5–27 (2010)
36. Marinakis, Y., Marinaki, M.: A Hybrid Genetic - Particle Swarm Algorithm for the Vehicle Routing Problem. Expert Systems with Applications 37, 1446–1455 (2010)
37. Marinakis, Y., Marinaki, M., Dounias, G.: A Hybrid Particle Swarm Optimization Algorithm for the Vehicle Routing Problem. Engineering Applications of Artificial Intelligence 23, 463–472 (2010)
38. Mester, D., Braysy, O.: Active guided evolution strategies for large scale capacitated vehicle routing problems. Computers and Operations Research 34, 2964–2975 (2007)
39. Migdalas, A.: Bilevel Programming in Traffic Planning: Models. Methods and Challenge Journal of Global Optimization 7, 381–405 (1995)
40. Migdalas, A., Pardalos, P.M.: Nonlinear Bilevel Problems With Convex Second Level Problem - Heuristics and Descent Methods. In: Du, D.-Z., et al. (eds.) Operations Research and its Application, pp. 194–204. World Scientific (1995)
41. Miller, T.: Hierarchical Operations and Supply Chain Planning. Springer, London (2001)
42. Min, H., Jayaraman, V., Srivastava, R.: Combined Location-Routing Problems: A Synthesis and Future Research Directions. European Journal of Operational Research 108, 1–15 (1998)
43. Nagy, G., Salhi, S.: Location-Routing: Issues, Models and Methods. European Journal of Operational Research 177, 649–672 (2007)

44. Or, I.: Traveling Salesman-Type Combinatorial Problems and their Relation to the Logistics of Regional Blood Banking. Ph. D. Thesis, Department of Industrial Engineering and Management Sciences, Northwestern University, Evanston, IL (1976)
45. Pereira, F.B., Tavares, J.: Bio-inspired Algorithms for the Vehicle Routing Problem. SCI, vol. 161. Springer, Heideberg (2008)
46. Perl, J., Daskin, M.S.: A Warehouse Location Routing Model. Transportation Research B 19, 381–396 (1985)
47. Pisinger, D., Ropke, S.: A general heuristic for vehicle routing problems. Computers and Operations Research 34, 2403–2435 (2007)
48. Poli, R., Kennedy, J., Blackwell, T.: Particle swarm optimization. An overview. Swarm Intelligence 1, 33–57 (2007)
49. Prins, C.: A simple and effective evolutionary algorithm for the vehicle routing problem. Computers and Operations Research 31, 1985–2002 (2004)
50. Prins, C., Prodhon, C., Calvo, R.W.: Solving the capacitated location-routing problem by a GRASP complemented by a learning process and a path relinking. 4OR 4, 221–238 (2006)
51. Prins, C., Prodhon, C., Calvo, R.W.: A Memetic Algorithm with Population Management (MA|PM) for the Capacitated Location-Routing Problem. In: Gottlieb, J., Raidl, G.R. (eds.) EvoCOP 2006. LNCS, vol. 3906, pp. 183–194. Springer, Heidelberg (2006)
52. Prins, C., Prodhon, C., Ruiz, A., Soriano, P., Calvo, R.W.: Solving the Capacitated Location-Routing Problem by a Cooperative Lagrangean Relaxation-Granular Tabu Search Heuristic. Transportation Science 41(4), 470–483 (2007)
53. Prins, C.: A GRASP $\times$ Evolutionary Local Search Hybrid for the Vehicle Routing Problem. In: Pereira, F.B., Tavares, J. (eds.) Bio-inspired Algorithms for the Vehicle Routing Problem. SCI, vol. 161, pp. 35–53. Springer, Heideberg (2008)
54. Reimann, M., Doerner, K., Hartl, R.F.: D-Ants: savings based ants divide and conquer the vehicle routing problem. Computers and Operations Research 31, 563–591 (2004)
55. Rochat, Y., Taillard, E.D.: Probabilistic diversification and intensification in local search for vehicle routing. Journal of Heuristics 1, 147–167 (1995)
56. Shi, Y., Eberhart, R.: A modified particle swarm optimizer. In: Proceedings of 1998 IEEE World Congress on Computational Intelligence, pp. 69–73 (1998)
57. Tarantilis, C.D.: Solving the vehicle routing problem with adaptive memory programming methodology. Computers and Operations Research 32, 2309–2327 (2005)
58. Tarantilis, C.D., Kiranoudis, C.T.: BoneRoute: an adaptive memory-based method for effective fleet management. Annals of Operations Research 115, 227–241 (2002)
59. Toth, P., Vigo, D.: The vehicle routing problem. Monographs on Discrete Mathematics and Applications. SIAM (2002)
60. Yu, V.F., Lin, S.W., Lee, W., Ting, C.J.: A simulated annealing heuristic for the capacitated location routing problem. Computers and Industrial Engineering 58, 288–299 (2010)

Chapter 4
CoBRA: A Coevolutionary Metaheuristic for Bi-level Optimization

François Legillon, Arnaud Liefooghe, and El-Ghazali Talbi

Abstract. This article presents CoBRA, a new parallel coevolutionary algorithm for bi-level optimization. CoBRA is based on a coevolutionary scheme to solve bi-level optimization problems. It handles population-based meta-heuristics on each level, each one cooperating with the other to provide solutions for the overall problem. Moreover, in order to evaluate the relevance of CoBRA against more classical approaches, a new performance assessment methodology, based on rationality, is introduced. An experimental analysis is conducted on a bi-level distribution planning problem, where multiple manufacturing plants deliver items to depots, and where a distribution company controls several depots and distributes items from depots to retailers. The experimental results reveal significant enhancements with respect to a more classical approach, based on a hierarchical scheme.

4.1 Introduction

Bi-level optimization problems allow to model a large number of real-life applications, with a hierarchical structure between two decision makers. It includes companies which have to face a legislator and security constraints [10], companies trying to predict consumer reaction [8], or a supply chain where a company has to predict its supplier reaction to determine the real cost of its decision [3].

Metaheuristics are a class of approximate algorithms focusing on finding good-quality solutions for large-size and complex problems, in a reasonable time [20]. While most of the existing literature about bi-level optimization focuses on

François Legillon
Tasker and INRIA Lille Nord Europe, France
e-mail: francois.legillon@inria.fr

Arnaud Liefooghe · El-Ghazali Talbi
University of Lille 1, CNRS, INRIA, France
e-mail: {arnaud.liefooghe, talbi}@lifl.fr

E.-G. Talbi (Ed.): *Metaheuristics for Bi-level Optimization*, SCI 482, pp. 95–114.
DOI: 10.1007/978-3-642-37838-6_4 © Springer-Verlag Berlin Heidelberg 2013

small-size linear problems (see for example [1, 9]), many real-life applications involve large-size instances and complex NP-hard problems, justifying the use of meta-heuristics. Meta-heuristics for bi-level optimization can be divided in two main classes. On the one hand, *hierarchical algorithms* try to solve the two levels sequentially, improving solutions on each level to get a good overall solution on both levels. Such algorithms include the repairing algorithm [12] , which considers the lower-level problem as a constraint and solve it during the evaluation step, or the constructing algorithm [13] which applies two improving algorithms on a population, one for each level, sequentially until meeting a stopping criterion. On the other hand, *coevolutionary algorithms* maintain two populations, one for each level, and try to improve it separately, while exchanging periodically information to keep an overall view on the problem, like in [16]. In cooperative coevolution, different sub-populations evolve a part of the decision variables, and complete solutions are built by means of a cooperative exchange of individuals from sub-populations [18].

This article focuses on a coevolutionary approach. Sub-problems involved in bi-level optimization can be tackled by meta-heuristics. Finding a good way to combine two meta-heuristics in order to solve a bi-level optimization problem would give a general methodology for bi-level optimization. First, we introduce a new algorithm, the *Co*evolutionary *B*i-level method using *R*epeated *A*lgorithms (CoBRA) . This co-evolutionary meta-heuristic is able to face general bi-level optimization problems, possibly involving complex large-size problems. Next, we introduce a new method for performance assessment, the rationality, able to more fully grasp the bi-level aspect of the problems than the Pareto efficiency. Rationality is based on the proximity from the optimum of the lower-level variables with the corresponding upper-level variables fixed. At last, to evaluate the performance of CoBRA against classical hierarchical approaches, we give an experimental analysis on a bi-level transportation problem involving a supply chain, the bi-level multiple depot vehicle problem introduced in [3]. This analysis includes the modeling of the problem, the instantiation of CoBRA on it and the study of the results with respect to the rationality metrics.

The paper is organized as follows. Section 4.2 gives the necessary background on bi-level optimization. Section 4.3 presents the new coevolutionary algorithm proposed in the paper for bi-level optimization, namely CoBRA. In Section 4.4, we discuss the issue of assessing the performance of approximate algorithms in bi-level optimization. The bi-level transportation problem under investigation in this paper is presented in Section 4.5, both in a single-objective and a multi-objective formulation. The experimental analysis of CoBRA is given Section 4.6. At last, the final section concludes the paper and gives directions for further research.

4.2 Bi-level Optimization

In this section we introduce a general bi-level optimization problem, and give a quick overview of state-of-the-art meta-heuristics for bi-level optimization.

4.2.1 General Principles of Bi-level Optimization

Bi-level optimization problems may be defined by the tuple (S,F,f) where S represents the set of feasible solutions, F the objective function(s) of the upper-level, and f the objective function(s) of the lower-level. For any $x \in S$ we separate the upper-level variables and the lower-level variables, respectively in x_u and x_l.

We define, for every x_u fixed, the set of rational reactions $R(x_u)$ as the set of x_l optimal in f.

$$R(S,f,x_u) = \begin{cases} \min_{x_l} f(x = (x_u,x_l)) = (f_1(x), f_2(x), \dots, f_n(x)) \\ \text{s.t. } x \in S \end{cases}$$

The bi-level problem consist in finding the solution $x \in S$ which is optimal with respect to f for x_u fixed and, respecting this constraint, optimal in F.

$$BP(\mathscr{S},\mathrm{F},\mathrm{f}) = \begin{cases} \min \mathrm{F}(x) \\ x \in \mathscr{S} \\ \text{s.t.} \begin{cases} x = (x_u,x_l) \\ x_l \in R(\mathscr{S},\mathrm{f},x_u) \end{cases} \end{cases}$$

Those problems induce a hierarchy between two decision makers:

- The *leader*, who chooses the upper part of the decision variables, x_u, and who tries to optimize $F(x)$.
- The *follower*, who chooses the lower part of the decision variables, x_l, and who tries to optimize $f(x)$.

The leader decides first. Then, the follower, knowing the leader decision, has to decide, in the view of optimizing its own objective function(s) f, without regarding the upper objective function(s) F. To optimize his choice, the leader then has to predict the follower *reaction*. This hierarchy can conduct to a higher complexity than both sub-problems. For instance, a NP-hard problem can be obtained from two linear problems [2].

This definition of bi-level optimization corresponds to the optimistic case, where the leader can "choose" the (x_u,x_l) couple in the set of $(x_u,x_l) \in S$ where $x_l \in R(x_u)$: the reaction has to be optimal, but if several reactions are optima (*i.e.* $|R(x_u)| > 1$) the leader has the last word . There exists a pessimistic case [14] which is not treated in this paper, where x_l is chosen as the leader worst case scenario in the set of rational responses.

4.2.2 Meta-heuristic Approaches for Bi-level Optimization

Meta-heuristics are approximate algorithms which allow to tackle large-size problem instances by delivering satisfactory solutions in reasonable time [20]. Due to

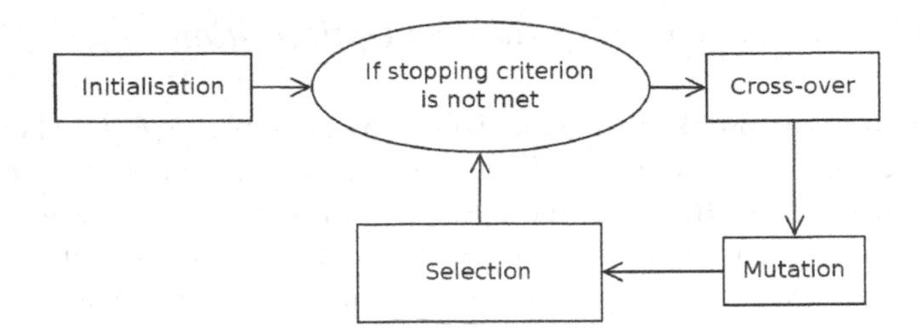

Fig. 4.1 General scheme of an evolutionary algorithm

their complexity, most bi-level optimization problems are tackled by approaches which involve a model reformulation masking the bi-level aspect of the problem (see [1, 9, 11, 15]), or involve meta-heuristics. Evolutionary algorithms are meta-heuristics mimicking the species evolution. We will use in this article several terms related to evolutionary algorithms: an *individual* is a feasible solution, a *population* is a set of individuals, a *mutation* is the creation of a new individual from an existing one, generally keeping some properties. A cross-over is the creation of individual(s), called *offspring*, from several other individuals called *parents*. The process of applying cross-over and mutation operators to a population in order to create a new population is called *generation*. On each generation, a *selection* step consists in selecting individuals to meet defined goals. Evolutionary algorithms consist in creating multiple generations and applying selections until a stopping criterion is met (Fig. 4.1). The reader is referred to [20] for more details about population-based meta-heuristics and evolutionary algorithms.

In this paper, we focus on coevolutionary approaches, a sub-group of meta-heuristics extending the evolutionary scheme. Coevolutionary algorithms consists in associating several evolutionary algorithms and applying transformations, such as mutation and cross-over, to distinct populations. A coevolution operator is then regularly applied between sub-populations to keep a global view on the whole problem. Oduguwa and Roy described BiGA [16], a coevolutionary algorithm to solve bi-level problems.

BiGA starts by initializing two distinct sub-populations using a heuristic, pop_u for the upper level and pop_l for the lower, then the upper part of the solutions is copied from pop_u to pop_l. Then during a parametrized number of generations, a selection process based on the respective level fitness values is applied on both sub-populations, followed by a mutation/crossover step. Then the sub-populations are evaluated, sorted, and coevolved, by copying the upper (resp. lower) variables to the lower (resp. upper) sub-population. At last, an archiving process occurs, before looping again to the selection step. The pseudo-code of BiGA is given in Algorithm 5.

Algorithm 5: BiGA

Data: initial population *pop*
$pop_l \leftarrow \text{selection}_{lower}(pop)$;
$pop_u \leftarrow \text{selection}_{upper}(pop)$;
Coevolution(pop_u, pop_l);
while *Stopping criterion not met* **do**
  Crossover(pop_u), crossover(pop_l);
  Mutation(pop_u), mutation(pop_l);
  Evaluation(pop_u), evaluation (pop_l);
  Elitist coevolution (pop_u, pop_l);
  Evaluation(pop_u), evaluation (pop_l);
  Archiving(pop_u), archiving (pop_l);
end
return *archive*

4.3 CoBRA, a Coevolutionary Meta-heuristic for Bi-level Optimization

In this section we introduce CoBRA, a new meta-heuristic to tackle bi-level problems.

4.3.1 General Principles

Most of literature works focus on linear bi-level problems (*ie*: formed with two linear sub-problems) or lower-level problems solvable in a reasonable amount of time. They use this property to discard the bi-level aspect of the problem. This article tries to define a more general methodology to solve bi-level optimization problems. The complexity of the considered problems lead us to consider the use of meta-heuristic, to obtain good-quality solutions in a reasonable amount of time.

We introduce a meta-heuristic, CoBRA, a *co*evolutionary *b*ilevel method using *r*epeated *a*lgorithms. Extending Oduguwa and Roy's BiGA [16], it is a coevolutionary meta-heuristic consisting in improving incrementally two different sub-populations, each one corresponding to one level, and periodically exchanging information with the other.

4.3.2 CoBRA Components

In order to instantiate CoBRA to solve a bi-level optimization problem, generic and problem-specific components have to be defined. Generic components, which can correspond to both sub-problems, consist in choosing the following:

- An improvement algorithm for each level, to improve the solutions on its level. We use, for single-objective levels, a classic evolutionary algorithm, and, for

multi-criterion levels, NSGA-II. Those algorithms are classic population-based meta-heuristic approaches [20].
- A coevolution strategy to decide how populations should exchange information.
- An archiving strategy to record the best solutions on every level, and to prevent the coevolution to change completely the sub-populations on a single generation.
- A stopping criterion to decide when the algorithm should stop.

Problem specific components still have to be designed to use CoBRA:

- Initialization operators, generally heuristics, which create a base population to begin the search process.
- Variation operators, level-specifics, which are then used by the improvements algorithms.
- Evaluation operators, corresponding to the f and F functions from the bi-level optimization model.

Figure 4.2 illustrates the outline of CoBRA.

4.3.3 General Algorithm

CoBRA is a coevolutionary algorithm using for each level a different population, and a different archive (Algo. 9.4). At each iteration, we apply the improvement algorithms, we archive the best solutions obtained, then we apply a selection operator to keep a constant size to the archive and to the populations. The final iteration step is then to coevolve the two sub-populations. Once the stopping criterion is met, CoBRA returns the lower-level archive.

Extending the BiGA approach, CoBRA involves several differences from the former:

1. The main difference is that CoBRA applies a complete algorithm, possibly iterating a certain number of generations, over each main algorithm iteration, instead of just applying variation operators. Evaluation process occurs during those improvement algorithms.

Algorithm 6: CoBRA

Data: initial population pop
$pop_u \leftarrow_{copie}$ pop;
$pop_l \leftarrow_{copie}$ pop;
while *Stopping criterion not met* **do**
  upper improvement (pop_u) and lower improvement (pop_l);
  upper archiving (pop_u) and lower archiving (pop_l);
  selection (pop_u) and selection (pop_l);
  coevolution(pop_u, pop_l);
  adding from upper archive (pop_u) and from lower archive (pop_l);
end
return *lower archive*

---

2. The coevolution process is not necessarily elitist: default coevolution strategy (Algo. 7) randomly coevolves solutions with each other.
3. The selection operations and the archives take place right after the improvement.

Algorithm 7: Random coevolution

Data: Populations *upPop* and *lowPop* of same size, *op* coevolution operator
Shuffle *upPop*;
foreach *i from 0 to size(upPop)* **do**
    op(upPop[i],lowPop[i]);

4.4 Performance Assessment and Bi-level Optimization

In this section, we introduce two new metrics for assessing the performance of heuristics on solving bi-level optimization problems.

4.4.1 Motivations

Being a problem with two different objective functions, a natural approach to tackle bi-level optimization problems would be to use a Pareto-based multi-objective approach . However bi-level optimization problems have a different structure. A good solution considering a similar problem approximating the Pareto frontier could be of bad quality in the bi-level way.

Bi-level optimization aim at identifying solutions in the form (x_u, x_l) which give good upper objective vectors, while being near the optimum regarding the lower objective for x_u fixed. This leads to the existence of good quality solutions not being on the Pareto frontier, and solutions on the Pareto frontier not necessarily being good quality solutions. Fig. 4.3 gives an example of objective functions giving a bi-level solution corresponding to a dominated solution in the Pareto sense. F and f are respectively the upper and the lower-level objective functions to be minimized, the leader chooses in {d,e,f} and the follower in {a,b}. The Pareto front would be composed of {(d,a),(f,a)} while the bi-level solution is (e,a).

We introduce the notion of *rationality* which correspond to the difficulty to improve a solution (x_u, x_l), with x_u fixed, according to the lower-level objective function. A rational solution is a solution where the follower reaction is rational, seeking for the optimality of its own objective function(s). We introduce two different rationality metrics, the direct one and the weighted one.

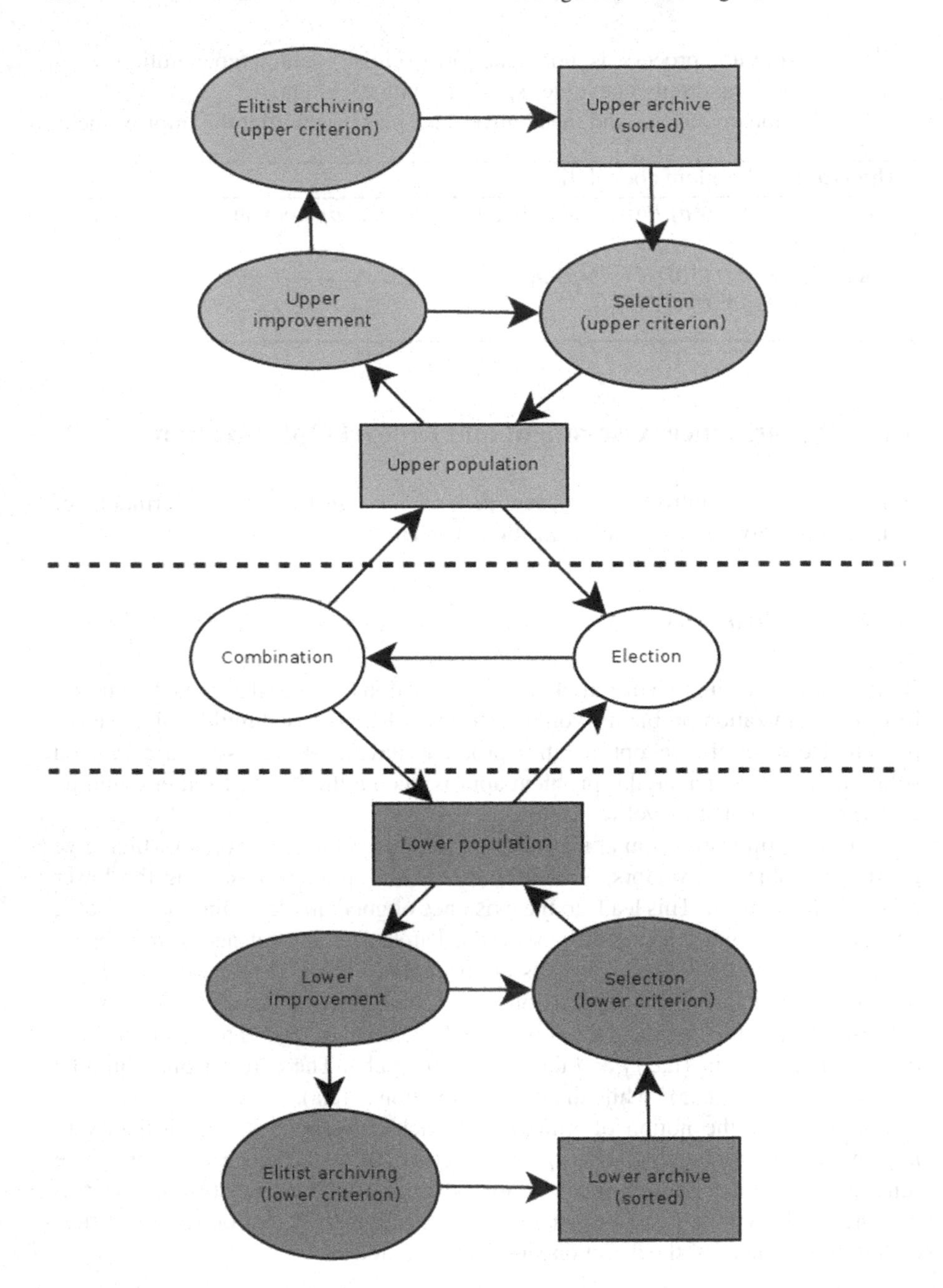

Fig. 4.2 CoBRA outline

F	a	b
d	0	1000
e	1	∞
f	300	∞

f	a	b
d	100	99
e	1001	∞
f	99	∞

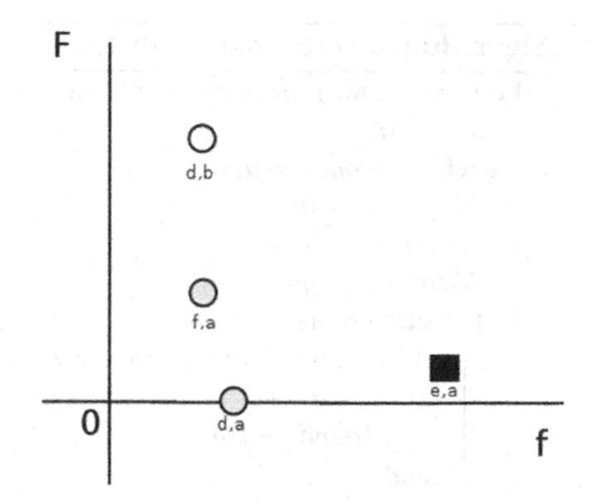

Fig. 4.3 Example of lower-level and upper-level objective functions whose optimal solution is dominated in terms of Pareto dominance

4.4.2 *Rationality*

4.4.2.1 Direct Rationality

The *direct rationality* measure corresponds to the difficulty of improving a solution without regarding the actual improvement: we simply consider the "improvability". To evaluate it for a population, we apply a parametrized number of time a "good" lower-level algorithm, and count how many times the algorithm did improve the solution (Algo. 8).

4.4.2.2 Weighted Rationality

The *weighted rationality* is another rationality measure working on the same principle as the direct rationality with the difference that, instead of counting how many times the algorithm was able to improve the solution, we also consider how much it was improved. Being able to improve a fitness by 0.001 or by 1000 does not give the same result to the rationality, whereas the direct approach would consider both as the same (Algo. 9). For bi-level optimization problems involving a multi-objective lower-level sub-problem, we used the multiplicative ε-indicator, an indicator to compare sets of objective vectors [21].

4.4.3 *Discussion*

The weighted rationality metric was introduced to compare results for a bi-level optimization problem composed with a hard lower-level problem. All the tested algorithms giving a bad direct rationality, we noticed that some algorithms were still doing better and were far nearer to the optimal on the lower-level than others. The weighted rationality is able to differentiate such algorithms.

Algorithm 8: Direct rationality test

```
Data: AlgoLow, pop, ni number of iterations
counter ← 0;
foreach gen from 1 to ni do
    neopop ← pop;
    found ← false;
    AlgoLow(neopop);
    foreach x in neopop do
        if (not found) and (x dominates an element of pop) then
            counter++;
            found ← false;
        end
    end
end
return counter/ni
```

Algorithm 9: Weighted rationality test

```
Data: AlgoLow, pop, ni number of iterations
ratio ← 0;
foreach gen from 1 to ni do
    neopop ← pop;
    AlgoLow(neopop);
    ratio=ratio+ε_ind(pop,neopop)/ni;
end
return ratio
```

We can note that those methods are not absolute, in the sense that we have to compare the algorithm using another algorithm, thus introducing a bias. Those measures compare the capacity of a meta-heuristic to use improvement algorithms, but do not actually compare the overall capacity to tackle the problem. To this end, we have to ensure that none of the tested algorithms is biased toward the improvement used by the rationality evaluation.

4.5 Application to Bi-level Transportation

In this section we define a bi-level transportation problem , involving two different companies in a supply chain: the leader transports goods from depots to retailers answering to the retailers demand, and a follower manages plants producing goods for the leader. The leader starts by deciding which depots should deliver goods, then the follower decides how to manufacture the goods, both decisions influencing the overall cost of solutions. Two variants of this problem are here considered, a single-objective one, and a multi-objective one.

4.5.1 A Bi-level Multi-depot Vehicle Routing Problem

The first problem, introduced by Calvete and Galé [3], consists of a bi-level problem where the leader controls a fleet of vehicles to deliver items from several depots to retailers, on the same principle as the classical multi-depot vehicle routing problem (MDVRP) . The follower controls a set of plants, and has to produce the items and deliver them to the depots according to the demand of the retailers it serves, thus answering a flow problem. The leader tries to minimize the total distance of his routes and the buying cost of the resources (depending on the lower-level decision). The follower minimizes the production cost and the distance traveled by the produced goods. The follower has to directly transport from plants to depots.

4.5.1.1 Problem Description

Let K, L, R and S denote the sets of plants, of depots, of retailers and of vehicles, respectively. Let E be the edge set between retailers and depots, b_r the demand of retailer r, $c^a_{i,j}$ the cost of transporting from depots or retailers i to j for the leader, $c^b_{k,l}$ the cost to buy and unload a unit produced in plant k into depot l for the leader, and $c^c_{k,l}$ the operational cost for plant k to produce and deliver to depot l for the follower.

The upper objective function is to minimize the sum of deliver costs from depots to retailers and buying from plants .

$$F(x,y) = \sum_{s \in S} \sum_{(i,j) \in E} c^a_{i,j} x^s_{i,j} + \sum_{k \in K} \sum_{l \in L} c^b_{k,l} y_{k,l}$$

with x the leader variables representing the routes chosen to deliver retailers, and y the follower variables representing the affectation of plants to depots. Then, the lower-level objective function is to minimize the sum of costs of producing items in plants and delivering it to depots.

$$f(x,y) = \sum_{k \in K} \sum_{l \in L} c^c_{k,l} y_{k,l}$$

The leader and follower follow a hierarchical order, where the leader choose routes, creating a demand for the depots corresponding to the retailers to be delivered, and where the follower has to respond to this new demand by associating a part of his plant production to depots.

$$\sum_{k \in K} y_{k,l} \geq \sum_{s \in S_l} \sum_{r \in R_s} b_r, \forall l \in L$$

Several other VRP-related constraints are omitted to improve readability. See [3] for more details about the problem.

4.5.1.2 Solution Representation

In the optic of doing an evolutionary algorithm, a solution representation was necessary. Using a generic bi-level representation, we had to decide a representation for each level. For the upper-level, we use a permutation: every retailer and every route (each route being associated to a depot) has an attributed number.

The route numbers in the permutation determine the routes start, and every retailers represent in order the actual route (Fig. 4.4). This representation facilitate the solution integrity, and suppress the need to check the number of routes and the "one visit per retailer" constraint. We use for the lower-level problem a more classical double matrix M, M_b^a representing the ratio of production sent from a to b. The quantity effectively sent is scaled down at the evaluation step if the sum of a column are over 1, and rounded down if not integer. This indirect representation permits to use classical algorithms without much adaptation work.

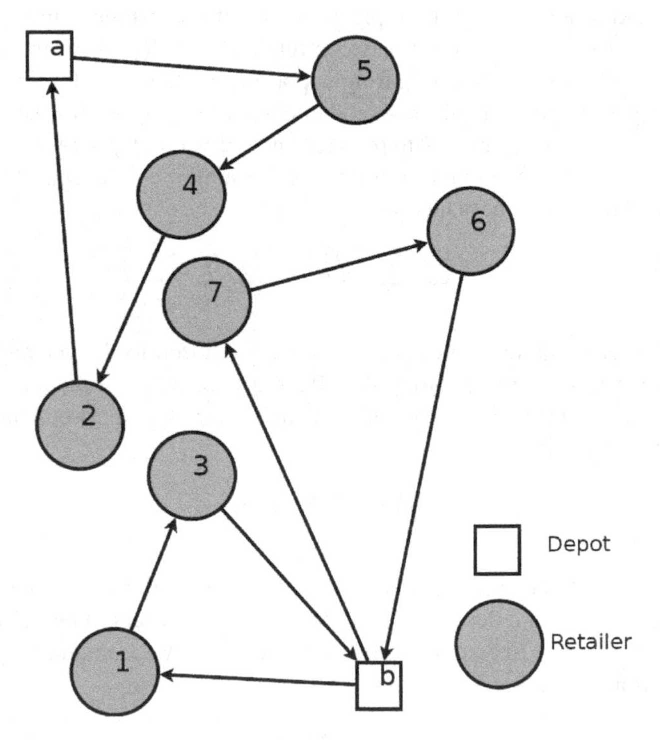

Fig. 4.4 Example of a VRP with 7 retailers, 2 depots, and 2 routes *per* depot, from the permutation $[5,4,2,9,7,6,10,1,3,8]$. Squares are for depots {a,b}, circles for retailers {1,2,3,4,5,6,7}.

4.5.1.3 Problem Instances

Two sets of instances[1] were generated to experiment the CoBRA efficiency. S_1 consist of instances created from MDVRP instances following the *modus operandi* described in [3]. We add as many plants as there are depots randomly located on the map. Then we set their maximal production to ensure that the instance is feasible. c^b and c^c follows a method described in [3]. Set S_1 contains 10 instances created from the 10 instances provided by Cordeau [4]. The second set S_2 consists of the same instances in which a higher fixed number of plants of 50 was added. Those instance parameters are described in Table 4.1.

Table 4.1 Description of S_1 and S_2 instances, R corresponding to the number of routes by depot

Instance	Depot	R	Plants (S_1)	Plants (S_2)	Retailer
bipr01	4	1	4	50	48
bipr02	4	2	4	50	96
bipr03	4	3	4	50	144
bipr04	4	4	4	50	192
bipr05	4	5	4	50	340
bipr06	4	6	4	50	288
bipr07	6	1	6	50	72
bipr08	6	2	6	50	144
bipr09	6	3	6	50	216
bipr10	6	4	6	50	288

4.5.2 A Multi-objective Bi-level Multi-depot Vehicle Routing Problem

The multi-objective bi-level multi-depot routing problem (M-BiMDVRP) is a variant of the BiMDVRP where the follower minimizes two costs instead of just one distance between plants and depots, aiming at finding a Pareto front approximation. The follower has to directly transport from plants to depots for this problem too. The lower-level objective function vector becomes:

$$f(x,y) = \left(\sum_{k \in K} \sum_{l \in L} c^c_{k,l} y_{k,l}, \sum_{k \in K} \sum_{l \in L} c^d_{k,l} y_{k,l} \right)$$

$c^d_{k,l}$ being another operational cost of plant k, to produce and delivering a unit of good to depot l, similar to c^c. While the leader still have to chose how to deliver

[1] Benchmark files are publicly available on the paradiseo website in the problems section at the following URL: `http://paradiseo.gforge.inria.fr/index.php?n=Problems.Problems`.

products from depots to retailers, the follower has to respond to a bi-objective problem , his goal being to find solutions which are Pareto efficient (see [5] for details on Pareto efficiency). We kept the same sets of instances as in BiMDVRP, to which we added the c^d cost independently generated on the same way as the c^c one.

4.6 Experimental Analysis

In order to evaluate the relevance of CoBRA for bi-level optimization, we conduct in this section an experimental analysis against a repairing algorithm, a classical approach which consider the lower-level optimality condition as a constraint, and simply try to find the best upper-level variable while "repairing" the lower-level one at the evaluation step.

4.6.1 Experimental Design

We conduct a two-part experimental analysis. In the first part, we apply the two algorithms on the bi-level multi-depot vehicle routing problem (BiMDVRP). We ran CoBRA and the repairing algorithm for BiMDVRP on S_1, and for M-BiMDVRP on S_1 and S_2. We run both of the algorithms 30 times with different seed values, since both algorithms use stochastic components.

Both algorithms use the same components (*i.e.* the improvement algorithms, the stopping criterion, the variation operators and the initializers). The reparation algorithm does not use any archiving or coevolution operator, and a different evaluation operator which apply a lower-level improvement algorithm before evaluating a solution. Once the stopping criterion is met, we evaluate the population with respect to three criteria:

- the population average upper-level fitness value,
- the direct rationality,
- the weighted rationality.

4.6.2 CoBRA instantiation for BiMDVRP an M-BiMDVRP

To use CoBRA on the BiMDVRP problem, several problem-specific components have to be chosen.

4.6.2.1 Upper-Level Problem-Related Components

For the MDVRP upper problem we use a combination of three variation operators:

RBX [19] is a cross-over operator copying routes from a parent, and then completing the offspring with routes from the other parents by removing visited retailers.

SBX [19] is a cross-over operator creating a new route, by taking half of a route starting from a single depot in each parents, keeping the order of each half, and then completing the offspring with the other routes and removing visited retailers.

Or-opt [17] is a mutation operator taking several retailers from a route and putting it in another. This operator changes the number of route which neither of the SBX and RBX can do.

Operators are applied on solutions uniformly chosen in the population.

4.6.2.2 Lower-Level Problem-Related Components

For the lower-level problem we use a combination of two operators:

UXover [6] is a crossover operator choosing elements uniformly for each parent solution matrix and putting it in the offspring.

Uniform mutation [7] is a mutation operator that add a parametrized real value $r_{lmut} \in [-0.5, 0.5]$ to each element of the solution matrix with a p_{lmut} probability.

4.6.2.3 Stopping Condition

The algorithm uses three stopping criteria, one for each improvement algorithm and one for the overall algorithm. Improvement algorithms use a generational stopping criterion which continue for a fixed number p_g of generations. The overall algorithm uses a lexical continuator which continue until no better solution is found for a fixed parameter p_l of generations, by using a lexical comparator (*i.e.* by comparing sequentially the objective values on each level).

4.6.2.4 Selection Operators

The algorithm uses three selection operators to choose which solution to keep from a generation to the next one, one for each improvement algorithm, and one for the overall algorithm. We use on both improvement algorithms a deterministic tournament, which randomly selects two solutions from the population and keep the best one. For the overall algorithm we use a survive-and-die replacement politic, which keeps a parametrized proportion of the best solutions n_{sad} from the last generation, and apply a deterministic tournament on the remaining part of the population in order to generate the next generation.

4.6.2.5 Archiving Strategy

The algorithm uses archives to keep record of the best solutions found over all generations. We define two different archive strategies depending on the number of lower-level objective function:

Single-objective lower-level strategy. We use a straight-forward archive that keeps the n best found solutions according the the level fitness value

Multi-objective lower-level strategy. The upper-level archive keeps the same strategy than in the single-objective case. The lower-level archive, at the insertion of a new individual i, starts by deleting any solution Pareto-dominated by i then inserts i if it's not dominated by any individual from the archive. If the archive size goes over n, we remove from the archive the worst elements according to the upper-level fitness values until the archive size returned under n.

4.6.2.6 Numerical Parameters

To use those components and CoBRA, the following parameters have to be set:

- n: the populations size, set to 100
- r_{lmut}: the uniform mutation adding parameter, set to 0.5
- p_{lmut}: the uniform mutation probability parameter, set to 0.1
- p_g: the number of generations each improvement generates, set to 10
- p_l: the number of generations CoBRA continues without improvement, set to 100
- n_{sad} the proportion of best solutions that are kept from the last generation, set to 0.8

4.6.3 Experimental Results

4.6.3.1 BiMDVRP

Table 4.2 shows numerical results for CoBRA and the repairing algorithm on instances from S_1. Here are displayed the average upper-level fitness value, and the best fitness value obtained in the lower-level archive, as well as the direct rationality metric value. Since direct rationality was enough to rank the algorithms, the weighted measure was not used.

CoBRA has a significantly better score for the rationality, on all the instances. For both algorithms, rationality is not related to the instance size. The repairing algorithm is doing better for the upper-level fitness value.

4.6.3.2 M-BiMDVRP

Tables 4.3 and 4.4 show the experimental results over the sets S_1 and S_2, respectively. The average and the best upper-level fitness values obtained in the lower archive, and the weighted rationality measure are given. Direct rationality did not permit to significantly decide between the coevolutionary and the hierarchical approach.

Both algorithms obtain similar upper-level fitness values. CoBRA is still having a better rationality. The rationality gap between CoBRA and the repairing algorithm increases with the instance size. The number of evaluations done by the algorithms are shown on Figure 4.5. The repairing algorithms needs a lot more evaluations, impairing the computational cost of the approach.

Table 4.2 Average upper-level fitness value, best upper-level fitness value and weighted rationality value for BiMDVRP instances from S_1

Instance	Averaged fitness		Best fitness		Direct rationality	
	CoBRA	Repair	CoBRA	Repair	CoBRA	Repair
bipr01	1883	**1848**	1676	**1626**	**0.6**	7.5
bipr02	4049	**3338**	3718	**2880**	**1.3**	5.4
bipr03	6058	**5849**	5604	**4712**	**2.7**	23.5
bipr04	**7172**	7368	6568	**6051**	**1.9**	19.9
bipr05	9750	**8535**	9493	**7179**	**0.6**	5.4
bipr06	15237	**11637**	14837	**9656**	**0.7**	5.9
bipr07	3165	**2917**	2851	**2453**	**0.9**	1.5
bipr08	7207	**5348**	6801	**4736**	**2.2**	22.9
bipr09	9825	**8326**	9343	**7042**	**2.0**	22.2
bipr10	14418	**12413**	13419	**12412**	**0.6**	13.5

Table 4.3 Average upper-level fitness value, best upper-level fitness value and weighted rationality value for M-BiMDVRP instances from S_1

Instance	Average Fitness		Best Fitness		Weighted rationality	
	CoBRA	Repair	CoBRA	Repair	CoBRA	Repair
mbipr01	**3151**	3570	**2930**	3002	**0.66**	21.50
mbipr02	**5980**	6559	**5729**	5792	**4.83**	140.50
mbipr03	**11459**	12369	**10887**	11230	**77.49**	562.76
mbipr04	**12985**	14346	**12568**	13158	**6.84**	195.84
mbipr05	**16067**	16872	**15317**	15982	**1.82**	52.89
mbipr06	**19408**	21291	**18523**	20079	**158.02**	839.89
mbipr07	**5195**	5790	4915	**4758**	**8.71**	253.39
mbipr08	**10566**	11691	**9943**	10543	**21.36**	106.70
mbipr09	**15948**	17519	**15330**	16247	**41.13**	727.62
mbipr10	**20849**	22798	**20361**	21523	**86.45**	1040.10

Table 4.4 Average upper-level fitness value, best upper-level fitness value and weighted rationality value for M-BiMDVRP instances from S_2

Instance	Averaged fitness		Best fitness		Weighted rationality	
	CoBRA	Repair	CoBRA	Repair	CoBRA	Repair
mbipr01	**3187**	3630	**2912**	3155	**16.69**	67.52
mbipr02	**6155**	6798	**5808**	6236	**25.61**	89.10
mbipr03	**11226**	12390	**10865**	11544	**31.55**	197.58
mbipr04	**13703**	14934	**13240**	14113	**27.18**	208.44
mbipr05	**15349**	16773	**14753**	16092	**111.10**	357.05
mbipr06	**19894**	21986	**19314**	21132	**45.41**	306.62
mbipr07	**5243**	5849	**4796**	5239	**18.86**	82.46
mbipr08	**10598**	11649	**10131**	10866	**15.02**	198.85
mbipr09	**15862**	17517	**15357**	16535	**21.61**	229.93
mbipr10	**20747**	22843	**20207**	22019	**39.43**	349.27

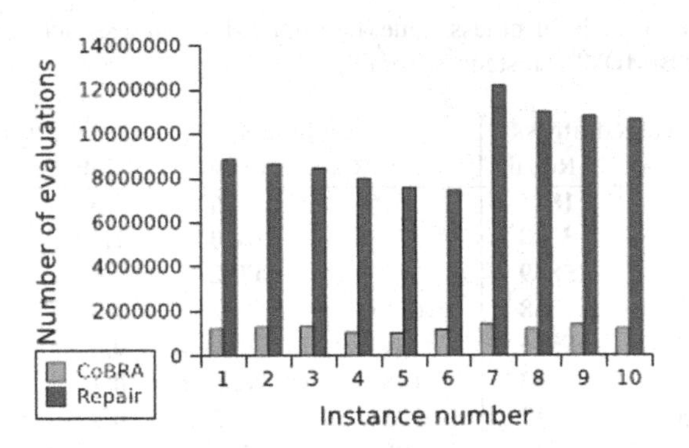

Fig. 4.5 Average number of evaluations required by CoBRA and the repairing algorithm on M-BiMDVRP instances from Set S_2

4.6.3.3 Discussion

CoBRA has a significant advantage in terms of rationality for all the runs we performed, while it does not always give a better upper-level fitness value. Rationality indicates the quality of the reaction predicted by the algorithm. A bad prediction is likely to lead to a bad solution: once applied to a real-life situation the follower will have greater chances to chose a better reaction for his own objective function(s), degrading the solution quality for the leader. Since CoBRA has a better rationality, we can better predict the outcome of the decisions. Thus we can conclude that CoBRA is more adapted to the bi-level aspect of the problem.

An explanation why the hierarchical algorithm does not select the more rational response would be that once an irrational solution $x = (x_u, x_l)$ is obtained, through a badly done reparation, which gives a better upper-level fitness value than the more rational response $x' = (x_u, x'_l)$, the overall algorithm will have a tendency to discard x' and keep x. We can conclude that the reparation approach needs either a good lower-level heuristic, an exact lower-level algorithm, or some properties over the problem (such as a strong correlation between the two levels) to be able to produce rational responses. This is the reason why the coevolution allows CoBRA to get a better rationality. We can conclude that the coevolutionary approach can give a significant enhancement for this problem.

4.7 Conclusions and Future Works

In this paper, we described *CoBRA*, a new general methodology to solve bi-level optimization problems. We introduced the concept of *rationality* for bi-level optimization problems and two new metrics to compare the performance of evolutionary algorithms for such purpose. Using those two metrics, we compared CoBRA against a classical hierarchical approach on a bi-level optimization problem of production/-transportation in its single-objective and multi-objective variants. Experimental results showed a significant advantage to the CoBRA approach in tackling the bi-level multiple depot problem against a classical hierarchical approach.

As future work, it would be interesting to look up a possible integration of diversification principles into CoBRA. Instead of considering the upper-level fitness values in the lower-level archive, it could be more efficient to keep a good diversity in the archive. This would give the opportunity for the algorithm to escape from local optima easier. Furthermore, the design of CoBRA is intrinsically parallel, since two sub-populations evolve independently,parallel computation would improve the performance in terms of computational time.

References

1. Anandalingam, G., White, D.: A solution method for the linear static stackelberg problem using penalty functions. IEEE Transactions on Automatic Control 35(10), 1170–1173 (1990)
2. Ben-Ayed, O., Blair, C.: Computational difficulties of bilevel linear programming. Operations Research 38(3), 556–560 (1990)
3. Calvete, H., Galé, C.: A Multiobjective Bilevel Program for Production-Distribution Planning in a Supply Chain. In: Multiple Criteria Decision Making for Sustainable Energy and Transportation Systems, pp. 155–165 (2010)
4. Cordeau, J., Gendreau, M., Laporte, G.: A tabu search heuristic for periodic and multi-depot vehicle routing problems. Networks 30(2), 105–119 (1997)
5. Deb, K.: Multi-Objective Optimization Using Evolutionary Algorithms. John Wiley & Sons, Inc., New York (2001)
6. Deb, K., Agrawal, R.: Simulated binary crossover for continuous search space. Complex systems 9(2), 115–148 (1995)
7. Deb, K., Goyal, M.: A combined genetic adaptive search (geneas) for engineering design. Computer Science and Informatics 26, 30–45 (1996)
8. Didi-Biha, M., Marcotte, P., Savard, G.: Path-based formulations of a bilevel toll setting problem. In: Optimization with Multivalued Mappings, pp. 29–50 (2006)
9. Eichfelder, G.: Multiobjective bilevel optimization. Mathematical Programming 123(2), 419–449 (2010)
10. Erkut, E., Gzara, F.: Solving the hazmat transport network design problem. Computers & Operations Research 35(7), 2234–2247 (2008)
11. Fliege, J., Vicente, L.: Multicriteria approach to bilevel optimization. Journal of optimization theory and applications 131(2), 209–225 (2006)
12. Koh, A.: Solving transportation bi-level programs with differential evolution. In: IEEE Congress on Evolutionary Computation, CEC 2007, pp. 2243–2250. IEEE (2008)
13. Li, X., Tian, P., Min, X.: A hierarchical particle swarm optimization for solving bilevel programming problems. In: Rutkowski, L., Tadeusiewicz, R., Zadeh, L.A., Żurada, J.M. (eds.) ICAISC 2006. LNCS (LNAI), vol. 4029, pp. 1169–1178. Springer, Heidelberg (2006)
14. Loridan, P., Morgan, J.: Weak via strong stackelberg problem: New results. Journal of Global Optimization 8, 263–287 (1996), doi:10.1007/BF00121269

15. Lv, Y., Hu, T., Wang, G., Wan, Z.: A penalty function method based on Kuhn-Tucker condition for solving linear bilevel programming. Applied Mathematics and Computation 188(1), 808–813 (2007)
16. Oduguwa, V., Roy, R.: Bi-level optimisation using genetic algorithm. In: 2002 IEEE International Conference on Artificial Intelligence Systems (ICAIS 2002), pp. 322–327 (2002)
17. Or, I.: Traveling salesman-type combinatorial problems and their relation to the logistics of regional blood banking. Northwestern University, Evanston (1976)
18. Potter, M.A., Jong, K.A.D.: Cooperative coevolution: An architecture for evolving coadapted subcomponents. Evolutionary Computation 8, 1–29 (2000)
19. Potvin, J., Bengio, S.: The vehicle routing problem with time windows part II: genetic search. INFORMS Journal on Computing 8(2), 165 (1996)
20. Talbi, E.-G.: Metaheuristics: from design to implementation. Wiley (2009)
21. Zitzler, E., Thiele, L., Laumanns, M., Fonseca, C.M., Grunert da Fonseca, V.: Performance assessment of multiobjective optimizers: An analysis and review. IEEE Transactions on Evolutionary Computation 7(2), 117–132 (2003)

Chapter 5
A Matheuristic for Leader-Follower Games Involving Facility Location-Protection-Interdiction Decisions

Deniz Aksen and Necati Aras

Abstract. The topic of this chapter is the application of a matheuristic to the leader-follower type of games—also called static Stackelberg games—that occur in the context of discrete location theory. The players of the game are a system planner (the leader) and an attacker (the follower). The decisions of the former are related to locating/relocating facilities as well as protecting some of those to provide service. The attacker, on the other hand, is interested in destroying (interdicting) facilities to cause the maximal possible disruption in service provision or accessibility. The motivation in the presented models is to identify the facilities that are most likely to be targeted by the attacker, and to devise a protection plan to minimize the resulting disruption on coverage as well as median type supply/demand or service networks. Stackelberg games can be formulated as a bilevel programming problem where the upper and the lower level problems with conflicting objectives belong to the leader and the follower, respectively. In this chapter, we first discuss the state of the art of the existing literature on both facility and network interdiction problems. Secondly, we present two fixed-charge facility location-protection-interdiction models applicable to coverage and median-type service network design problems. Out of these two, we focus on the latter model which also involves initial capacity planning and post-attack capacity expansion decisions on behalf of the leader. For this bilevel model, we develop a matheuristic which searches the solution space of the upper level problem according to tabu search principles, and resorts to a CPLEX-based exact solution technique to tackle the lower level problem. In addition, we also demonstrate the computational efficiency of using a hash function, which helps to uniquely identify and record all the solutions visited, thereby avoids cycling altogether throughout the tabu search iterations.

Deniz Aksen
College of Admin. Sciences and Economics, Koç University, İstanbul, Turkey
e-mail: dakseN@ku.edu.tr

Necati Aras
Dept. of Industrial Engineering, Boğaziçi University, İstanbul, Turkey
e-mail: arasn@boun.edu.tr

E.-G. Talbi (Ed.): *Metaheuristics for Bi-level Optimization*, SCI 482, pp. 115–151.
DOI: 10.1007/978-3-642-37838-6_5 © Springer-Verlag Berlin Heidelberg 2013

5.1 Introduction and Background

5.1.1 Man-Made Attacks as a Source of Disruption

The malicious acts of terror syndicates around the world have been threatening the public and governments in the last 20 years. The massive and well-prepared attacks in New York 1993 and 2001, in İstanbul and Morocco 2003, in Madrid 2004, and in London, Bali and Sharm el-Sheikh 2005 point to the increased sophistication in the planning of such terrorist deeds . This emerging trend is highlighted also in Perl's report of 2006 [41] presented to the U.S. Congress. The Internet age amenities help modern-day terrorists to collect the necessary information for tactical planning prior to any attack. Thanks to the availability of information in the public domain, criminals can utilize their resources in the best possible way to inflict maximum harm. In response to this trend, governments and security officials have been fostering the protection of critical facilities against such attacks. The protection of hard targets, however, heightened the attractiveness of soft targets for terrorist attacks [14]. Taliban's attack on the telecommunication towers in Afghanistan [42] is a significant example, where the attacker's primary aim was disrupting service rather than killing people. Another example is the assault on an ambulance station in Northern Ireland [8] where the station was shut down after the attack causing a major disruption in emergency services. As can be seen in these examples, critical infrastructure comprises certain physical assets of a system the loss of which leads to significant disruption in the system's operational and functional capabilities. Examples are given in [16] as transportation linkages such as bridges, viaducts, tunnels and railroads; public service facilities; government buildings; power plants; critical stockpiles; key personnel; and even national landmarks whose loss would severely impact the morale of the public.

5.1.2 Preliminary Interdiction Models

Security planning and protection of critical infrastructure has been motivating the OR community to develop a wide variety of interdiction models in the last years. The terms interdiction and attack are used interchangeably in the literature. They both refer to the deliberate act of attempting to destroy or damage one or more components of an infrastructure or service system in order to impair its overall performance. In a recent encyclopedia article, Smith [50] argues that interdiction models and algorithms can effectively identify critical components in a complex network without resorting to exhaustively enumerating worst-case scenarios, and can be coupled with fortification models to assess the benefits of protecting a network. In this regard, the identification of vulnerabilities is realized from the perspective of the attacker to anticipate the extent of the maximal possible or worst-case disruption in service provision. This analysis pinpoints the particular facilities that would be targeted by the attacker in the worst-case scenario. Once they have been identified, the system planner can devise a protection plan to minimize the worst-case disruption.

The scope of interdiction models can be divided into two major lines: network interdiction and facility interdiction. In the latter, the target of disruption by the attacker is facility or supply nodes offering some service to a set of customer nodes. Arcs providing the linkage in between the two entities are not considered. However, the former was studied much earlier and more extensively in the literature. In this line, the seminal work belongs to Wollmer [55]. He described the ever first network interdiction model where the objective is to minimize the maximum possible flow between a pair of source and sink nodes by the removal of a certain number of arcs from the graph. Complete and partial interdiction of arc capacities in maximum flow networks under budget and cardinality constraints was later studied in-depth by Wood [56]. Attacks with probabilistic outcome were introduced into network interdiction models for the first time by Cormican et al. [18]. The objective of their model is the minimization of the expected maximum flow through a graph where the attacker's arc interdictions follow a Bernoulli process with success probability $(1-p)$.

The first models in the line of facility interdiction are found in [17], where the authors deal with disruptions both in coverage and median type supply/demand networks. They formulated two models from an attacker's viewpoint given that there are p existing facilities serving the customers. In the r-interdiction median problem (RIM), the objective is to maximize the demand-weighted total distance by attacking r out of p facilities where the customers of the disrupted facilities have to be reassigned to undamaged facilities to get service. In the r-interdiction covering problem (RIC), the goal of the attacker is to determine a subset of r facilities among the set of p existing ones, which if destroyed will yield the greatest reduction in covered customer demand. It is not difficult to see that RIM and RIC are the antitheses of the well-known p-median and maximal covering problems, respectively. Church et al. [17] presented also an exhaustive survey of interdiction models published before 2004.

5.1.3 Recent Network and Power Grid Interdiction Models

The shortest path network interdiction model based on the unsolicited elongation of arc lengths was first introduced by Fulkerson and Harding [20]. Later, Israeli and Wood [29] adopted this model to analyze the impact of complete destruction, i.e., removal of arcs from the shortest path graph. Lim and Smith [33] studied the multi-commodity flow network version where arc capacities can be reduced either partially or completely. The attacker's objective is to minimize the maximum profit that can be obtained from shipping commodities across the network. The case of asymmetric arc length information held by the evader and the attacker of a shortest path network was discussed in Bayrak and Bailey [7]. Finally, Khachiyan et al. [31] revisited Israeli and Wood's shortest path interdiction model, and looked into two subcases: *total limited interdiction*, where a fixed number k of arcs can be removed from the network, and *node-wise limited interdiction*, where for each node v of the network a fixed number $k(v)$ of outgoing arcs can be removed. In all five papers, the attacker is assumed to have some budget for interdicting arcs, and each arc is associated with a positive interdiction expense.

The stochastic version of the shortest path interdiction model has been studied from the attacker's perspective by Hemmecke et al. [28], Held et al. [27], and Held and Woodruff [26]. The objective in the stochastic version is to maximize the expected minimum distance between a pair of source and sink on a network with uncertain characteristics. A variation of this model is the case where the attacker's objective is to maximize the probability that the shortest path length exceeds a particular threshold value.

On the front of node-based connectivity and flow interdiction, one can cite Murray et al.'s paper [38] among recent works. Their paper presents an optimization approach for identifying connectivity and/or flow interdiction bounds with respect to the origin-destination pairs of a network. Royset and Wood [43] proposed a deterministic, single level and bi-objective arc interdiction model in maximum flow networks. In this model, the interdictor seeks to destroy some arcs beyond repair so as to minimize both the total interdiction cost and maximum flow.

Real-world applications of bilevel network interdiction models have been lately presented by Arroyo and Galiana [6], Motto et al. [37], and Salmerón et al. [44]. In these papers, an electric power grid involving different vulnerable components such as generators, transmission lines, transformers and substations is threatened by a disruptive agent, e.g., a group of terrorists. The disruptive agent in the upper level problem tries to maximize the costs arising by power generation and intentionally-engineered power outage (load shedding). In turn, the grid operator in the lower level problem tries to minimize either this or a different total cost function by taking corrective actions while meeting power requests of consumers.

5.1.4 Protection-Interdiction Models

Governments cannot afford to safeguard all critical assets of a country so as to make them 100 percent immune. Murray et al. [38] point to fiscal constraints that may limit the scope of applicable protective measures. Yet, protecting or fortifying a subset of an infrastructure or service system against disruptive action can be a viable alternative to the complete redesign of that system to minimize the aftermath of interdiction. Scaparra and Church [47] count the addition of security/guards, perimeter fencing, surveillance cameras, and strengthened telecommunications among possible measures of protection against an interdictor. Protection can reduce the probability of the loss of a facility or network component (critical asset), or can prevent it altogether. A broad range of models investigating this issue within the context of network and facility interdiction can be found in Snyder et al. [53]. The ensuing protection-interdiction (or fortification-interdiction) models are used to identify the critical assets whose protection sustains the post-attack system functionality as much as possible. The amount of protection is explicitly restricted either through a budget constraint or by the cardinality of assets that can be hardened against interdiction.

The presence of two players in these models, namely a system planner and an attacker, gives rise to a bilevel programming structure. Another commonly used name for bilevel programs is leader-follower or static Stackelberg games . A bilevel programming (BP) problem is a special case of multilevel optimization with two levels or two parties, one of whom takes the leader's position, and the other one is the follower making his or her plan based on the leader's decision. In a bilevel protection-interdiction problem, protection and location or network design decisions are modeled in the upper level problem, whereas the lower level interdiction model is used to assess the impact of the most damaging disruption for a given protection and location or network design strategy. This modeling prototype can be found in the following articles which are listed in chronological order:

- O'Hanley et al. [39] dealt with the problem of budget-constrained protection of critical ecological sites . There is a conservation planner who reserves (protects) ecological sites against a hypothetical adversary who destroys a subset of unprotected (nonreserve) sites. The authors developed a mixed-integer bilevel programming model based on the RIC of Church et al. [17] to minimize the maximum species losses following the worst-case loss of a restricted subset of nonreserve sites.
- Church and Scaparra [16] incorporated the protection of facilities into the RIM model to obtain the interdiction median problem with fortification (IMF).
- Scaparra and Church [45] proposed another formulation for the IMF referred to as the maximal covering problem with precedence constraints (MCPC).
- Scaparra and Church [46] solved a BP formulation of the r-interdiction median problem with fortification (RIMF) based on an implicit enumeration algorithm performed on a search tree.
- Smith and Lim [51] reviewed the ongoing research in the area of three-stage network interdiction problems in which the network operator fortifies the network by increasing capacities, reducing flow costs, or defending network elements before the interdictor takes action.
- Scaparra and Church [47] studied another version of the RIMF with capacitated facilities as a trilevel programming problem .
- Aksen et al. [2] solved the budget-constrained protection of facilities with expandable capacities in a median type supply/demand network.
- Losada et al. [34] dealt with the partial protection of uncapacitated facilities with nonzero post-attack recovery times in a median type service network over a multi-period planning horizon.
- Liberatore et al. [32] studied the stochastic RIMF.
- Cappanera and Scaparra [13] solved a multi-level interdiction problem to determine the optimal allocation of protective resources in a shortest path network so as to maximize its robustness to an attacker who is capable of ruining unprotected arcs and nodes of that network.

5.1.5 Location-Interdiction and Network Design-Interdiction Models

In this type of interdiction models, the network or system designer (the defender) assumes the role of the leader in the static Stackelberg game, and decides on the structure of the underlying system which may be a network infrastructure or a median or coverage-type supply/demand system. The attacker—in the role of the follower—makes his plan in order to inflict as much damage as possible to this system. Research on this type of interdiction models is quite new and rare. To be cited on the network design-interdiction front is the paper by Smith et al. [52], which considers budget constrained arc construction and partial arc interdiction in a multicommodity flow network. The authors built on the work of Lim and Smith [33] by incorporating an additional design layer into the linear BP problem. The resulting model is a three-level defender-attacker game. The defender first constructs a network each arc of which has a fixed construction cost, a maximum capacity, and a per-unit flow profit. Next, the attacker interdicts a set of arcs, and in the third level the defender determines the set of flows through the surviving network to maximize her post-interdiction profit. The objective of the upper level problem includes a weighted combination of flow profits before and after interdiction minus arc construction costs. The distinguishing merit of Smith et al.'s model is that the defender's and attacker's objective functions are not identical. The former includes a weighted combination of net profits before and after interdiction, while the latter minimizes the net profits after interdiction only.

In the category of shortest path networks , the first network design-interdiction problem was defined in Berman and Gavious [11], where the leader and the follower of the underlying Stackelberg game are the State and the terrorist, respectively. The State determines K sites of emergency response facilities on a shortest path network comprising multiple cities, and decides the total amount of protective resources outside the network. This amount dictates the probability that a terrorist attack on a city will succeed. The higher the resources, the lower an attack's success probability. The terrorist is assumed to have exact information about the response facility sites chosen by the State. Following the terrorist strike on a city, resources are sent from the closest response facility over the shortest path in the network between that facility and the attacked city. The objective of the terrorist is to maximize the loss which is represented by the product of the delay in sending the required resources and the expected damage at the attacked city. The State's objective, on the other hand, is the minimization of the sum of this loss and the total cost of opening response facilities and installing protective resources. The problem is solved first for the case of single response facility ($K = 1$), then for multiple facilities. In a recent follow-up paper by Berman et al. [12], the assumption of the terrorist being perfectly knowledgeable about the response facility sites is relaxed. This leads to a simultaneous move game between the two players for which Nash equilibria can be found numerically. Due to the intractability of the resulting game-theoretical problem , solutions are found only for the case of $K = 1$.

On the front of facility location-interdiction , we are aware of only two published papers. The first one is due to O'Hanley and Church [40], which considers a maximal coverage type supply/demand system. The defender of this system has to decide on the locations of at most p facilities by choosing from a set of candidate sites which are all exposed to disruptive actions by an intelligent attacker. This bilevel model is inspired by the maximal covering problem (MCP) , first introduced by Church and ReVelle [15]. The authors couple MCP with RIC, which means that facility locations are determined while anticipating their effect on the most disruptive interdiction pattern of the attacker. This way, a pre-attack layout of facility sites can be obtained that is more robust to worst-case losses inflicted by the attacker. The resulting problem is called the maximal covering location-interdiction problem (MCLIP) , and formulated as a bilevel mixed-integer program (BMIP). A bilevel decomposition based algorithm is applied to the BMIP, where two separate problems, i.e. the upper level master problem and the lower level subproblem derived by decoupling the original BMIP are solved sequentially.

The second paper dealing with a facility location-interdiction problem was written by Berman et al. [10]. The authors studied a defensive p-median maximal covering problem with a single link (arc) interdiction. The attacker deliberately severs one of the network links to reduce the coverage of customer nodes as much as possible. The defender, on the other hand, tries to maximize coverage following the destruction of that link. Rather than using a bilevel programming framework, the leader's and follower's problems are solved separately with three heuristics.

5.1.6 The Triple Problem of Facility Location-Protection-Interdiction

A defender-attacker game revolving around a facility location-interdiction problem that also involves the defender's facility protection decisions is found in Aksen et al. [3] for a p-median problem, in Aksen and Aras [1] for a fixed-charge facility location problem, and in Keçici et al. [30] for a maximal coverage type service network with fixed-charge facilities. In the first two papers, the defender has to plan also for the initial capacity acquisition and post-attack capacity expansion at the facilities as necessary in order to accommodate all customer demands in the event of the worst-case interdiction by the attacker. We can say that the defender's facility location decisions bring about an additional location layer to the protection-interdiction problem in Aksen et al. [2]. In all three papers, both protection and interdiction decisions are implemented on an all-or-none basis. This means that neither partial protection nor partial interdiction is possible. While [3] and [30] put a budget limit on the sum of facility protection expenditures, in [1] it is added directly to the objective function of the defender in the upper level problem.

We may treat the generic facility location-protection-interdiction problem as a service network design problem (SND). In this chapter, we revisit the models of Keçici et al. [30] and Aksen and Aras [1] that are proposed to capture the triple SND of facility location-protection-interdiction in a BP framework. We then develop an efficient tabu search based matheuristic with hashing (TSH) which is capable of producing high-quality solutions to this SND both on median and coverage-type service networks. We show the results of extensive computations performed with TSH on a set of randomly generated median-type networks with capacity acquisition and expansion costs. For future research we suggest several extensions of this SND for which TSH can still prove an effective method to solve the upper level problem.

The remainder of the chapter is organized as follows. In Sect. 7.3 we give two location-protection-interdiction models formulated as bilevel integer programs. The first one is for a coverage-type network, while the second one is for a median-type network. The solution approach, which is a matheuristic based on tabu search, is outlined in detail in Sect. 7.4. Section 7.5 includes the computational results on randomly generated test instances. Finally, Sect. 7.6 concludes the chapter.

5.2 Two Service Network Design Models

5.2.1 A Location-Protection-Interdiction Model for Coverage-Type Networks

In this problem setting it is assumed that there exist $|J_1|$ operational facilities providing service to customers that are located at $|I|$ different zones with demand d_i. Since an attacker aims to minimize the service coverage by interdicting r unprotected facilities, the system planner's objective is to maximize the post-attack service coverage by taking one or more of the following actions: relocating existing facilities, opening new facilities at some of the $|J_2|$ candidate sites, and protecting some of the existing and new facilities to render them invulnerable to any attack. A different cost is incurred associated with each of these actions: g_{ij} is the cost of relocating an existing facility from site $j \in J_1$ to site $k \in J_2$, f_j is the cost of opening a new facility at site $j \in J_2$, and h_j is the protection cost of an existing or new facility at site j. The amount of the total cost cannot exceed a predetermined level of budget given as b.

The service coverage can be modeled using the *gradual coverage* concept introduced in [9], where the coverage of a customer zone is full when the distance between the customer zone and a facility is smaller than a lower limit R_1, and it decreases gradually to zero when the distance becomes larger than an upper limit $R_2 > R_1$. As pointed out in [9], the gradual coverage idea might be a good approximation in some real-life applications as opposed to the "all or nothing" property of the classical maximal covering model developed in [15]. According to this model, the coverage is full at a distance smaller than or equal to a threshold value R, and suddenly drops and remains at zero beyond this value. In the present model, a linear decay function between R_1 and R_2 is adopted, where the level of partial service coverage a_{ij} is given as follows when the distance between customer zone i and the facility at site j is equal to c_{ij}.

$$a_{ij} = \begin{cases} 1 & \text{if } c_{ij} \leq R_1 \\ \frac{R_2 - c_{ij}}{R_2 - R_1} & \text{if } R_1 < c_{ij} \leq R_2 \\ 0 & \text{if } c_{ij} > R_2 \end{cases}$$

It is assumed that when a customer zone is partially covered by multiple facilities, the final coverage can be obtained by taking the maximum of the partial coverage value of each facility. This implies that the resulting coverage is achieved by the closest facility to a customer zone when R_1 and R_2 are the same for each facility.

Since the system planner makes his decisions with the anticipation of the attacker's decision about the destruction of some of the facilities, the situation can be formulated as a BP model. The upper level problem (ULP) corresponds to the system planner who is the leader of the game, while the lower level problem (LLP) belongs to the attacker who is the follower. We assume that the system planner has complete information about the attacker's problem, i.e., the objective function and the constraints of the LLP. Similarly, the attacker is assumed to be fully knowledgeable about the protection status of the facilities so that he will not waste his resources by attacking protected facilities.

Before presenting the BP model with binary variables at both levels, we give the notation used in the formulation.

Index Sets:

I = set of customer zones,
J_1 = set of existing facility sites,
J_2 = set of candidate facility sites,
$J = J_1 \bigcup J_2$ = set of existing and candidate facility sites.

Parameters:

d_i = demand of customer zone $i \in I$,
f_j = fixed cost of opening a new facility at site $j \in J_2$,
g_{jk} = cost of relocating an existing facility from site $j \in J_1$ to site $k \in J_2$,
h_j = fixed cost of protecting a facility at site $j \in J$,
b = available budget of the system planner,
r = the maximum number of facilities the attacker can interdict,
a_{ij} = partial coverage of customer zone $i \in I$ by the facility at site $j \in J$.

Decision Variables:

$$Y_j = \begin{cases} 1 & \text{if there is a facility at site } j \in J, \\ 0 & \text{otherwise.} \end{cases}$$

$$X_{jk} = \begin{cases} 1 & \text{if an existing facility at site } j \in J_1 \text{ is relocated to site } k \in J_2, \\ 0 & \text{otherwise.} \end{cases}$$

$$S_j = \begin{cases} 1 \text{ if facility at site } j \in J \text{ is destroyed by the attacker,} \\ 0 \text{ otherwise.} \end{cases}$$

$$P_j = \begin{cases} 1 \text{ if facility at site } j \in J \text{ is protected,} \\ 0 \text{ otherwise.} \end{cases}$$

V_i = fraction of customer zone i's demand served after the attack.

Note that if there is an existing facility at site $j \in J_1$, or a new facility is opened at site $j \in J_2$ or an existing facility is relocated to site $j \in J_2$, then $Y_j = 1$. The BP model corresponding to the coverage-type location-protection-interdiction problem described above is given as follows:

$$\max_{\mathbf{X},\mathbf{Y},\mathbf{P}} \; Z_{\text{sys}} = \sum_{i \in I} d_i V_i \tag{5.1}$$

s.t.

$$(1 - Y_j) = \sum_{k \in J_2} X_{jk} \qquad j \in J_1 \tag{5.2}$$

$$\sum_{j \in J_1} X_{jk} \le Y_k \qquad k \in J_2 \tag{5.3}$$

$$\sum_{j \in J_2} f_j (Y_j - \sum_{k \in J_1} X_{kj}) + \sum_{j \in J_1} \sum_{k \in J_2} g_{jk} X_{jk} + \sum_{j \in J} h_j P_j \le b \tag{5.4}$$

$$P_j \le Y_j \qquad j \in J \tag{5.5}$$

$$Y_j, P_j \in \{0,1\} \qquad j \in J \tag{5.6}$$

$$X_{jk} \in \{0,1\} \qquad j \in J_1, k \in J_2 \tag{5.7}$$

where **X**, **Y**, and **P** solve:

$$\min_{\mathbf{S},\mathbf{V}} \; Z_{\text{att}} = \sum_{i \in I} d_i V_i \tag{5.8}$$

s.t.

$$a_{ij}(Y_j - S_j) \le V_i \qquad i \in I, j \in J \tag{5.9}$$

$$S_j \le Y_j - P_j \qquad j \in J \tag{5.10}$$

$$\sum_{j \in J} S_j \le r \tag{5.11}$$

$$0 \le V_i \le 1 \qquad i \in I \tag{5.12}$$

$$S_j \in \{0,1\} \qquad j \in J \tag{5.13}$$

In the above formulation (5.1)–(5.7) represent the system planner's ULP, while (5.8)–(5.13) constitute the attackers LLP. In the ULP, the objective function (5.1) is the maximization of the total post-attack service coverage. Constraints (5.2) and (5.3) establish the relationship between location variables Y_j and relocation variables X_{jk}. In particular, the first set of constraints guarantee that if a facility is relocated from site $j \in J_1$ to any other site $k \in J_2$, then the location variable Y_j must be equal to zero. The second set of constraints state that if a facility is relocated to site $k \in J_2$, then $Y_k = 1$ must hold true. Note that these constraints are valid in the case that no facility is relocated to site $k \in J_2$ (i.e., the left-hand side is zero) and a new facility is opened at site $k \in J_2$ (i.e., the right-hand side is one). Constraint (5.4) makes sure that the total expenditure of the system planner does not exceed the available budget. The first term on the left-hand side of this inequality represents the cost of new facility openings, while the second and third terms constitute the cost of the relocations and protections, respectively. Constraints (5.5) suggest that a facility cannot be protected if it is not opened. Binary restrictions are imposed on location variables Y_j and protection variables P_j in constraints (5.6), and on relocation variables X_{jk} in constraints (5.7).

In the LLP, the objective function (5.8) is the same as that of the system planner, but the sense of optimization is opposite. Namely, the attacker aims at reducing the total post-attack service coverage as much as possible. Constraints (5.9) ensure the coverage of each customer zone by the closest open facility which is not interdicted by the attacker. Note that if customer zone i can be served from multiple facilities, the final coverage represented by variable V_i is set to the maximum of the partial coverage due to each facility. In other words, this customer zone is served from the closest facility. Constraints (5.10) are logical conditions, which prevent the attacker from interdicting facilities either not existing at all or opened/relocated but also protected by the system planner. Constraint (5.11) states that the attacker can interdict at most r facilities. Constraints (5.12) ensure that coverage variables V_i are between zero and one, and constraints (5.13) are binary restrictions on interdiction variables S_j.

There are $2|J| + |J_1||J_2|$ binary variables and $2|J| + 1$ constraints in the ULP of the BP model defined in (5.1)–(5.13). Its LLP contains $|I|$ continuous and $|J|$ binary variables and $(|I| + 1)|J| + |I| + 1$ constraints.

5.2.2 A Location-Protection-Interdiction Model for Median-Type Networks

In this subsection, we consider a unified bilevel fixed-charge location problem called BFCLP. The problem is defined on a median-type fixed-charge service network. We introduce a BP model for this problem. In the upper level (leader's problem), a system planner is the decision maker who decides about the following issues: which facilities should be opened, which of them should be protected, and what should their initial capacity levels be so as to minimize the sum of the costs incurred before and after the interdiction attempt of an attacker. Here, the number of facilities to be opened is a decision variable as is the case in the well-known uncapacitated facility location problem. Each customer goes to the nearest opened facility, which cannot deny service. Thus, the system planner has to assure the necessary capacity acquisition at each opened facility to meet the total demand of customers who will choose that facility for service. We assume that a protected facility becomes immune to any attack, hence cannot be interdicted by the attacker. As opposed to earlier facility protection models IMF [16] and RIMF [45] which use a cardinality constraint on the number of protected facilities and BCRIMF-CE [2] which uses a budget limit on the total cost of protection, BFCLP explicitly includes the cost of protection in the system planner's objective function.

As mentioned earlier, the system planner's objective includes costs that are realized both before the interdiction (BI) and after the interdiction (AI). BI costs include the following components.

BI-1: The total fixed cost of opening facilities where each facility may have a different fixed cost.
BI-2: The total fixed cost of protecting the opened facilities where each facility may have a different protection cost.
BI-3: The sum of capacity acquisition costs at the opened facilities, which may vary from one facility to another depending on the unit capacity acquisition cost and the total customer demand met.
BI-4: The sum of demand-weighted traveling costs from customer locations to the respective nearest facilities opened before interdiction.

Given the decisions of the system planner in the upper level, the attacker in the lower level (follower's problem) chooses at most r facilities to destroy. As is the case with the previous model, the attacker has perfect information about the protection status of the facilities. This means that he never hits a protected facility. As a consequence of the attack on a facility, all its customers have to be reallocated to the nearest non-interdicted facilities since no customer must be left out even after the attack. This, in turn, leads to a necessary expansion at the facilities that are still operational. In our model the capacity expansion cost is incurred at a unit rate, which may be different for each facility. In summary, AI costs consist of the following components.

AI-1: The sum of capacity expansion costs incurred by the non-interdicted facilities due to the reallocation of the customer demand originally satisfied by the interdicted facilities.
AI-2: The sum of extra demand-weighted traveling costs arising due to the reallocation of the customers from the facilities severed by the attacker.
AI-3: The sum of post-attack demand-weighted traveling costs between all customers and their respective nearest facilities surviving the attack.

Note that AI-2 is included in AI-3. The system planner pursues the minimization of the sum of BI-1 through BI-4, AI-1 and AI-2. BI costs should be included in his objective function Z_{sys} since they constitute the setup and operating costs in an undisrupted service network. The probability of disruption due to malicious acts may be fairly small. Thus, ignoring BI costs could lead to a facility location plan which turns out to be very cost-ineffective in the absence of interdiction. Another issue is the inclusion of AI-2 instead of AI-3 in Z_{sys}. Its rationale is to avoid the double-counting of accessibility costs for those customers whose post-attack facility assignments do not differ from their pre-attack assignments.

The attacker's objective is to maximize the total of AI-1 and AI-3. His objective function Z_{att} should include AI-1 because capacity expansion due to customer reallocation is a direct outcome of facility interdiction. Not only is it highly expensive in many real situations, but also impractical due to spatial restrictions or even undoable due to lack of personnel or equipment. The inclusion of AI-1 in Z_{att} indicates the attacker's intent to cause this sort of inconvenience in the system. Finally, AI-3 should be added to Z_{att} as well, since the attacker tries to maximize the accessibility costs for all customers, not just for those who are reallocated to another facility in the wake of an interdiction.

Index Sets:

I = set of customer zones
J = set of candidate facility sites

Parameters:

c_{ij} = shortest distance between customer node $i \in I$ and facility site $j \in J$,
α = customers' traveling cost per unit distance per unit demand,
d_i = demand at customer node $i \in I$,
f_j = fixed cost of opening a facility at site $j \in J$,
h_j = fixed cost of protecting a facility at site $j \in J$,
e_j = unit capacity acquisition / expansion cost for the facility at site $j \in J$,
r = the maximum number of facilities the attacker can interdict.

Decision Variables:

$$Y_j = \begin{cases} 1 \text{ if there is a facility opened at site } j \in J, \\ 0 \text{ otherwise.} \end{cases}$$

$$P_j = \begin{cases} 1 \text{ if the facility at site } j \in J \text{ is protected,} \\ 0 \text{ otherwise.} \end{cases}$$

$$S_j = \begin{cases} 1 \text{ if facility at site } j \in J \text{ is destroyed by the attacker,} \\ 0 \text{ otherwise.} \end{cases}$$

$$B_{ij} = \begin{cases} 1 \text{ if customer } i \in I \text{ is assigned to the facility at site } j \in J \text{ before the attack,} \\ 0 \text{ otherwise.} \end{cases}$$

$$A_{ij} = \begin{cases} 1 \text{ if customer } i \in I \text{ is assigned to the facility at site } j \in J \text{ after the attack,} \\ 0 \text{ otherwise.} \end{cases}$$

We will also use an auxiliary set called $\mathscr{F}_{ij}$ which is defined as the subset of candidate facility sites that are at least as close as site j is to customer i, i.e., $\mathscr{F}_{ij} = \{k \in J | c_{ik} \leq c_{ij}\}$. The mathematical model of the BFCLP is given as follows.

$$\min_{\mathbf{B},\mathbf{P},\mathbf{Y}} \; Z_{\text{sys}} = \sum_{j\in J} f_j Y_j + \sum_{j\in J} h_j P_j + \sum_{i\in I}\sum_{j\in J}(e_j + \alpha c_{ij}) d_i B_{ij}$$
$$+ \sum_{i\in I}\sum_{j\in J}(e_j + \alpha c_{ij}) d_i (1 - B_{ij}) A_{ij} \qquad (5.14)$$

s.t.

$$\sum_{j\in J} B_{ij} = 1 \qquad i \in I \qquad (5.15)$$
$$\sum_{i\in I} B_{ij} \le n Y_j \qquad j \in J \qquad (5.16)$$
$$\sum_{k\in \mathscr{F}_{ij}} B_{ik} \ge Y_j \qquad i \in I, j \in J \qquad (5.17)$$
$$P_j \le Y_j \qquad j \in J \qquad (5.18)$$
$$B_{ij}, P_j, Y_j \in \{0,1\} \qquad i \in I, j \in J \qquad (5.19)$$

where **B**, **P**, and **Y** solve:

$$\max_{\mathbf{S},\mathbf{A}} \; Z_{\text{att}} = \sum_{i\in I}\sum_{j\in J} e_j d_i (1 - B_{ij}) A_{ij} + \sum_{i\in I}\sum_{j\in J} d_i \alpha c_{ij} A_{ij} \qquad (5.20)$$

s.t.

$$\sum_{j\in J} A_{ij} = 1 \qquad i \in I \qquad (5.21)$$
$$\sum_{j\in J} S_j \le r \qquad (5.22)$$
$$S_j \le Y_j - P_j \qquad j \in J \qquad (5.23)$$
$$\sum_{i\in I} A_{ij} \le n Y_j (1 - S_j) \qquad j \in J \qquad (5.24)$$
$$\sum_{k\notin \mathscr{F}_{ij}} A_{ik} \le 1 + S_j - Y_j \qquad i \in I, j \in J \qquad (5.25)$$
$$A_{ij} \ge B_{ij}(1 - S_j) \qquad j \in J \qquad (5.26)$$
$$S_j, A_{ij} \in \{0,1\} \qquad i \in I, j \in J \qquad (5.27)$$

In the BFCLP formulation above, (5.14)–(5.19) represent the upper level problem and (5.20)–(5.27) correspond to the LLP of BFCLP. Expression (5.14) shows the objective function Z_{sys} of the system planner. The first component of this objective is the fixed cost of opening facilities (BI-1). The second component is the facility protection cost (BI-2). The third component is equivalent to (BI-3+BI-4). We do not have to account for BI-3 if $e_j = e$ for all $j \in J$. The fourth objective component in (1) represents (AI-1+AI-2). The multiplication of the post-attack assignment variable A_{ij} with $(1 - B_{ij})$ ensures that the system planner will incur no additional accessibility and capacity expansion costs for those customers who keep going to the same closest facilities as before the attack. Constraints (5.15) require that each customer be assigned to exactly one facility before the interdiction. Constraints (5.16) prohibit illegal assignment of customers to a facility that is not opened. The set of constraints (5.17) are closest assignment (CA) constraints, which—together with the constraints (5.16) and binary decision variables Y_j—explicitly enforce the assignment of each and every customer to its closest facility put in service. For any pair of customer i and site j, if no facility is opened at j ($Y_j = 0$), then the constraint has no effect. If a facility is opened at site j ($Y_j = 1$), then customer i will go to the facility at site j or to another opened facility at the same or shorter distance than j. Constraints (5.18) suggest that a facility cannot be protected if it is not opened. Binary constraints on the decision variables shown in (5.19) conclude the system planner's problem.

The values of the variables B_{ij}, Y_j, and P_j obtained in the ULP are parsed as input parameters to the lower level problem. The objective function Z_{att} in (5.20) is equal to (AI-1+AI-3). The first constraint of the LLP given in (5.21) requires that each customer be assigned to exactly one facility after the attack. The next constraint in (5.22) states that the attacker can interdict at most r facilities. Constraints (5.23) are logical conditions, which prevent the attacker from interdicting facilities either not opened at all or opened, but also protected by the system planner. These inequalities provide at the same time a linkage between the upper and lower level problems. Their right-hand side value can never be negative due to the upper level constraints (5.18). Thus, the interdiction variables S_j are never forced to become less than zero. Constraints (5.24) ensure the logic that no customer is assigned to an interdicted facility or to a facility not opened earlier by the system planner. The right-hand side of the inequality in (5.24) is a linear expression for a given value of Y_j. So, there is no need to linearize the multiplication of $Y_j(1 - S_j)$.

Constraints (5.25) enforce the post-attack assignment of customers to the closest non-interdicted facilities. Their working mechanism is as follows. First, observe that given a pair of customer i and site j, the summation on the left-hand side of (5.25) is over all other sites k that are farther from i than j. Now if the facility at j is lost ($S_j = 1$), then (5.25) has no effect. If facility j is not opened by the system planner, hence cannot be hit by the attacker either ($Y_j = S_j = 0$), then (5.25) is again ineffective. However, if facility j is opened but not hit during the attack ($Y_j = 1, S_j = 0$), then the right-hand side of (5.25) equals zero. With the facility j being available in this case, (5.25) now prevents customer i from being assigned to any facility site k farther than j. The reader is referred to Aksen et al. [2] for further discussion of the CA constraints used in the formulation of BFCLP.

In constraints (5.26) we use the values of the binary decision variables B_{ij} as constants imported from the upper level solution. Therefore, constraints (5.26) like (5.24) are actually linear constraints. They state that a customer who has been assigned to the closest of the opened facilities in the upper level problem should still stay with the same facility in the lower level problem unless it is lost due to interdiction. In other words, if facility j is not lost ($S_j = 0$) and if customer i used to go to j before the attack ($B_{ij} = 1$), then he/she will go again to j after the attack ($A_{ij} = 1$).

This constraint is crucial in the presence of two or more equidistant closest facilities around some customer node. If neglected, the attacker's optimal solution unnecessarily reassigns such a customer from her pre-attack facility to another facility at the same distance in order to maximize the objective component AI-1 without violating the CA constraint in (5.25).

Finally, binary constraints on the decision variables S_j and A_{ij} are provided in (5.27) to conclude the lower level problem. One may ask why constraint (5.22) is written as an inequality rather than equality. Its main reason is to account for the trivial situation when $\sum_{j\in J} Y_j < r$ (the number of opened facilities is below r), or when the system planner decides to protect more than $(\sum_{j\in J} Y_j - r)$ facilities against interdiction.

Note that BFCLP has $2|I||J|+3|J|$ binary decision variables where $|I||J|+2|J|$ variables are in the ULP and $|I||J|+|J|$ variables are in the LLP. It has $2|I||J|+5|J|+2|I|+1$ constraints $|I||J|+2|J|+|I|$ of which are in the ULP and the remaining $|I||J|+3|J|+|I|+1$ ones in the LLP.

5.3 A Tabu Search Based Matheuristic for the BFCLP

In their working paper, Scaparra and Church [47] stated that the introduction of capacity constraints within a protection-interdiction model greatly increases the complexity of the problem and requires the formulation of a three-level program. For our unified facility location-protection-interdiction problem BFCLP we proposed in the previous section a pure integer linear bilevel programming formulation which has binary variables in both the upper and lower level problems. Moore and Bard [36] showed that mixed-integer bilevel programming problems (MIBPPs) are *NP*-hard. The authors observed that—unlike the single-level mixed-integer problems—when the integrality constraints of a given MIBPP are relaxed, the solution of the relaxed problem does not provide a valid lower bound on the global optimum of the MIBPP even if that solution is integral. They developed a branch-and-bound type of enumerative solution algorithm to tackle the MIBPP, and experimented with different versions of their algorithm on a test bed of 50 randomly generated MIBPP instances. The number of integer variables in either the leader's or the follower's problem is at most 10, which admits the restricted applicability of Moore and Bard's algorithm.

More recently, Gümüş and Floudas [24] introduced a comprehensive methodology to solve a variety of mixed-integer nonlinear bilevel programming problems including the pure integer linear BPP. It capitalizes on a novel deterministic global optimization framework combined with a reformulation/ linearization technique (RLT) originally developed by Sherali and Adams [48] and further extended by Sherali et al. [49] to exploit special structures found in linear mixed-integer 0-1 problems. RLT transforms the mixed-integer inner problem constraint set of the given BPP into the continuous domain. Provided that the inner problem is bounded, the transformation results in a polytope with all vertices defined by binary values.

This polytope is actually equivalent to the convex hull of all integer feasible solutions of the respective inner problem. RLT converts the inner problem to a linear programming problem with respect to inner variables. The inner problem can be then replaced with the set of equations that define its necessary and sufficient Karush-Kuhn-Tucker (KKT) optimality conditions . In this way, the original BPP is reduced into a single level optimization problem to be solved to global optimality. This final job is accomplished by Cplex provided that the problem is a mixed-integer linear problem. If there are continuous nonlinear terms in the problem, but all integer variables are binary, linear and separable, the so-called global optimization procedure SMIN-αBB can be used. Otherwise, another procedure called GMIN-αBB solves the single level optimization problem to optimality. Both procedures are discussed in detail in the monograph of Floudas [19].

The literature is not so rich in heuristic methods proposed for MIBPPs. A recent example of such methods is due to Hecheng and Yuping [25], where an exponential-distribution based genetic algorithm (GA) has been proposed for two classes of MIBPPs: one is where the follower's objective function and constraints are separable with respect to the follower's variables, and the other is where they are convex when the follower's variables are not restricted to integers. The GA essentially fixes the leader's decision variables, and obtains a proven optimal solution to the linear relaxation of the follower's mixed-integer problem by using a simplified branch-and-bound method. The algorithm proceeds according to the principles of genetic search until it converges to the best possible values of the leader's decision variables.

No matter how effective the methodology of Gümüş and Floudas [24] is, it may be impractical even for a 5-facility and 50-customer instance of the BFCLP, since such a modest instance would require as many as 515 binary variables and 626 constraints. The use of efficient heuristic methods to obtain quality solutions to the BFCLP in a reasonable computational time becomes inevitable as the problem size increases towards realistic values. Motivated by this fact, we first propose a tabu search based matheuristic called TSH (*T*abu *S*earch with *H*ashing) . It involves the search for the best facility locations using tabu search principles. When the locations of the facilities are fixed, i.e., when a solution of the system planner's problem (ULP) is provided, the attacker's problem (LLP) can be solved to optimality conditional on the protection plan of the system planner. It yields an interdiction scheme that is in the best interest of the attacker. The post-attack customer-facility assignments in that solution are used to calculate the system planner's objective value in the present BFCLP instance. We utilize the commercial mixed-integer programming solver Cplex 11.2 to solve the LLP to optimality. Cplex is a contemporary benchmark in the field of optimization software. We can classify TSH as a matheuristic due to the involvement of Cplex as an exact solution technique in it. Matheuristics are optimization algorithms which consist of a metaheuristic merged with mathematical programming methodologies . The reader is referred to an edited volume by Maniezzo et al. [35] dedicated to the topic of matheuristics.

5.3.1 *Background of Tabu Search*

Tabu search (TS) is a celebrated metaheuristic algorithm that has been widely applied to many difficult combinatorial optimization problems known in the literature or encountered in real life. An in-depth TS study can be found in Glover and Laguna [22], among others. The latest advances in this methodology alongside its key aspects have been presented in Glover et al. [23]. We acknowledge an excellent short tutorial on the fundamental concepts of TS by Gendreau [21]. There have been numerous successful TS applications for the p-median and fixed charge facility location problems and their extensions (see, for example, Sun [54]; Aras and Aksen [4]; Aras et al. [5]).

5.3.2 *Key Features of the Tabu Search Algorithm TSH*

5.3.2.1 Initial Solution of TSH

If $r < |J|$, we randomly select $p = \max\{r+1, [|\log_2|J||]$ facilities and open them in the unprotected mode. The operator $[|\cdot|]$ in this formula rounds off its argument to the nearest integer number. Should $|J|$ be smaller than $(r+1)$, we open all of the candidate facilities in the protected mode. This initial solution is guaranteed to be feasible regardless of how big r is. If $r < |J|$, the attacker's problem for the initial solution can be solved by Cplex as a RIM problem since none of the p facilities will be protected in that situation.

5.3.2.2 Neighborhood Structure

TSH capitalizes on a large-scale neighborhood structure comprising five types of moves. The advantage of large-scale neighborhood structures is that they allow a more thorough search of the solution space. The types of moves executed in each TSH iteration are explained below.

1-Drop: One of the open facilities will be closed.
1-Add: A facility either in the protected or unprotected mode will be opened at one of the candidate sites without a facility yet.
1-Flip: The protection status of an open facility will be switched on or off.
1-Swap-Int: Two facilities opened in opposite protection modes will swap their modes.
1-Swap-Ext: A facility will be opened at one of the candidate sites without a facility yet, and one of the currently open facilities will be closed. The new facility will be opened either in the protected or unprotected mode.

Let p and π respectively denote the numbers of opened and protected facilities in the current solution. The number of available candidate sites is then $(m-p)$, and the number of opened, but unprotected facilities equals $(p-\pi)$. Using this extra notation, we give in Table 5.1 the properties of the move types in TSH. The fourth column of the table shows the maximum number of neighborhood solutions that can be generated for the current solution σ_t using the respective move type. This number for the 1-Swap-Ext move is divided further by a granularity coefficient called ratio of neighborhood size *RNS* ($RNS \geq 1$), which shrinks the actual search space explored. If $p > 1$ and $RNS > 1$ as well, then less-than-possible 1-Swap-Ext moves are performed on σ_t. The selection of which ones to perform is random. Yet we guarantee that in each TSH iteration as many distinct moves as the applying maximum neighborhood size will be executed during the exploration of the 1-Swap-Ext neighborhood.

Table 5.1 Properties of the move types used in TSH

Move Types	Affects p	Affects π	Max. Neigh. Size
1-Add	Yes	Possible	$2(\lvert J\rvert - p)$
1-Drop	Yes	Possible	p
1-Flip	No	Yes	p
1-Swap-Int	No	No	$\pi(p-\pi)$
1-Swap-Ext	No	Possible	$\begin{cases} \lceil 2p(\lvert J\rvert - p)/RNS \rceil & \text{if } p > 1, \\ 2(\lvert J\rvert - 1) & \text{otherwise.} \end{cases}$

Throughout the TSH iterations, we do not allow any move to create an infeasible solution for the BFCLP. An infeasible solution is where the facility location-protection plan of the system planner is too weak to secure the sustainability of the public service system in the wake of the worst-case attacks. To put it in other words, a facility location-protection plan of the system planner is infeasible if $p \leq r$ and $\pi = 0$. In response to such a plan, the attacker would interdict all facilities opened, thereby paralyze the whole system.

5.3.2.3 Solution Representation and Hash Values

We represent any solution to the upper level problem in (5.14)–(5.19) as a unique string made up of ternary digits (*trits*) 0, 1, and 2. The trits 0, 1, 2 correspond to a closed (not opened), an opened (but not protected), and a protected facility, respectively. Figure 7.1 illustrates a sample solution string for $|J| = 10$ where the opened facilities are 1, 3, 5, and 8 among which 5 and 8 have been protected. The ternary string of a given solution can be treated as a ternary number, which converts to a unique decimal number as shown in Fig. 7.1. The ternary-to-decimal conversion is a "one-to-one" and "onto" mapping between all feasible or infeasible facility location-protection plans of the system planner and positive integer numbers in the range of $[0, 3^{|J|} - 1]$. Let σ be the ternary string representation of a particular upper level solution in which the status of the jth facility is indicated by $trit_j$ for

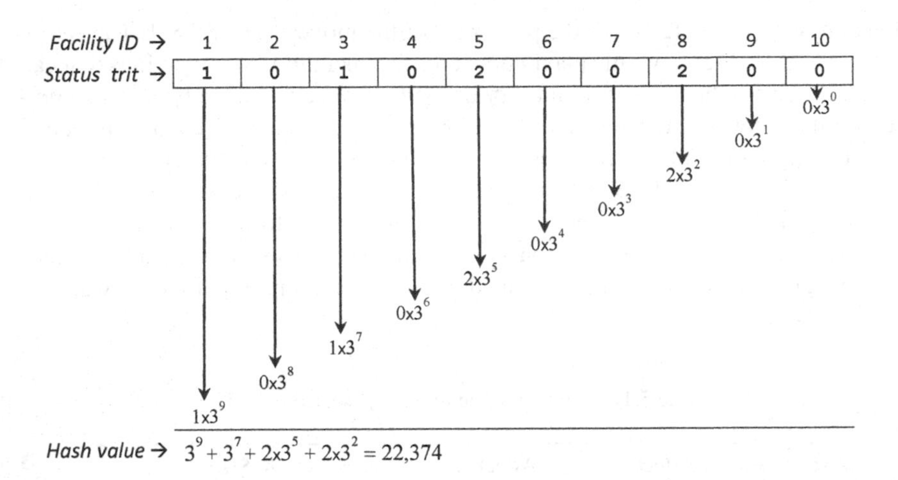

Fig. 5.1 The unique ternary string representation and hash value of a sample BFCLP solution σ

$j = 1, \ldots, |J|$. The unique decimal integer number associated with this solution is referred to as hash value and is given by the following hash function:

$$Hash(\sigma) = \sum_{j=1}^{|J|} trit_j \times 3^{|J|-j} \tag{5.28}$$

The function $Hash(\sigma)$ in (5.28) takes in a facility location-protection plan (a solution vector) and returns a unique integer to be stored in a hash list. Hash lists reduce the memory requirements and the computational burden associated with checking whether a newly constructed solution is already stored in the explicit memory. The hash value of a new neighborhood solution ($hash_{\text{neigh}}$) can be computed in $O(1)$ time if we know the current solution's hash value ($hash_{\text{curr}}$). Table 5.2 lists the formulae of this computation for each move type of the neighborhood structure in TSH.

Table 5.2 Computation of the hash value of a neighborhood solution in TSH

Move Type	ID of Fac.1 added or flipped	ID of Fac.2 dropped or flipped	Formula of $hash_{\text{neigh}}$ given the current solution σ and its hash value $hash_{\text{curr}}$
1-Add	j	–	$(hash_{\text{curr}} + trit_j \times 3^{\lvert J\rvert-k})$
1-Drop	–	k	$(hash_{\text{curr}} - trit_k \times 3^{\lvert J\rvert-j})$
1-Flip	j	–	$(hash_{\text{curr}} + 3^{\lvert J\rvert-j})$ if $trit_j$ was 1 in σ, $(hash_{\text{curr}} - 3^{\lvert J\rvert-j})$ otherwise.
1-Swap-Int	j	k	$(hash_{\text{curr}} + 3^{\lvert J\rvert-j} - 3^{\lvert J\rvert-k})$ where $trit_j$ was 2 in σ
1-Swap-Ext	j	k	$(hash_{\text{curr}} + trit_j \times 3^{\lvert J\rvert-j} - trit_k \times 3^{\lvert J\rvert-k})$

5.3.2.4 Prevention of Cycling by Means of a Hash List

Cycling can be defined as the endless or exclusive execution of the same sequence of moves until a stopping condition is satisfied, which results in revisiting the same set of solutions in the algorithmic loop. A search algorithm should be prevented from falling back to a recently visited solution; otherwise it can never escape local optima (Woodruff and Zemel [57]). In tabu search this is typically ensured by the use of a tabu list that records the recent history of the search. The tabu principle states that if a particular move/move attribute/solution attribute was previously placed in the tabu list, i.e., declared tabu, and if it is still tabu active, then it cannot be reversed or undone. Tabus are useful to help the search move away from previously visited portions of the search space, hence perform more extensive exploration (Gendreau [21]).

The simplest scheme for mitigating cycling is to keep a certain number of most recently visited solutions in an explicit tabu list, and forbid them at the current iteration. Actually, cycling is completely avoided by storing in the tabu list all the previous solutions accepted as the then-current solutions. Woodruff and Zemel [57] have argued that this approach can be computationally very expensive. Instead of recording complete solutions, they have proposed to compute the hash values of newly found solutions by using a hash function, and put them in a hash list. A new hash value can be compared with the previously recorded values to avoid cycling. Hash values and hash lists nullify the need to set tabu restrictions and tabu list sizes, which are otherwise indispensable features of an attributive memory.

5.3.2.5 Minimization of the Number of Cplex Calls

Woodruff and Zemel [57] have listed the following three goals to be met by an effective pair of a hash function and hash list:

1. Computation and update of the hash values should be as easy as possible.
2. The integers generated should be in a range that requires reasonable storage and comparison effort.
3. The probability of collision (also known as hashing error) should be low. A collision occurs when the hash function returns the same hash value for two different solutions.

The hash function we use for TSH, namely $Hash(\sigma)$ in (5.28), and the sorted doubly linked *HashList* illustrated in Fig. 7.2 meet these goals. The update of the hash value is accomplished in $O(1)$ time as described in Table 5.2. The maximum possible value attained by $Hash(\sigma)$ is $(3^{|J|}-1)$, which can be stored in the computer memory as a 17-digit precision number of type `long double`. This way, the probability of collision drops to zero as long as $|J| \leq 34$ since $(3^{34}-1)$ is a 17-digit integer, and the compiler of a computer programming language can precisely perform basic arithmetic operations with it.

Moreover, the number of Cplex calls is minimized by preventing the attacker's problem from being solved more than once for the same solution vector σ of the

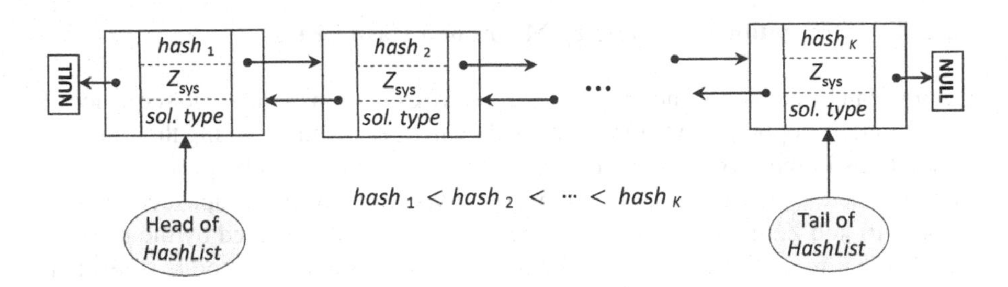

Fig. 5.2 The doubly linked *HashList* to store the hash values in ascending order

system planner. This is achieved by recording in the *HashList* not only the hash value, but also the objective value $Z_{sys}(\sigma)$ and the type (*current* or *neighbor*) of each solution constructed during the TSH iterations. The hash value of each new solution is searched in the sorted *HashList* starting from both the head and the tail of the list. If it is found, the associated $Z_{sys}(\sigma)$ is retrieved and compared with the objective value of the so far best neighborhood solution. If not found, the objective value Z_{sys} of the new solution is computed first. Next, its hash value is inserted together with the associated Z_{sys} at the appropriate position of the sorted list. After the neighborhood search has been completed, the best neighborhood solution becomes the new current solution, and its type field in the *HashList* is changed from *neighbor* to *current*. The preservation of the ascending order of hash values in the *HashList* is critical to the efficiency of this bilinear search procedure.

5.3.2.6 Stopping Conditions

We use two stopping conditions. The first one is completing the maximum number of iterations (*Max_Iter*), which is set equal to 200. The second one is reaching the maximum number of successive iterations during which the incumbent does not improve (*Max_Nonimp_Iter*). This number is set equal to 100. TSH is stopped as soon as either of these conditions is satisfied.

5.3.2.7 Probabilistic Nature of the Algorithm

TSH is probabilistic in the sense that it involves randomness which comes from the random selection of an initial feasible facility location-protection plan, and from the granularity of the 1-Swap-Ext neighborhood. Depending on the random number seed, TSH starts with a different initial solution in each run. Hence, it needs to be restarted multiple times with different seeds in order to get the best solution overall. Likewise, for a given current solution σ_{curr}, the particular 1-Swap-Ext neighborhood that will be constructed and evaluated depends on the random number seed and on the value of the granularity coefficient *RNS* given in Table 5.1. We adopt these two elements of randomness as a diversification mechanism of TSH.

5.3.2.8 Borderline Cases

Attention to the following borderline cases can be time-saving during the execution of TSH iterations.

- If $p = \pi$, then there is no need to solve the LLP of the attacker. The system planner's post-attack objective value will be equal to the pre-attack level. Neither will the post-attack customer-facility assignments differ from the pre-attack assignments.
- If $p - \pi = 1$, then the trivial solution to the attacker's problem is to interdict the only opened, but not protected facility. It is unnecessary to call Cplex to solve the associated LLP.

The outline of the algorithm TSH is presented below with the following additional notation.

t	the iteration counter,
τ	the counter of successive iterations during which the incumbent does not improve,
σ_t	the current solution at iteration t,
σ_t^{neigh}	a solution in the neighborhood of σ_t,
σ_t^{best}	the best solution identified in the neighborhood of σ_t,
$MOVES$	the set of move types comprising the neighborhood structure of TSH,
Obj_t	the system planner's objective value (Z_{sys}) in the current solution σ_t,
Obj_{neigh}	the system planner's objective value in a neighborhood solution σ_t^{neigh},
Obj_{best}	the system planner's objective value in the best neighborhood solution σ_t^{best},
Obj^*	the system planner's objective value in the *Incumbent*.

5.3.3 *Partial Validation of TSH Solutions with Exhaustive Search*

More evidence is needed to assess the solution quality performance of the proposed matheuristic TSH. Only then can one justify its competence in solving the BFCLP. For this purpose, we have developed a conditional exhaustive search algorithm called ESV-p. We use this algorithm to partially validate and assess the quality of the solutions produced by TSH. It works conditionally on the value of p^* stands for the number of opened (protected as well as unprotected) facilities in the best TSH solutions yielding $(Z^*_{\text{sys}})_{\min}$ values. In other words, it solves a given BFCLP instance under the condition that the total number of facilities to be opened in either mode will be fixed to p^*. Unlike in the p-median version, this number is unknown in the BFCLP. It must be nonzero to secure the provision of public service to the customers in the system, but it can be any positive integer less than or equal to $|J|$. Therefore, one can actually guarantee global optimality by running ESV-p $|J|$ times for each possible value of p^* in the interval $[1, |J|]$. Yet, for large $|J|$ this would be

```
1: while t ≤ Max_Iter and τ ≤ Max_Nonimp_Iter do
2:    Obj_best ← ∞
3:    for each move type k ∈ MOVES do
4:       /* Pay attention to the granularity of 1-Swap-Ext neighborhood. */
5:       for each feasible neighborhood solution σ_t^neigh of type k do
6:          Compute hash_neigh according to Table 2 using hash_curr
7:          if σ_t^neigh was previously recorded in the HashList as current then
8:             Bypass it and continue with the next σ_t^neigh
9:          else if σ_t^neigh was previously recorded in the HashList as neighbor then
10:            Retrieve Obj_neigh
11:         else
12:            Solve the attacker's problem CRIM associated with σ_t^neigh to optimality
               using Cplex
13:            Based on the optimal CRIM solution, calculate Obj_neigh for σ_t^neigh
14:            Record σ_t^neigh as neighbor with its hash_neigh and Obj_neigh in the HashList
15:         end if
16:         if Obj_neigh < Obj_best then
17:            Obj_best ← Obj_neigh and σ_t^best ← σ_t^neigh
18:         end if
19:         if Obj_neigh < Obj* then
20:            Obj* ← Obj_neigh and Incumbent = σ_t^neigh
21:         end if
22:      end for
23:   end for
24:   if the Incumbent has improved in this iteration then
25:      τ ← 1
26:   else
27:      Increment τ
28:   end if
29:   σ_{t+1} = σ_t^best and change the solution type of σ_{t+1} recorded in the HashList to
      current
30:   Increment t
31: end while
32: Return the Incumbent and Obj* as the best feasible solution to BFCLP
```

virtually impossible because of the exponential time complexity of the algorithm. ESV-p explores in an outer loop $\binom{|J|}{p}$ site combinations to open p facilities. For each combination $J_p = j_1, \ldots, j_p$ it has to check all subsets of sites to be protected. If q out of p facilities are protected in some combination J_p and if $\rho = \min\{r, p-q\}$, then as many as $\sum_{k=1}^{\rho} \binom{p-q}{k}$ possible interdiction patterns of the attacker must be examined for that J_p. In our experiments ESV-p tackles BFCLP instances in somewhat tolerable times for $p^* \leq 8$.

To validate TSH, we exploit ESV-p in the following way: first, we check the p^* value found in the best TSH solution. If $p^* \leq 8$, we run ESV-p for $p = p^*$. Secondly, if $p^* > 1$, we run ESV-p for $p = p^* - 1$. Thirdly, ESV-p is run again to solve the given BFCLP instance for $p = p^* + 1$ provided that $p^* \leq 7$. Each time we record $Z_{\text{sys}}^{\text{ESV}}(p)$, the objective value found by ESV-p. We then compare the lowest of these denoted by $(Z_{\text{sys}}^{\text{ESV}})_{\min}$ against the best TSH objective value $(Z_{\text{sys}}^*)_{\min}$. In the end, if $Z_{\text{sys}}^{\text{ESV}}(p^*) = (Z_{\text{sys}}^*)_{\min}$ and if $(Z_{\text{sys}}^{\text{ESV}})_{\min} = Z_{\text{sys}}^{\text{ESV}}(p^*)$, the partial validation of the TSH solution concludes affirmatively.

5.4 Computational Results

In this section we present extensive test results obtained with TSH. We validate TSH solutions partially, where possible, with an exhaustive search algorithm called ESV-p, which is described in the sequel. All coding was done in C with Microsoft Visual Studio 2005 using Cplex 11.2 Callable Library. We measured CPU times on an Intel Xeon X5460 3.16 GHz Quad-Core server equipped with 16 GB RAM.

5.4.1 Random Generation of Test Instances

Our test bed comprises a total of 60 randomly generated BFCLP instances, which vary in m (the number of candidate sites), r (the number of interdictions in the worst case), h_j (cost of protecting a facility at site j), and α (customers' traveling cost per

Table 5.3 Random problem generation template employed in the computational study

Parameters	Values
m	10, 15, 20, 25, 30
n	$10m$
r	$1, 2, 3$
(R, L)	(1000, 1500)
$(\text{cx}_i, \text{cy}_i)$	Let $R_i = R \times U(0,1), \theta_i = 2\pi \times U(0,1)$. Then, $\text{cx}_i = [\lvert R_i \cos\theta_i \rvert]$ and $\text{cy}_i = [\lvert R_i \sin\theta_i \rvert]$
$(\text{fx}_j, \text{fy}_j)$	$\text{fx}_j = -0.5L + \frac{L}{m} \times U[0,m]$ and $\text{fy}_j = -0.5L + \frac{L}{m} \times U[0,m]$
d_i	$10 + 5 \times U[0,18]$
f_j	$10{,}000 + 1{,}250 \times U[0,8]$
$(h_j)_{\text{low}}$	$0.5f_j$
$(h_j)_{\text{high}}$	$3.0f_j$
e_j	$10 + 2.5 \times U[0,4]$
α_{low}	0.10
α_{high}	0.20

unit distance per unit demand). The instances are named to be indicative of the m, r, and h_j values. For example, the name [10-1-lo] implies that the system planner can open a facility at 10 candidate sites, the attacker can hit at most one facility, and h_j values are low, i.e., equal to half of the fixed costs of opening.

In each instance $n = |I|$ (the number of customers) is set to $10m = 10|J|$. Customer nodes are uniformly distributed over a circular area centered at the origin (0,0) with a radius $R = 1000$. Candidate sites are uniformly dispersed on $(m+1)$ equidistant horizontal and vertical lines which hypothetically dice a square centered at the origin (0,0) with a side length of 1500. All coordinates are rounded to the nearest integer. In the random instance generation template given in Table 5.3, the operator $[|\cdot|]$ rounds off its argument to the nearest integer; (cx_i, cy_i) and (fx_j, fy_j) stand for the coordinates of customer node i and facility site j, respectively; $U(0,1)$ stands for a uniform random number in the interval [0, 1); and finally $U[lb, ub]$ symbolizes a random integer number between and inclusive of a lower bound lb and an upper bound ub.

5.4.2 *Preliminary Analysis of the Neighborhood Structure*

First, we investigated whether TSH would do better with fewer move types. For this investigation, we left out the move types in Table 5.1 one at a time, and re-performed the TSH iterations with four instead of five move types. Five different reduced neighborhood structures were applied this way. However, none of them helped improve the best objective value in any of the 60 test instances. This preliminary analysis convinced us of the necessity of five move types in the neighborhood structure. We recorded the selection frequencies of each move type responsible for the progress from the current to the next solution during the algorithmic loop of TSH. The frequencies were averaged over 150 runs of 30 test instances each with low and high unit distance costs. We observed that 30.7%, 29.7%, 2.8%, 1.3%, and 35.6% of the current-to-next solution updates were attributed to the 1-Add, 1-Drop, 1-Flip, 1-Swap-Int, and 1-Swap-Ext moves, respectively, in case of low unit distance costs. In the test instances with α_{high}, these frequency percentages were calculated as 33.4%, 30.5%, 4.8%, 5.2%, and 26.1%. We also experimented with different values of the granularity coefficient *RNS*, which shrinks the size of the 1-Swap-Ext neighborhood as was explained in Table 5.1 before. We observed that the best CPU time and incumbent quality trade-off was achieved at $RNS = 5$; hence we adopted this coefficient in our experiments.

5.4.3 *Test results of TSH*

Due to the probabilistic nature of TSH, we ran the code of the algorithm in a multi-start scheme by solving each instance five times with five different random number seeds as was explained in Sect. 5.3.2.7. Objective values and CPU times attained by TSH in 30 test instances with α_{low} and α_{high} are shown in Tables 5.4 and 5.5, respectively, alongside the number of times Cplex was called to solve the lower level problem. The minimum and average objective values Z^*_{sys} found by TSH in five runs are reported in the second and third columns of the tables. Column 4 shows the average CPU times consumed by TSH in those five runs. The columns labeled as p^* and π^* show respectively the numbers of opened and protected facilities in the best TSH solutions which yield the $(Z^*_{\text{sys}})_{\min}$ values. Finally, the last column gives the number of Cplex calls performed throughout the TSH iterations in each instance.

The computing platform employed in our study requires 1,007 (1,124) seconds on the average to perform a single run of TSH on the test instances for which the α value is equal to 0.10 (0.20). The main reason for these considerably long CPU times of the algorithm TSH is the number of calls to Cplex. Although we use a hash list as shown in Figure 7.2 to store the objective values of all visited solutions, and although this absolutely eliminates duplicate objective value calculations for the same solution during the TSH iterations, Cplex is still called 6,346 times per instance in the average of $5 \times 30 = 150$ runs on test instances with α_{low}. The average CPU time of TSH increases to 1,124 seconds per run from 1,007 seconds when α_{high} is used. Also the number of Cplex calls per run rises to 9,360 averaged over 30 instances each solved five times. This is due to the fact that the share of the demand-weighted traveling cost in the system planner's objective function increases as the customers' unit distance traveling cost rises from 0.10 to 0.20 per unit demand, which makes opening more facilities than before viable. This, in turn, renders the search for the best facility configuration by TSH more difficult, thus more time-consuming.

Moreover, TSH opens and protects more and more facilities as m goes from 10 to 25. This trend is not sustained when m increases from 25 to 30, but also in that case the average π^* ascends although the total number of opened facilities goes down.

Another interesting observation that can be drawn from Table 5.5 is that among 60 test instances there are just three in which the system planner opens a mixture of protected and unprotected facilities to counter the attacker's disruptive threats. Only in the instances "20-r2-hi", "20-r3-hi", and "30-r3-hi" with α_{high}, the system planner opens 10, 10, and 15 facilities, respectively and protects one of them. Otherwise, the system planner's best strategy found by TSH is to open facilities either in the protected or unprotected mode exclusively. At most six facilities are opened and protected by the system planner. This situation is observed in six particular instances where m is 25 or 30, $\alpha = 0.20$, and h_j values are low. Once h_j increases from $0.5 f_j$ to $3.0 f_j$ the system planner prefers to open many more facilities in the unprotected mode in lieu of any protected facility. The only exception is "30-r3-hi" with α_{high}. While the system planner opens and protects six facilities in the low h_j version of this instance, the best defense policy in the high h_j version suggests that 15 facilities be opened and one of them be protected.

Table 5.4 Objective values and CPU times attained by TSH for α_{low}

Problem	$(Z^*_{\text{sys}})_{\text{min}}$	$(Z^*_{\text{sys}})_{\text{avg}}$	CPU_{avg}	p^*	π^*	N_{Cplex}
10-r1-lo	383,583	383,583	0.6	1	1	110
10-r1-hi	465,833	468,419	20.6	1	1	2,534
10-r2-lo	383,583	383,583	0.5	1	1	100
10-r2-hi	465,833	488,041	14.3	1	1	949
10-r3-lo	383,583	383,583	1.5	1	1	136
10-r3-hi	465,833	465,833	24.9	1	1	1,293
Averages	*424,708*	*428,840*	*10.4*	*1.0*	*1.0*	*854*
15-r1-lo	554,693	554,693	5.9	2	2	790
15-r1-hi	656,982	658,659	143.1	6	0	6,700
15-r2-lo	554,693	554,693	7.8	2	2	709
15-r2-hi	749,255	749,255	353.0	7	0	6,800
15-r3-lo	554,693	554,693	16.4	2	2	924
15-r3-hi	729,693	819,743	435.8	2	2	4,524
Averages	*633,335*	*648,622*	*160.4*	*3.5*	*1.3*	*3,408*
20-r1-lo	670,595	706,431	66.9	3	3	5,736
20-r1-hi	804,407	804,492	267.5	5	0	7,654
20-r2-lo	670,595	670,595	20.6	3	3	1,613
20-r2-hi	904,898	914,619	985.8	8	0	8,488
20-r3-lo	670,595	670,595	9.8	3	3	650
20-r3-hi	998,940	1,021,150	1246.8	10	0	8,737
Averages	*786,671*	*797,980*	*432.9*	*5.3*	*1.5*	*5,480*
25-r1-lo	917,001	934,455	138.4	4	4	5,715
25-r1-hi	1,032,114	1,048,461	1670.9	8	0	15,592
25-r2-lo	917,001	917,001	227.9	4	4	3,269
25-r2-hi	1,153,094	1,153,094	4775.2	10	0	15,956
25-r3-lo	917,001	917,001	237.9	4	4	2,283
25-r3-hi	1,240,523	1,244,311	10513.5	11	0	15,709
Averages	*1,029,456*	*1,035,721*	*2,927.3*	*6.8*	*2.0*	*9,754*
30-r1-lo	947,879	981,942	249.0	5	5	12,815
30-r1-hi	1,066,937	1,073,992	2148.8	7	0	15,572
30-r2-lo	947,879	947,879	172.7	5	5	6,079
30-r2-hi	1,181,813	1,232,369	4386.8	11	0	19,924
30-r3-lo	947,879	947,879	139.7	5	5	5,144
30-r3-hi	1,178,930	1,282,046	1919.9	2	2	13,880
Averages	*1,045,219*	*1,077,684*	*1,502.8*	*5.8*	*2.8*	*12,236*
Grand Avg.	783,878	797,770	1,006.8	4.5	1.7	6,346

5.4.4 Comparison of the Best TSH Solutions with the ESV-p Results

The partial validation of the best TSH solutions through the use of ESV-p is presented, where possible, in Table 5.6 and Table 5.7 for low and high distance costs,

Table 5.5 Objective values and CPU times attained by TSH for α_{high}

Problem	$(Z^*_{sys})_{min}$	$(Z^*_{sys})_{avg}$	CPU_{avg}	p^*	π^*	N_{Cplex}
10-r1-lo	625,715	625,715	4	3	3	804
10-r1-hi	706,085	736,985	18	5	0	1,797
10-r2-lo	625,715	625,715	1	3	3	267
10-r2-hi	843,632	853,590	53	7	0	3,268
10-r3-lo	625,715	625,715	3	3	3	396
10-r3-hi	746,416	911,121	54	1	1	2,625
Averages	*695,546*	*729,807*	*22.1*	*3.7*	*1.7*	*1,526*
15-r1-lo	849,962	849,962	43	4	4	4,844
15-r1-hi	952,103	955,308	205	8	0	6,135
15-r2-lo	849,962	893,152	53	4	4	2,051
15-r2-hi	1,073,465	1,089,345	356	9	0	5,598
15-r3-lo	849,962	849,962	15	4	4	862
15-r3-hi	1,087,123	1,139,347	430	2	2	5,680
Averages	*943,763*	*962,846*	*183.7*	*5.2*	*2.3*	*4,195*
20-r1-lo	1,018,362	1,053,347	172	5	5	9,748
20-r1-hi	1,154,365	1,154,603	591	7	0	8,303
20-r2-lo	1,018,362	1,116,695	211	5	5	8,429
20-r2-hi	1,262,446	1,263,197	1,010	10	1	11,049
20-r3-lo	1,018,362	1,117,974	177	5	5	6,618
20-r3-hi	1,363,394	1,366,011	1,236	10	1	11,936
Averages	*1,139,215*	*1,178,638*	*566.1*	*7.0*	*2.8*	*9,347*
25-r1-lo	1,425,903	1,499,490	1,123	6	6	16,262
25-r1-hi	1,517,865	1,539,055	1,613	10	0	13,048
25-r2-lo	1,425,903	1,547,308	504	6	6	11,884
25-r2-hi	1,680,009	1,706,407	3,120	13	0	15,236
25-r3-lo	1,425,903	1,508,689	1,269	6	6	10,486
25-r3-hi	1,803,353	1,869,447	11,287	15	0	17,572
Averages	*1,546,489*	*1,611,733*	*3,152.5*	*9.3*	*3.0*	*14,081*
30-r1-lo	1,414,479	1,485,410	380	6	6	15,363
30-r1-hi	1,504,678	1,566,019	1,844	11	0	19,747
30-r2-lo	1,414,479	1,494,753	569	6	6	17,315
30-r2-hi	1,679,823	1,706,480	3,524	11	0	22,010
30-r3-lo	1,414,479	1,414,479	182	6	6	7,411
30-r3-hi	1,829,376	1,898,706	3,675	15	1	24,048
Averages	*1,542,886*	*1,594,308*	*1,695.7*	*9.2*	*3.2*	*17,649*
Grand Avg.	1,173,580	1,215,466	1,124.0	6.9	2.6	9,360

respectively. The last column labeled as "Gap" in both tables shows the relative percent gaps between $(Z^{ESV}_{sys})_{min}$ and $(Z^*_{sys})_{min}$. A negative value would indicate that ESV-p yields a lower objective value for the system planner than TSH; however, this is never the case in the tables. The only nonzero gap is observed for the instance [15-2-hi] in Table 5.7, which is actually due the unavailability of $Z^{ESV}_{sys}(p^*)$ since the p^* value in the best TSH solution is larger than eight.

Table 5.6 Partial validation of the best TSH solutions for α_{low}

	TSH over five runs		ESV-p results for p^*-1, p^*, and p^*+1				
Problem	$(Z^*_{\text{sys}})_{\min}$	p^*	$Z^{\text{ESV}}_{\text{sys}}(p^*\text{-}1)$	$Z^{\text{ESV}}_{\text{sys}}(p^*)$	$Z^{\text{ESV}}_{\text{sys}}(p^*\text{+}1)$	$(Z^{\text{ESV}}_{\text{sys}})_{\min}$	*Gap* (%)
10-r1-lo	383,583	1	–	383,583	401,501	383,583	0.00
10-r1-hi	465,833	1	–	465,833	562,647	465,833	0.00
10-r2-lo	383,583	1	–	383,583	401,501	383,583	0.00
10-r2-hi	465,833	1	–	465,833	562,647	465,833	0.00
10-r3-lo	383,583	1	–	383,583	401,501	383,583	0.00
10-r3-hi	465,833	1	–	465,833	562,647	465,833	0.00
15-r1-lo	554,693	2	645,245	554,693	559,948	554,693	0.00
15-r1-hi	656,982	6	660,602	656,982	668,598	656,982	0.00
15-r2-lo	554,693	2	645,245	554,693	559,948	554,693	0.00
15-r2-hi	749,255	7	760,103	749,255	753,183	749,255	0.00
15-r3-lo	554,693	2	645,245	554,693	559,948	554,693	0.00
15-r3-hi	729,693	2	733,995	729,693	810,938	729,693	0.00
20-r1-lo	670,595	3	693,422	670,595	687,205	670,595	0.00
20-r1-hi	804,407	5	836,278	804,407	805,159	804,407	0.00
20-r2-lo	670,595	3	693,422	670,595	687,205	670,595	0.00
20-r2-hi	904,898	8	914,219	904,898	–	904,898	0.00
20-r3-lo	670,595	3	693,422	670,595	687,205	670,595	0.00
20-r3-hi	998,940	10	–	–	–	–	–
25-r1-lo	917,001	4	948,170	917,001	935,585	917,001	0.00
25-r1-hi	1,032,114	8	1,054,522	1,032,114	–	1,032,114	0.00
25-r2-lo	917,001	4	948,170	917,001	935,585	917,001	0.00
25-r2-hi	1,153,094	10	–	–	–	–	–
25-r3-lo	917,001	4	948,170	917,001	935,585	917,001	0.00
25-r3-hi	1,240,523	11	–	–	–	–	–
30-r1-lo	947,879	5	964,121	947,879	965,728	947,879	0.00
30-r1-hi	1,066,937	7	1,091,886	1,066,937	1,071,259	1,066,937	0.00
30-r2-lo	947,879	5	964,121	947,879	965,728	947,879	0.00
30-r2-hi	1,181,813	11	–	–	–	–	–
30-r3-lo	947,879	5	964,121	947,879	965,728	947,879	0.00
30-r3-hi	1,178,930	2	1,303,965	1,178,930	1,227,813	1,178,930	0.00

The results in Table 5.6 and Table 5.7 do not prove the global optimality of the best TSH solutions since ESV-p has been run only for a limited number of p values, namely p^*-1, p^*, and p^*+1. However, they serve as further evidence for the solution quality of TSH. We remark that CPU times of the runs have been purposely omitted, since ESV-p is uncompetitive in that venue. Especially in the test instances with $m=25$ and $m=30$, we observed extravagant solution times. When m equals 25, the algorithm consumes approximately 33 min, 4 h, and 12.5 h on average for $p=6$, $p=7$, and $p=8$, respectively. These times jump to 2 h, 36.5 h, and 77 h when m equals 30. In conclusion, ESV-p cannot be a viable alternative to TSH.

Table 5.7 Partial validation of the best TSH solutions for α_{high}

	TSH over five runs		ESV-p results for p^*-1, p^*, and p^*+1				
Problem	$(Z^*_{\text{sys}})_{\min}$	p^*	$Z^{\text{ESV}}_{\text{sys}}(p^*\text{-}1)$	$Z^{\text{ESV}}_{\text{sys}}(p^*)$	$Z^{\text{ESV}}_{\text{sys}}(p^*\text{+}1)$	$(Z^{\text{ESV}}_{\text{sys}})_{\min}$	*Gap* (%)
10-r1-lo	625,715	3	642,589	625,715	647,729	625,715	0.00
10-r1-hi	706,085	5	746,048	706,085	739,591	706,085	0.00
10-r2-lo	625,715	3	642,589	625,715	647,729	625,715	0.00
10-r2-hi	843,632	7	871,743	843,632	861,852	843,632	0.00
10-r3-lo	625,715	3	642,589	625,715	647,729	625,715	0.00
10-r3-hi	746,416	1	–	746,416	806,132	746,416	0.00
15-r1-lo	849,962	4	874,083	849,962	853,016	849,962	0.00
15-r1-hi	952,103	8	957,569	952,103	–	952,103	0.00
15-r2-lo	849,962	4	874,083	849,962	853,016	849,962	0.00
15-r2-hi	1,073,465	9	1,085,828	–	–	1,085,828	1.14
15-r3-lo	849,962	4	874,083	849,962	853,016	849,962	0.00
15-r3-hi	1,087,123	2	1,225,366	1,087,123	1,137,833	1,087,123	0.00
20-r1-lo	1,018,362	5	1,020,485	1,018,362	1,026,877	1,018,362	0.00
20-r1-hi	1,154,365	7	1,175,580	1,154,365	1,155,872	1,154,365	0.00
20-r2-lo	1,018,362	5	1,020,485	1,018,362	1,026,877	1,018,362	0.00
20-r2-hi	1,262,446	10	–	–	–	–	–
20-r3-lo	1,018,362	5	1,020,485	1,018,362	1,026,877	1,018,362	0.00
20-r3-hi	1,363,394	10	–	–	–	–	–
25-r1-lo	1,425,903	6	1,433,096	1,425,903	1,432,375	1,425,903	0.00
25-r1-hi	1,517,865	10	–	–	–	–	–
25-r2-lo	1,425,903	6	1,433,096	1,425,903	1,432,375	1,425,903	0.00
25-r2-hi	1,680,009	13	–	–	–	–	–
25-r3-lo	1,425,903	6	1,433,096	1,425,903	1,432,375	1,425,903	0.00
25-r3-hi	1,803,353	15	–	–	–	–	–
30-r1-lo	1,414,479	6	1,448,031	1,414,479	1,416,260	1,414,479	0.00
30-r1-hi	1,504,678	11	–	–	–	–	–
30-r2-lo	1,414,479	6	1,448,031	1,414,479	1,416,260	1,414,479	0.00
30-r2-hi	1,679,823	11	–	–	–	–	–
30-r3-lo	1,414,479	6	1,448,031	1,414,479	1,416,260	1,414,479	0.00
30-r3-hi	1,829,376	15	–	–	–	–	–

5.4.5 *Sensitivity of the TSH Solutions to Problem Parameters*

Our next goal is to conduct an analysis with regard to three parameters in the model: the protection cost (h_j), the maximum number of facilities that can be interdicted by the attacker (r), and the unit distance traveling costs per unit demand (α). We recorded the $(Z^*_{\text{sys}})_{\min}$ values of the best TSH solutions averaged over the parameters h_j, r, and α in Table 5.8. $\overline{p^*}$ and $\overline{\pi^*}$ indicate, respectively, the average numbers of opened and protected facilities in the best TSH solutions. The columns of Table 5.8 corresponding to $(h_j)_{\text{high}}$ reveal that as r increases $(Z^*_{\text{sys}})_{\min}$ takes on larger values. This is a consequence of the demand-weighted traveling costs increased by extra interdictions. However, that is true only when the protection costs are high and the cost of protecting added facilities exceeds the cost of capacity expansions induced by extra interdictions. When the protection costs are low, on the other hand, the system planner can afford to protect all opened facilities, and the value of r does not have any effect on the average $(Z^*_{\text{sys}})_{\min}$. Moreover, $(Z^*_{\text{sys}})_{\min}$ increases as expected when α doubles.

Table 5.8 Change in $(Z^*_{\text{sys}})_{\min}$, p^*, π^* with respect to problem parameters

	$\alpha_{\text{low}} = 0.10$				$\alpha_{\text{high}} = 0.20$				Grand
	$(h_j)_{\text{low}}$	$(\overline{p^*}, \overline{\pi^*})$	$(h_j)_{\text{high}}$	$(\overline{p^*}, \overline{\pi^*})$	$(h_j)_{\text{low}}$	$(\overline{p^*}, \overline{\pi^*})$	$(h_j)_{\text{high}}$	$(\overline{p^*}, \overline{\pi^*})$	Avg.
$r=1$	694,750	(3.0, 3.0)	805,254	(5.4, 0.2)	1,066,884	(4.8, 4.8)	1,167,019	(8.2, 0.0)	933,477
$r=2$	694,750	(3.0, 3.0)	890,979	(7.4, 0.2)	1,066,884	(4.8, 4.8)	1,307,875	(10.0, 0.2)	990,122
$r=3$	694,750	(3.0, 3.0)	922,784	(5.2, 1.0)	1,066,884	(4.8, 4.8)	1,365,933	(8.6, 1.0)	1,012,588

As the attacker's threat level is elevated in problems with $(h_j)_{\text{high}}$, p^* exhibits a unimodal pattern. The first increase from $r = 1$ to $r = 2$ causes additional facilities to be opened, but π^* remains more or less the same due to the fact that h_j values are three times higher than f_j. When r increases to three, however, the system planner cannot afford opening his facilities in the unprotected mode anymore. He needs to protect one facility on average against interdiction. In return, he tries to offset the considerably high protection cost of that facility by opening a lesser number of facilities in the unprotected mode.

5.4.6 *Effect of the Protection Costs on the Best Facility Configuration*

In our final experiment, we picked the problem instance $(m = 20, r = 3)$ with α_{high} to test the effect of h_j on the best facility configuration. The reason for selecting this particular instance is that its p^* and π^* values show significant variation between low and high protection costs. Moreover, its cumulative CPU time over five runs is less than $1\frac{3}{4}$ h so that multiple replications can be performed in a reasonable amount of time. We started h_j's at zero and increased them in multiples of $0.5f_j$ until they reached $3.0f_j$ which is equal to $(h_j)_{\text{high}}$. The results including the objective values and the numerals of the opened, protected, and interdicted facilities can be seen in Table 5.9 for each test. We observe that when $h_j = 0$, seven facilities are opened all of which are protected; thus no interdiction occurs. A slight increase in the protection costs induces three facilities (#6, #16, #18) to leave and one facility (#15) to enter the initial configuration. All of them are still opened in the protected mode. When h_j values double becoming equal to f_j, three facilities in the previous configuration (#8, #15, #17) are replaced by one facility (#6) again in the protected mode. This configuration remains the same until the highest level of the protection costs, namely $3.0f_j$. At this level the system planner cannot endure the cost of protecting more than one facility; thus he opens one protected and nine unprotected facilities three of which are subsequently targeted in worst-case interdiction by the attacker.

Table 5.9 Effect of the increasing protection costs

h_j	$(Z^*_{sys})_{min}$	Fac. opened in unprotected mode	Fac. opened in protected mode	Facilities interdicted
0.0	918,726	none	6, 8, 10, 16, 17, 18, 20	none
$0.5f_j$	1,018,362	none	8, 10, 15, 17, 20	none
$1.0f_j$	1,085,638	none	6, 10, 20	none
$1.5f_j$	1,137,888	none	6, 10, 20	none
$2.0f_j$	1,190,138	none	6, 10, 20	none
$2.5f_j$	1,242,388	none	6, 10, 20	none
$3.0f_j$	1,363,394	4, 6, 8, 12, 13, 15, 16 17, 18	10	12, 13, 16

5.5 Conclusions

In this chapter we review the state of the art of the network and facility interdiction literature by differentiating between interdiction problems that include either an extra protection or network design/facility location dimension. Of interest to us is the unified problem of facility location-protection- and interdiction which was discussed in only three recent references before. This triple facility interdiction problem can be modeled as a static Stackelberg game between a system planner and a potential attacker. We revisit two versions of this problem: The first one is defined as a bilevel maximal coverage problem where the system planner has to make new facility opening, existing facility relocation, and facility protection decisions. The second one is defined as a bilevel fixed-charge facility location problem. Here, the system planner has to decide about the locations of new facilities, the protection status and initial capacity acquisition at the opened facilities as well as post-attack capacity expansions to remedy capacity reductions caused by the attacker's worst-case interdiction. In both versions the system planner is free to open arbitrarily many facilities; he is restricted only by the number of available candidate sites. After presenting both models, we focus on the solution of the latter which we refer to as the bilevel fixed-charge location problem (BFCLP). The leader of the static Stackelberg game in this problem is the planner of a public supply/demand system whose clientele always goes to the closest facility available to get service. The follower of the game is an attacker with the objective to inflict the maximum possible disruption to this public service system. To achieve his objective, the attacker destroys up to a fixed number of facilities beyond repair unless they are opened in the protected mode at additional site-specific costs. Thus, the system planner needs to decide the protection status of the opened facilities too.

We elaborate a matheuristic, called TSH, to solve the BFCLP. It is a probabilistic tabu search algorithm capitalizing on a hash list which records the objectives and hash values of all solutions explored. The use of hashing helps avoid cycling and boosts the efficiency of the tabu search by minimizing the number of objective function evaluations. TSH searches the space of candidate facility locations to determine the best feasible configuration of protected and unprotected facilities. For each such configuration, a binary programming problem is solved to optimality using the commercial solver Cplex 11.2 to find the best interdiction plan of the attacker. The attacker's best interdiction plan is synonymous with the worst-case interdiction faced by the system planner. We also develop a conditional exhaustive search and validation algorithm called ESV-p by means of which the quality of each TSH solution is tested and confirmed. Experiments with both solution approaches demonstrated the suitability of the proposed matheuristic TSH for producing high-quality solutions to the BFCLP in tolerable CPU times.

There can be two extensions of this work. One is formulating the same BFCLP model such that partial interdiction of facilities is allowed. This means that facilities will not be rendered totally out of service in the wake of an attack, but will continue to provide service with less-than-full capacity depending on the degree of interdiction. Another extension could be to incorporate into the model the partial protection of facilities alongside their partial interdiction. This approach, albeit much more intractable than complete protection and interdiction, could disclose a quite different service network design and interdiction strategy on behalf of the system planner and the attacker enabling the former to better utilize his protective resources and the latter to do so with his limited offensive resources.

References

1. Aksen, D., Aras, N.: A bilevel fixed charge location model for facilities under imminent attack. Comp. Oper. Res. 39(7), 1364–1381 (2012)
2. Aksen, D., Piyade, N., Aras, N.: The budget constrained r-interdiction median problem with capacity expansion. Central European Journal of Operations Research 18(3), 269–291 (2010)
3. Aksen, D., Aras, N., Piyade, N.: A bilevel p-median model for the planning and protection of critical facilities. Journal of Heuristics (2011), doi: 10.1007/s10732-011-9163-5
4. Aras, N., Aksen, D.: Locating collection centers for distance- and incentive-dependent returns. Int. J. Prod. Econ. 111(2), 316–333 (2008)
5. Aras, N., Aksen, D., Tanuğur, A.G.: Locating collection centers for incentive-dependent returns under a pick-up policy with capacitated vehicles. Eur. J. Oper. Res. 191(3), 1223–1240 (2008)
6. Arroyo, J.M., Galiana, F.D.: On the solution of the bilevel programming formulation of the terrorist threat problem. IEEE Trans. Power Systems 20(2), 789–797 (2005)
7. Bayrak, H., Bailey, M.D.: Shortest path network interdiction with asymmetric information. Networks 52(3), 133–140 (2008)
8. BBC News UK, Arsonists target ambulance station. BBC Online Network (1999), `http://news.bbc.co.uk/2/hi/uk_news/378254.stm` (cited November 12, 2011)

9. Berman, O., Krass, D., Drezner, Z.: The gradual covering decay location problem on a network. Eur. J. Oper. Res. 151, 474–480 (2003)
10. Berman, O., Drezner, T., Drezner, Z., Wesolowsky, G.O.: A defensive maximal covering problem on a network. International Transactions in Operational Research 16(1), 69–86 (2009)
11. Berman, O., Gavious, A.: Location of terror response facilities: a game between state and terrorist. European Journal of Operational Research 177(2), 1113–1133 (2007)
12. Berman, O., Gavious, A., Huang, R.: Location of response facilities: a simultaneous game between state and terrorist. International Journal of Operational Research 10(1), 102–120 (2011)
13. Cappanera, P., Scaparra, M.P.: Optimal allocation of protective resources in shortest-path networks. Transportation Science 45(1), 64–80 (2011)
14. Chalk, P., Hoffman, B., Reville, R., Kasupski, A.: Trends in terrorism. RAND Center for Terrorism Risk Management Policy (2005), `http://www.rand.org/content/dam/rand/pubs/monographs/2005/RAND_MG393.pdf` (cited November 12, 2011)
15. Church, R.L., ReVelle, C.: The maximal covering location problem. Papers in Regional Science 32(1), 101–118 (1974)
16. Church, R.L., Scaparra, M.P.: Protecting critical assets: the r-interdiction median problem with fortification. Geographical Analysis 39(2), 129–146 (2007)
17. Church, R.L., Scaparra, M.P., Middleton, R.S.: Identifying critical infrastructure: the median and covering facility interdiction problems. Annals of the Association of American Geographers 94(3), 491–502 (2004)
18. Cormican, K.J., Morton, D.P., Wood, R.K.: Stochastic network interdiction. Oper. Res. 46(2), 184–197 (1998)
19. Floudas, C.A.: Deterministic Global Optimization: Theory, Methods and Applications. Nonconvex Optimization and Its Applications, vol. 37. Kluwer Academic Publishers, Dordrecht (2000)
20. Fulkerson, D.R., Harding, G.C.: Maximizing the minimum source-sink path subject to a budget constraint. Math Prog. 13(1), 116–118 (1977)
21. Gendreau, M.: An introduction to tabu search. In: Glover, F., Kochenberger, G.A. (eds.) Handbook of Metaheuristics, pp. 37–54. Kluwer Academic Publishers, Boston (2003)
22. Glover, F., Laguna, M.: Tabu search. Kluwer Academic Publishers, Dordrecht (1997)
23. Glover, F., Laguna, M., Martí, R.: Principles of Tabu Search. In: Gonzalez, T. (ed.) Handbook on Approximation Algorithms and Metaheuristics. Chapman & Hall/CRC, Boca Raton (2007)
24. Gümüş, Z.H., Floudas, C.A.: Global optimization of mixed-integer bilevel programming problems. Computational Management Science 2(3), 181–212 (2005)
25. Hecheng, L., Yuping, W.: Exponential distribution-based genetic algorithm for solving mixed-integer bilevel programming problems. Journal of Systems Engineering and Electronics 19(6), 1157–1164 (2008)
26. Held, H., Woodruff, D.L.: Heuristics for multi-stage interdiction of stochastic networks. Journal of Heuristics 11(5–6), 483–500 (2005)
27. Held, H., Hemmecke, R., Woodruff, D.L.: A decomposition algorithm applied to planning the interdiction of stochastic networks. Naval Res. Logistics 52(4), 321–328 (2005)
28. Hemmecke, R., Schultz, R., Woodruff, D.L.: Interdicting stochastic networks with binary interdiction effort. In: Woodruff, D.L. (ed.) Network Interdiction and Stochastic Integer Programming. Operations Research/Computer Science Interfaces Series, vol. 22, pp. 69–84. Kluwer, Boston (2003)

29. Israeli, E., Wood, R.K.: Shortest-path network interdiction. Networks 40(2), 97–111 (2002)
30. Keçici, S., Aras, N., Verter, V.: Incorporating the threat of terrorist attacks in the design of public service facility networks. Optimization Letters (2011), doi:10.1007/s11590-011-0412-1
31. Khachiyan, L., Boros, E., Borys, K., Elbassioni, K., Gurvich, V., Rudolf, G., Zhao, J.: On short paths interdiction problems: total and node-wise limited interdiction. Theory of Computing Systems 43(2), 204–233 (2008)
32. Liberatore, F., Scaparra, M.P., Daskin, M.S.: Analysis of facility protection strategies against an uncertain number of attacks: The stochastic R-interdiction median problem with fortification. Comp. Oper. Res. 38(1), 357–366 (2011)
33. Lim, C., Smith, J.C.: Algorithms for discrete and continuous multicommodity flow network interdiction problems. IIE Transactions 39(1), 15–26 (2007)
34. Losada, C., Scaparra, M.P., O'Hanley, J.R.: Optimizing system resilience: a facility protection model with recovery time. Eur. J. Oper. Res. 217(3), 519–530 (2012)
35. Maniezzo, V., Stützle, T., Voß, S. (eds.): Matheuristics: Hybridizing Metaheuristics and Mathematical Programming. Annals of Information Systems, vol. 10. Springer (2009)
36. Moore, J.T., Bard, J.F.: The mixed-integer linear bilevel programming problem. Oper. Res. 38(5), 911–921 (1990)
37. Motto, A.L., Arroyo, J.M., Galiana, F.D.: MILP for the analysis of electric grid security under disruptive threat. IEEE Transactions on Power Systems 20(3), 1357–1365 (2005)
38. Murray, A.T., Matisziw, T.C., Grubesic, T.H.: Critical network infrastructure analysis: interdiction and system flow. Journal of Geographical Systems 9(2), 103–117 (2007)
39. O'Hanley, J.R., Church, R.L., Gilless, K.: Locating and protecting critical reserve sites to minimize expected and worst-case losses. Biological Conservation 134(1), 130–141 (2007)
40. O'Hanley, J.R., Church, R.L.: Designing robust coverage to hedge against worst-case facility losses. Eur. J. Oper. Res. 209(1), 23–36 (2011)
41. Perl, R.: Trends in terrorism: Congressional Research Service Report for Congress. The Library of Congress, Order Code: RL33555 (2006), `http://www.dtic.mil/cgi-in/GetTRDoc?AD=ADA464744&Location=U2&doc=GetTRDoc.pdf` (cited November 12, 2011)
42. Radio Free Europe-Radio Liberty, Afghanistan Report: March 8, 2008. Afghanistan: Mobile-Phone Towers are Taliban's New Target (2008), `http://www.rferl.org/content/article/1347757.html` (accessed November 12, 2011)
43. Royset, J.O., Wood, R.K.: Solving the bi-objective maximum-flow network interdiction problem. INFORMS J. Computing 19(2), 175–184 (2007)
44. Salmerón, J., Wood, K., Baldick, R.: Worst-case interdiction analysis of large-scale Electric power grids. IEEE Trans. Power Systems 24(1), 96–104 (2009)
45. Scaparra, M.P., Church, R.L.: A bilevel mixed integer program for critical infrastructure protection planning. Comp. Oper. Res. 35(6), 1905–1923 (2008a)
46. Scaparra, M.P., Church, R.L.: An exact solution approach for the interdiction median problem with fortification. Eur. J. Oper. Res. 189(1), 76–92 (2008b)
47. Scaparra, M.P., Church, R.L.: Protecting supply systems to mitigate potential disaster: A model to fortify capacitated facilities. Kent Business School Working Paper No.209, University of Kent, Canterbury, UK (2010)
48. Sherali, H.D., Adams, W.P.: A hierarchy of relaxations between the continuous and convex hull representations for zero-one programming problems. SIAM Journal on Discrete Mathematics 3(3), 411–430 (1990)

49. Sherali, H.D., Adams, W.P., Driscoll, P.J.: Exploiting special structures in constructing a hierarchy of relaxations for 0-1 mixed integer problems. Oper. Res. 46(3), 396–405 (1998)
50. Smith, J.C.: Basic interdiction models. In: Cochran, J. (ed.) Wiley Encyclopedia of Operations Research and Management Science (EORMS), Wiley, New York (2010), `http://eu.wiley.com/WileyCDA/Section/id-380764.html` (accessed: November 12, 2011)
51. Smith, J.C., Lim, C.: Algorithms for network interdiction and fortification games. In: Chinchuluun, A., Pardalos, P.M., Migdalas, A., Pitsoulis, L. (eds.) Pareto Optimality, Game Theory and Equilibria, pp. 609–644. Springer, New York (2008)
52. Smith, J.C., Lim, C., Sudargho, F.: Survivable network design under optimal and heuristic interdiction scenarios. Journal of Global Optimization 38(2), 181–199 (2007)
53. Snyder, L.V., Scaparra, M.P., Daskin, M.S., Church, R.L.: Planning for disruptions in supply chain networks. In: Greenberg, H.K. (ed.) TutORials in Operations Research, pp. 234–257. INFORMS, Baltimore (2006)
54. Sun, M.: Solving the uncapacitated facility location problem using tabu search. Comp. Oper. Res. 33(9), 2563–2589 (2006)
55. Wollmer, R.: Removing arcs from a network. Oper. Res. 12(6), 934–940 (1964)
56. Wood, R.K.: Deterministic network interdiction. Mathematical and Computer Modelling 17(2), 1–18 (1993)
57. Woodruff, D.L., Zemel, E.: Hashing vectors for tabu search. Annals of Oper. Res. 41(2), 123–137 (1993)

Chapter 6
A Metaheuristic Framework for Bi-level Programming Problems with Multi-disciplinary Applications

Andrew Koh

Abstract. Bi-level programming problems arise in situations when the decision maker has to take into account the responses of the users to his decisions. Several problems arising in engineering and economics can be cast within the bi-level programming framework. The bi-level programming model is also known as a Stackleberg or leader-follower game in which the leader chooses his variables so as to optimise his objective function, taking into account the response of the follower(s) who separately optimise their own objectives, treating the leader's decisions as exogenous. In this chapter, we present a unified framework fully consistent with the Stackleberg paradigm of bi-level programming that allows for the integration of meta-heuristic algorithms with traditional gradient based optimisation algorithms for the solution of bi-level programming problems. In particular we employ Differential Evolution as the main meta-heuristic in our proposal. We subsequently apply the proposed method (DEBLP) to a range of problems from many fields such as transportation systems management, parameter estimation and game theory. It is demonstrated that DEBLP is a robust and powerful search heuristic for this class of problems characterised by non smoothness and non convexity.

6.1 Introduction

This paper introduces a meta-heuristic framework for solving the Bi-level programming Problem (BLPP) with a multitude of applications [26, 74]. As a historical footnote, the term "bi-level programming" was first coined in a technical report by Candler and Norton in [20] who were concerned with general multilevel programming problems. The BLPP is a special case of a multilevel programming problem restricted to two levels. Prior to that time, the BLPP was known simply as a mathematical program with an optimisation problem in the constraints [18] but had

---

Andrew Koh
Institute for Transport Studies, University of Leeds, Leeds, LS2 9JT, United Kingdom
e-mail: a.koh@its.leeds.ac.uk

E.-G. Talbi (Ed.): *Metaheuristics for Bi-level Optimization*, SCI 482, pp. 153–187.
DOI: 10.1007/978-3-642-37838-6_6 © Springer-Verlag Berlin Heidelberg 2013

already found military applications [19]. In economics and game theory, a BLPP is a Stackleberg [106] or "leader-follower" game (see Fig. 6.1) in which the leader chooses his variables so as to optimise his objective but continues to take into account the response of the follower(s) who when independently optimising their separate objectives, treat the leader's decisions as an exogenous input [72].

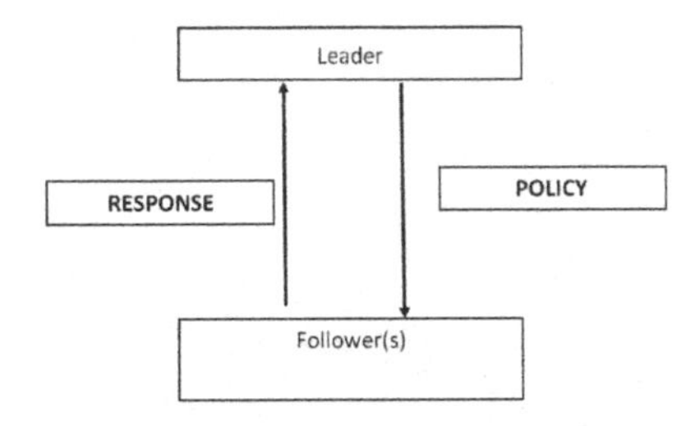

Fig. 6.1 Pictoral Representation of a BLPP

BLPPs possess in common the following three characteristics [86, 117]:

- The decision-making units are interactive and exist within a hierarchical structure.
- Decision making is sequential from higher to lower level. The lower level decision maker executes its policies after decisions are made at the upper level.
- Each unit independently optimises its own objective functions but is influenced by actions taken by other units.

The BLPP has been a subject of intense research and several notable volumes have been published to date [7, 32, 72, 87]. At the same time applications of BLPP can be found in fields as diverse as chemical engineering [49], robot motion planning and control [72], production planning [8] occurring in a multitude of disciplines [7]. In tandem, there has been much work on the development of solution methodologies (see [26, 32] for a review of these).

This chapter is structured as follows. In Section 7.3 we formally introduce the BLPP and provide a brief and by not means exhaustive review of solution methodologies for the BLPP. In Section 7.4 we discuss the Differential Evolution for Bi-Level Programming (DEBLP) meta-heuristic introduced in [61] and apply it to problems in transportation systems management in Section 7.5. To emphasise the multidisciplinary applications to which DEBLP is applicable, Section 7.6 provides examples of BLPPs arising from parameter estimation problems. Section 7.7 introduces a method for handling constraints integrating recent developments in evolutionary algorithms to propose a method to handle constraints in the leader's problem. Having set up this framework, we are ready in Section 6.7 to apply DEBLP to solving Generalised Nash Equilibrium Problems (GNEP) which when formulated as BLPPs are in fact characterised by constraints in the leader's objective. Section 6.8 summarises and provides extensive directions for further research on problems related to the topic of BLPPs.

6.2 The Bi-level Programming Problem

6.2.1 A General BLPP

We can write a generic BLPP as the system of equations in Eq. 6.1. The unique feature of Eq. 6.1 is that the constraint region is implicitly determined by yet another optimisation problem[1]. This constraint is always active. The upper level problem denoted as Program U, is given in Eq. 6.1a,

$$\text{Program } U \begin{cases} \min\limits_{\mathbf{x}\in X} U(\mathbf{x},\mathbf{y}) \\ \text{subject to} \\ G(\mathbf{x},\mathbf{y}) \leq 0 \\ E(\mathbf{x},\mathbf{y}) = 0 \end{cases} \tag{6.1a}$$

where for given $\mathbf{x}$, $\mathbf{y}$ is the solution to the lower level program (Program L) in 6.1b:

$$\text{Program } L \begin{cases} \min\limits_{\mathbf{y}\in Y} L(\mathbf{x},\mathbf{y}) \\ \text{subject to} \\ g(\mathbf{x},\mathbf{y})) \leq 0 \\ e(\mathbf{x},\mathbf{y})) = 0 \end{cases} \tag{6.1b}$$

In the formulation in Eq. 6.1 we define the following mappings: $U, L : \mathbb{R}^{n_1} \times \mathbb{R}^{n_2} \rightarrow \mathbb{R}^1, G : \mathbb{R}^{n_1} \times \mathbb{R}^{n_2} \rightarrow \mathbb{R}^{q_1}, g : \mathbb{R}^{n_1} \times \mathbb{R}^{n_2} \rightarrow \mathbb{R}^{q_2}, E : \mathbb{R}^{n_1} \times \mathbb{R}^{n_2} \rightarrow \mathbb{R}^{r_1}, e : \mathbb{R}^{n_1} \times \mathbb{R}^{n_2} \rightarrow \mathbb{R}^{r_2}$. In the general case the objectives and constraints at both levels are non-linear. The sets X and Y representing the search domains for 6.1a and 6.1b respectively are defined as follows: $X = \left\{ (x_1, x_2, ..., x_{n_1})^{\mathsf{T}} \in \mathbb{R}^{n_1} \middle| x_i^l \leq x_i \leq x_i^u, i = 1, ..., n_1 \right\}$ and $Y = \left\{ (y_1, y_2, ..., y_{n_2})^{\mathsf{T}} \in \mathbb{R}^{n_2} \middle| y_j^l \leq y_j \leq y_j^u, j = 1, ..., n_2 \right\}$ with T denoting the transpose. Arising from the "leader-follower" analogy of BLPPs, we use the terms leader's variables and upper level variables interchangeably when referring to $\mathbf{x}$.

6.2.2 Mathematical Programs with Equilibrium Constraints

We also define a class of BLPPs known as the Mathematical Programs with Equilibrium Constraints (MPECs) . MPECs are BLPPs where the lower level problem consists of a variational inequality (VI) [26].

$$\text{Program } U \begin{cases} \min\limits_{\mathbf{x}\in X} U(\mathbf{x},\mathbf{y}) \\ \text{subject to} \\ G(\mathbf{x},\mathbf{y}) \leq 0 \\ E(\mathbf{x},\mathbf{y}) = 0 \end{cases} \tag{6.2a}$$

[1] Hence the original name of mathematical programs with optimisation problems in the constraints chosen by Bracken and McGill in [18].

where for given $\mathbf{x}$, $\mathbf{y}$ is the solution of the VI in Program L 6.2b:

$$L(\mathbf{x},\mathbf{y})^{\mathsf{T}}(\mathbf{y}-\mathbf{y}^{*}) \geq 0, \forall \mathbf{y} \in \Upsilon(x) \tag{6.2b}$$

Another class of problems closely related to MPECs are Mathematical Programs with *Complementarity Constraints* [67] which feature in place of a VI, a Complementarity Problem instead in Program L. However since the VI is a generalization of the Complementarity Problem [56, 82], we will treat these two categories as synonymous for the purposes of this chapter and neglect the theoretical distinctions. We return in Section 6.4 to give an example of MPECs that arise naturally in transportation systems management.

6.2.3 Solution Algorithms for the BLPP

When all functions (both objectives and constraints) at both levels are linear and affine , this class of problems is known as the linear-BLPP . However even in this deceptively "simple" case the problem is still nondeterministic polynomial time hard [11]. Even when both the upper level and the lower level are convex programming problems , the resulting BLPP itself can be non-convex [12] . Non convexity suggests the possibility of multiple local optima. Ben-Ayed and Blair [11] demonstrated the failure of both the Parametric Complementarity Pivot Algorithm [13] and the Grid Search Algorithm [5] to locate the optimal solution . Since then, progress has been made in solving the linear-BLPP and techniques including implicit enumeration [21], penalty based methods [2] and methods based on Karush -Kuhn-Tucker (KKT) conditions [41] have been developed. (See [117] for a detailed review of the algorithms available for the linear-BLPP).

Turning to solution algorithms for the general BLPP, several intriguing attempts have been proposed to solve it. One early proposal was the Iterative Optimisation Algorithm (IOA) [3, 107]. This method involved solving the Program U for fixed $\mathbf{y}$ and using the solution thus obtained to solve the lower level problem, Program L, and repeatedly iterating between the two programs until some convergence criteria is met. However the IOA was shown to be an exact method for solving a Cournot Nash game [40, 42] rather than the Stackleberg game that the BLPP reflects. The IOA implicitly assumes that the leader is myopic as he does not take into account the follower's reaction to his policy [42]. To be consistent with the Stackleberg model, the leader must be modelled as endowed with knowledge of the follower's reaction function which the leader knows the follower will obey.

The primary difficulty with solving MPECs is that they fail to satisfy certain technical conditions (known as constraint qualifications) at any feasible point [23, 102]. The penalty interior point algorithm (PIPA) was proposed in [72]. Unfortunately a counterexample in [67] demonstrates that PIPA can converge to a nonstationary point. Subsequent research has led to the development of many other techniques to solve the MPEC such as the piecewise sequential quadratic programming in [72], branch-and-bound [6], nonsmooth approaches [32, 87] and smoothing methods [38].

Wrapping up this section, we summarise briefly the use of methods based on meta-heuristics. Meta-heuristics including stochastic optimisation techniques are recognised as useful tools for solving problems such as the BLPPs which do not necessarily satisfy the classical optimisation assumptions of continuity, convexity and differentiability. Techniques include Simulated Annealing (SA) [1], Tabu Search (TS) [48], Genetic Algorithms (GA) [47], Ant Colony Optimisation (ACO) [34], Particle Swam Optimisation (PSO) [57] and Differential Evolution (DE) [94, 95, 108].

SA was used to optimise a chemical process plant layout design problem formulated as a BLPP in [100] and a Network Design Problem formulated as an MPEC [43]. ACO techniques for BLPPs are found in [93]. GAs have been used to solve BLPPs in *inter alia* [73, 86, 111, 114, 122]. PSO was applied to BLPPs in e.g. [126]. DE was used for BLPPs in [61] where an example demonstrated the inability of the TS method implemented in [96] to locate the global optima of a test function. Despite their reported successes in tackling very difficult problems, it must be emphasised that heuristics provide no guarantee of convergence to even a local optimum. Despite this heuristics have been succesfully used to solve a variety of difficult problems such as the BLPP.

6.3 Differential Evolution for Bi-Level Programming (DEBLP)

Differential Evolution for Bi-Level Programming (DEBLP) was initially proposed in [61] to tackle BLPPs arising in transportation systems management. It is developed from the GA Based Approach proposed in [111, 122] but substitutes the use of binary coded GA strings with real coded DE [95] as the meta-heuristic instead.

DE is a simple algorithm that utilises perturbation and recombination to optimise multi-modal functions and has already shown remarkable success when applied to the optimisation of numerous practical engineering problems [94, 95, 108]. On the other hand, many years of research have resulted in the development of a plethora of robust gradient based algorithms for tackling many operations research questions posed as non-linear programming problems (NLP) [9, 70, 85]. If we momentarily ignore the upper level problem, then for fixed $\mathbf{x}$, Eq. 6.1b is effectively an NLP[2] which can be tackled by dedicated NLP tools such as sequential quadratic programming [9, 70, 85]. Such considerations motivated the development of the DEBLP meta-heuristic which sought to synergise DE's well-documented global search capability to optimise the upper level problem with the dedicated NLP tools focused on solving the lower level problem. More importantly, as we shall emphasise later, DEBLP continues to maintain the crucial "leader-follower" paradigm upon which the BLPP is founded.

[2] Note that for fixed $\mathbf{x}$, the lower level problem in the MPEC in Eq. 6.2b can also be solved using deterministic methods. See e.g. [82] for a review of deterministic solution algorithms for VIs.

In the rest of this section, we provide an overview of the operation of the DEBLP algorithm and discuss some of its limitations. However we temporarily neglect consideration of the $q_1 + r_1$ upper level constraints in Program U. Our discussion of the procedure used to ensure satisfaction of the upper level inequality and/or equality constraints is postponed till later (see Section 7.7).

6.3.1 Differential Evolution

Conventional deterministic optimisation methods generally operate on a single trial point, transforming it using search directions computed based on first (and possibly, second) order conditions until some criteria measuring convergence to a stationary point is satisfied [9, 70, 85]. On the other hand, population based meta-heuristics such as DE operate with a population of trial points instead. The idea here is that of improving each member throughout the operation of the algorithm by way of an analogy with Darwin's theory of evolution[3].

Let there be π members in such a population of trial points. Specifically we denote the population at iteration *it* as $\mathscr{P}^{it}$. An illustration of such a population is given in Eq. 6.3. Each member of $\mathscr{P}^{it}$ representing a single trial point $\mathbf{x}_k^{it} = (x_{k,1}^{it}, \ldots, x_{k,n_1}^{it}), k = \{1, \ldots, \pi\}$, also known as an individual, is a n_1 dimensional vector that represents the upper level variables (see Eq. 6.1a). To avoid notational clutter, we drop the *it* superscript as long as it does not lead to confusion. Without loss of generality, we will assume minimization. The DEBLP algorithm is outlined in Algorithm 10 which we elaborate upon in the ensuing paragraphs of this section.

$$\mathscr{P}^{it} = \begin{pmatrix} \mathbf{x}_1^{it} \\ \vdots \\ \mathbf{x}_k^{it} \\ \vdots \\ \mathbf{x}_\pi^{it} \end{pmatrix} = \begin{pmatrix} x_{1,1}^{it} & x_{1,2}^{it} & \cdots & x_{1,n_1}^{it} \\ \vdots & \vdots & \ddots & \vdots \\ x_{k,1}^{it} & x_{k,2}^{it} & \cdots & x_{k,n_1}^{it} \\ \vdots & \vdots & \ddots & \vdots \\ x_{\pi,1}^{it} & x_{\pi,2}^{it} & \cdots & x_{\pi,n_1}^{it} \end{pmatrix} \tag{6.3}$$

6.3.1.1 Generate Parent Population

When the algorithm begins, real parameters in each dimension i of each member k of $\mathscr{P}$, that comprise the parent population, are randomly generated within the lower and upper bounds of the domain of the BLPP as in Eq. 6.4.

$$x_{k,i} = rand(0,1)(x_i^u - x_i^l) + x_i^l, k \in \{1, ..., \pi\}, i \in \{1, ..., n_1\}. \tag{6.4}$$

In Eq. 6.4 $rand(0,1)$ is a pseudo random number generated from an uniform distribution between 0 and 1.

[3] Hence some of these methods are sometimes referred to as evolutionary algorithms in the literature.

Algorithm 10: Differential Evolution for Bi-Level Programming (DEBLP)

1. Randomly generate parent population $\mathscr{P}$ of π individuals.
2. Evaluate $\mathscr{P}$

set iteration counter $it = 1$

3. While stopping criterion not met, do:

   For each individual in $\mathscr{P}^{it}$, do:

   (a) Mutation and Crossover to create a single child from individual.
   (b) Evaluate the child using a hierarchical strategy.
   (c) Selection: If the child is fitter than the individual, the child replaces the parent. Otherwise, the child is discarded.

   End For

   $it = it + 1$

   End While

6.3.1.2 Evaluation

The evaluation process to determine the fitness[4] of a trial point in the population has to be developed within the Stackleberg model [106] since we have to specifically model the leaders taking into account the response (reaction) of the followers to his strategy $\mathbf{x}$. One way to accomplish this is via a "two stage" or hierarchical strategy which is achieved as follows.

In the first stage, for each individual k vector of the leader's decision variables $\mathbf{x}_k$, we solve Program L i.e. Eq. 6.1b to obtain $\mathbf{y}$ by using deterministic methods such as linear programming or sequential quadratic programming [9, 70, 85]. With $\mathbf{y}$ so obtained, we are then able to carry out the second stage which involves computing the value of the upper level objective U, corresponding to each individual vector of the leader's decision variables input in the first stage.

It is worth highlighting that this procedure is *different* from the IOA described earlier in Section 7.3 as DEBLP obviates any iteration between the two levels. Instead, entirely consistent with the "leader-follower" paradigm, the leader's vector $\mathbf{x}_k$ being manipulated by DE is offered as an exogenous input to the lower level program to be solved in the first stage. One obvious drawback of doing this is the resulting increase in computational burden which has been significantly reduced by advances in computing power.

6.3.1.3 Mutation and Crossover

The objective of mutation and crossover is to produce a child vector $\mathbf{w}_k$ from the parent. This is accomplished by stochastically adding to the parent vector the

[4] The term "fitness" used in such evolutionary meta-heuristics is borrowed from its analogy with evolution where Darwin's concept of survival of the fittest is a conerstone. In minimization problems, when comparing two individuals, the fitter individual is the one that evaluates to a *lower* upper level objective.

factored difference of two other randomly chosen vectors from the population as shown in Eq. 6.5.

$$w_{k,i} = \begin{cases} x_{s1,i} + \lambda(x_{s2,i} - x_{s3,i}) & \text{if } rand(0,1) < \chi \text{ or } i = intr(1,n_1) \\ x_{k,i} & \text{otherwise} \end{cases} \quad (6.5)$$

In Eq. 6.5, $s1$,$s2$ and $s3 \in \{1,2,\ldots,\pi\}$ are randomly chosen population indices distinct from each other and also distinct from the current population member index k. $rand(0,1)$ is a pseudo random real number between 0 and 1 and $intr(1,n_1)$ is a pseudo random integer between 1 and n_1. The mutation factor $\lambda \in (0,2)$ is a parameter which controls the magnitude of the perturbation and $\chi \in [0,1]$ is a probability that controls the ratio of new components in the offspring. The or condition in Eq. 6.5 ensures that the child vector $\mathbf{w}_k$ will differ from its parent $\mathbf{x}_k$ in at least one dimension.

We stress that the mutation and crossover strategy shown in Eq. 6.5 is not the only possible strategy available though this is the one used in this work. Other strategies are found in [94, 95, 108]. Nevertheless all the strategies of DE reflect a common theme: the creation of the child vector $\mathbf{w}_k$ via the arithmetic recombination of randomly chosen vectors along with addition of difference vector(s) typified in Eq. 6.5.

6.3.1.4 Enforce Bound Constraints

Mutation and crossover can however produce child vectors that lie outside the bounds of the original problem specification. There are several ways to ensure satisfaction of these constraints. One could set the parameter equal to the limit exceeded or regenerate it within the bounds. Alternatively, following [94], we reset out of bound values in each dimension i half way between its pre-mutation value and the bound violated as shown in Eq. 6.6.

$$w_{k,i} = \begin{cases} \frac{x_{k,i}+x_i^l}{2} \text{if } w_{k,i} < x_i^l \\ \frac{x_{k,i}+x_i^u}{2} \text{if } w_{k,i} > x_i^u \\ w_{k,i} \text{ otherwise} \end{cases} \quad (6.6)$$

6.3.1.5 Selection

Once the hierarchical evaluation process is carried out on the child vector $\mathbf{w}_k$ produced, we can compare the fitness obtained with that of its parent $\mathbf{x}_k$. This means that comparison is against the same k parent vector[5] on the basis of whichever of the two gives a lower value for Program U. Assuming minimization the one that produces a lower value survives to become a parent in the following generation as shown in Eq. 6.7.

[5] This is sometimes referred to as "one to one" comparison in [94, 95, 108].

$$\mathbf{x}_k^{it+1} = \begin{cases} \mathbf{w}_k^{it} & \text{if } U(\mathbf{w}_k^{it}, L(\bullet)) \leq U(\mathbf{x}_k^{t}, L(\bullet)) \\ \mathbf{x}_k^{it} & \text{otherwise} \end{cases} \tag{6.7}$$

These steps are repeated until some user specified termination criteria is met, and this is usually when it reaches the maximum number of iterations, although other criteria are possible [95].

6.3.2 Control Parameters of DE

Unless otherwise stated, for all experiments reported throughout this chapter we used a Mutation Factor, λ, of 0.9 and a Probability of Crossover, χ, of 0.9. The population size, π, and the maximum number of iterations allowed varied for each of the BLPPs we investigated and these will be clearly stated in the relevant sections. Because DEBLP is a stochastic meta-heuristic, we always carry out 30 independent runs with different random seeds. All numerical experiments were conducted using MATLAB™ 7.8 running on a 32 bit Windows™ XP machine with 4 GB of RAM.

6.3.3 Implicit Assumptions of DEBLP

Through the rest of this paper we will demonstrate in examples from various disciplines that DEBLP is a powerful and robust solution methodology for handling a variety of problems formulated as BLPPs. However we are cognizant at the outset two key limitations of our approach:

1. DEBLP is a heuristic: with its strength arising from it avoiding reliance on the objective functions being differentiable and/or satisfying convexity properties and hence able to handle a large class of intrinsically non smooth problems. However it should recognised that for this very reason, it is not generally possible to establish convergence of the algorithm to even a local optimum.
2. DEBLP implicitly assumes that the Program L is convex for fixed $\mathbf{x}$ and can be solved to global optimality by deterministic methods and that failure to solve the lower level problem to global optimality does not affect the solution of Program U.

This section has focused on defining the motivation for, and outlining, the DEBLP meta-heuristic which sought to synergise the exploratory power of DE with robust deterministic algorithms focused on solving the lower level problem. Recognizing its limitations, in the next section, we apply DEBLP to control problems arising from Transportation Systems Management formulated as BLPPs where the lower level program is shown to be convex for a given tuple of the leader's variables.

6.4 Applications to Transportation Systems Management

In this section, we study two problems in transportation systems management. In applications, the leader in Program U could be thought of as a regulatory authority applying control strategies (policy) that influence the travel choices of the followers who are the highway users on the road network. It will be shown under certain assumptions, the followers problem can be established as a VI thus the problems under consideration are MPECs.

6.4.1 The Lower Level Program in Transportation

In the transportation systems management literature, Program L has an interpretation in that it is the mathematical formulation representing the follower's (road user's) route choice [15] on a highway network. This is often referred to as the Traffic Assignment Problem (TAP). Traffic assignment aims to determine the number of vehicles and the travel time on different road sections of a traffic network, given the travel demand between different pairs of origins and destinations [60].

Definition 1. *[115] The journey times on all the routes actually used are equal, and not greater than those which would be experienced by a single vehicle on any unused route.*

The TAP is founded on the behavioral premise of Wardrop's User Equilibrium as given in Definition 1. In effect this states that user equilibrium is attained when no user can decrease his travel costs by unilaterally changing routes. The TAP provides the link flow vector (v) when user equilibrium is attained.

To facilitate exposition of Program L, consider a transportation network represented as a graph with N nodes and A links/arcs, and let:

P: the set of all paths/routes in the network,

H: the set of all Origin Destination (OD) pairs in the network,

P_h: the set of paths connecting an OD pair $h, h \in H$,

F_p: the flow on route/path $p, p \in P$,

v_a: the link flow on link $a\, \mathbf{v} = [v_a], a \in A$,

$c_a(v_a)$: the travel cost of utilising the link a, as a function of link flow v_a on that link only, $\mathbf{c}(\mathbf{v}) = [c_a(v_a)], a \in A$

c_p: the travel cost of path $p, p \in P$,

δ_{ap} : a dummy variable that is 1 if the path $p, p \in P$ uses link a $, a \in A$, 0 otherwise and

Ω: the set of feasible flows and demands.

On the demand side, we assume that there is an amount of demand $d_h, h \in H$ ($d_h \geq 0$) wishing to travel between OD pair h and μ_h is the minimum travel cost that OD pair $h, h \in H$.

6.4.1.1 TAP as a Variational Inequality

Lemma 6.1. *Wardrop's Equilibrium Condition of route choice implies that at equilibrium the following conditions are simultaneously satisfied:*

$$
\begin{aligned}
&F_{p\in P_h} \geq 0 \Leftrightarrow c_{p\in P_h} = \mu_h \quad \forall h \in H, \forall p \in P;\\
&F_{p\in P_h} = 0 \Leftrightarrow c_{p\in P_h} \geq \mu_h \quad \forall h \in H, \forall p \in P;\\
&d_h = \sum_{p\in P_h} F_p \quad \forall h \in H, \forall p \in P;
\end{aligned}
$$

Lemma 6.1 states that path p connecting OD pair h will be used by the travellers if and only if the cost of travelling on this route is the minimum travel cost between that OD pair. The Variational Inequality (VI) in Eq. 6.8 restates Wardrop's Equilibrium Condition.

$$
\text{Find } \mathbf{v}^* \in \Omega \text{ such that } \mathbf{c}(\mathbf{v}^*)^{\mathsf{T}}(\mathbf{v}-\mathbf{v}^*) \geq 0, \forall \mathbf{v} \in \Omega \tag{6.8}
$$

Proposition 6.1. *The solution of the Variational Inequality defined in Eq. 6.8 results in a vector of link flows demands ($\mathbf{v}^* \in \Omega$) that satisfies Wardrop's Equilibrium Condition of route choice given by Lemma 6.1.*

Proof. For a proof of Proposition 6.1, see [28, 105]. □

6.4.1.2 Convex Optimisation Reformulation

In the particular instance (and in the cases considered in this chapter) when the travel cost of using a link is dependent only on its own flow[6], there exists an equivalent convex optimisation program for the VI (Eq. 6.8) as shown in Eq. 6.9.

$$
\min_{\mathbf{v}} L = \sum_{\forall a} \int_0^{v_a} c_a(z)dz \tag{6.9a}
$$

Subject to:

$$
\sum_{p\in P_h} F_p = d_h \, , h \in H \tag{6.9b}
$$

$$
v_a = \sum_{p\in P} F_p \delta_{ap} \, , a \in A \tag{6.9c}
$$

$$
F_p \geq 0, p \in P. \tag{6.9d}
$$

The objective of the program in Eq. 6.9 is a mathematical construct, with no behavioral interpretation, employed to solve for the equilibrium link flows that satisfies Wardrop's Equilibrium Condition [103]. In this program, the first constraint states that the flow on each route used by each OD pair is equal to the total demand for that OD pair. The second constraint is a definitional constraint which stipulates that

[6] This is known as the separability assumption.

the flow on a link comprises flow on all routes that use that link. The last constraint restricts the equilibrium flows and demands to be non negative. These linear constraints define Ω. Since Ω is closed and convex, the equilibrium link flows $\mathbf{v}^* \in \Omega$ are unique [15]. In practice, it is usually the case that traffic assignment algorithms (see examples in texts such as [89, 103]) are used to solve Program L.

6.4.2 Continuous Optimal Toll Pricing Problem (COTP)

The continuous optimal toll pricing problem involves selecting an optimal toll level for each predefined tolled link in the network [11]. With a view to controlling congestion, there has been renewed interests by transportation authorities globally to study this "road pricing " problem (e.g. Singapore, London, Stockholm).

6.4.2.1 Model Formulation

In addition to the notation defined at the start of this section, we introduce the following notation to describe the COTP. Let:

$t_a(v_a)$: the travel time on link a, as a function of link flow v_a on that link only,
T: the set of links that are tolled $T \subseteq A$
τ: the vector of tolls, $\tau = [\tau_a],\ a \in T$
$\tau_a^{\max}, \tau_a^{\min}$: the upper and lower bounds of toll charge on link a, $a \in T$

Total travel cost, conventionally measured as the sum product of the travel times and traffic flows on all links in the network, may be interpreted as the social cost of the transport sector and acts as a proxy for the resource cost to the economy of the highway system. The objective of the upper level decision maker in the COTP is to minimise this by encouraging more efficient routing of traffic by levying tolls on the road users in the network. The upper level program is the system in Eq. 6.10.

$$\min_{\tau} U = \sum_{a \in A} v_a t_a(v_a) \tag{6.10a}$$

Subject to:

$$\begin{aligned} &\tau_a^{\min} \le \tau_a \le \tau_a^{\max}, \quad a \in T \\ &\tau_a = 0, \quad a \notin T \end{aligned} \tag{6.10b}$$

Note however that $\mathbf{v}$ can only be obtained by solving Program L in Eq. 6.9. Thus in terms of Figure 6.1, the policy variables $\mathbf{x}$ is the toll vector τ and the follower's response is the traffic routing that manifests in the vector of link flows on the road network $\mathbf{v}$ that in turns affect the leader's objective.

Recall that in defining the lower level program in Eq. 6.9, the road user was assumed to consider the travel cost of utilising an arc a, $a \in A$. Eq. 6.11 maps the travel time $t_a(v_a)$ on an arc a, into the equivalent travel costs $c_a(v_a)$.

$$c_a(v_a) = \begin{cases} t_a(v_a) + \tau_a & \text{if } a \in T \\ t_a(v_a) & \text{otherwise} \end{cases} \tag{6.11}$$

6.4.2.2 Previous Work on the COTP

Various solution algorithms have been proposed for the COTP. Yang and Lam proposed a linearisation based method that uses derivative information to form approximations to the upper level objective [118] known as a sensitivity based analysis algorithm (SAB). However it has been pointed out [122] the global optimality of the SAB algorithm is not assured and that obtaining a local optimum is indeed possible. Another derivative-based method was derived from constraint accumulation [66]. A review of algorithms for the COTP is found in [111].

6.4.2.3 Example

We illustrate the use of DEBLP to solve the COTP with an example from [118]. Fig. 6.2 shows the network which has 6 nodes and 7 links. Link numbers are written above the links and node numbers are indicated accordingly. There are two OD pairs between nodes 1 and 3 and between 2 and 4 of 30 trips each. The rest of the nodes represent junction/intersections of the road network and travel is in the direction indicated by the arrows. The link travel times $t_a(v_a)$ take the explicit function forms as given in Eq. 6.12.

$$t_a(v_a) = t_a^0(1 + 0.15(\frac{v_a}{Cap_a})^4) \tag{6.12}$$

In Eq. 6.12, t_a^0 is the free flow travel time of the link a and Cap_a is the capacity of link a. The parameter details for the network and the upper bound on tolls $\tau_a^{\max}$ are found in [118] and given in Table 6.1. Note that $\tau_a^{\min} = 0, \forall a \in T$.

For this example, we use a population size, π, of 20 and allowed a maximum of 50 iterations in each of 30 runs. Table 6.2 compares the results of DEBLP with that of two deterministic algorithms (direct from [118] and our implementation of the algorithm of [66] together with a GA based method from [122]. UPO refers to the value of (Upper level) Objective in Eq. 6.10. It can be seen from Table 6.2 that the four different algorithms provided different tolls underlying the multimodal nature of this problem. However the upper level objective function values are the same in all cases. This bears testimony to the multimodal nature of the COTP where many different toll vector tuples could potentially result in attaining the same upper level objective function value.

6.4.3 *Continuous Network Design Problem*

The continuous network design problem (CNDP) aims to determine the optimal capacity enhancements of existing facilities of a traffic network [43]. Care has to be taken when solving the CNDP because additional capacity can counter productively

Table 6.1 Network Parameters for COTP Example

Link a	t_a^0	Cap_a	$\tau_a^{\max}$
1	8	20	5
2	9	20	5
3	2	20	2
4	6	40	2
5	3	20	2
6	3	25	2
7	4	25	2

Table 6.2 Comparison of existing against DEBLP results for COTP Example

Method	Deterministic		Stochastic	
Tolls	[118]	Method of [66]	[122]	DEBLP
Link 1	3.82	2.667	4.324	3.824
Link 2	4.265	3.548	4.976	3.92
Link 3	0.472	0.038	0.035	0.564
Link 4	0.476	0.154	1.759	0.462
Link 5	0.294	0.116	0.016	0.145
Link 6	0.472	0.038	0.127	0.396
Link 7	0.294	0.116	0.013	0.111
UPO	628.6	628.6	628.6	628.6

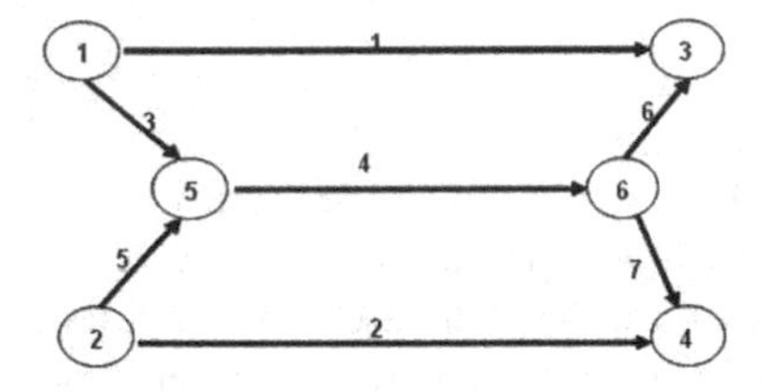

Fig. 6.2 Network for COTP Example [118]

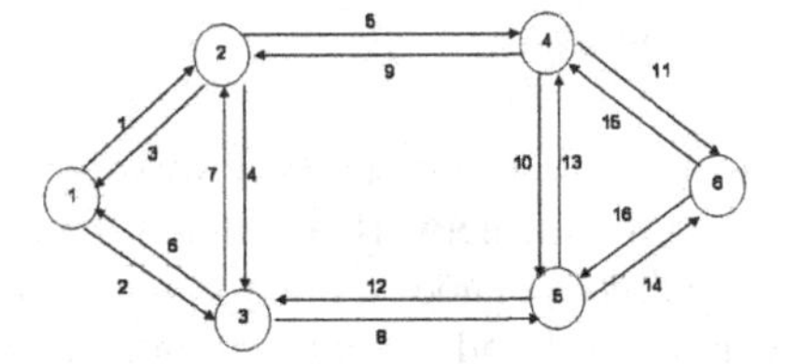

Fig. 6.3 Network for CNDP Example 1 [27]

increase the total network travel time and this is a phenomenon is known as Braess's paradox [17]. Braess's paradox has been known to occur in transportation [17] and telecommunication networks [63].

6.4.3.1 Model Formulation

To proceed with this example, we introduce additional notation as follows (others as previously defined):

κ: the set of links that have their individual capacities enhanced, $\kappa \subseteq A$.

β: the vector of capacity enhancements, $\beta = [\beta_a]$, $a \in \kappa$

$\beta_a^{\max}$,$\beta_a^{\min}$: the upper and lower bounds of capacity enhancements, $a \in \kappa$.

d_a: the monetary cost of capacity increments per unit of enhancement, $a \in \kappa$.

Cap_a^0: existing capacity of link a, $a \in A$.

θ: conversion factor from monetary investment costs to travel cost units.

In the CNDP, the regulator aims to minimise the sum of the total travel times and investment costs with constraints on the amount of capacity additions while Program L determines the user's route choice, for a given β, once again based on Wardrop's principle of route choice as mentioned previously. Hence the CNDP seeks a $|\kappa|$ dimension vector of capacity enhancements optimal to the following BLPP in Eq. 6.13:

$$\min_{\tau} U = \sum_{a \in A} v_a t_a(v_a) + \sum_{\forall a \in K} \theta d_a \beta_a \tag{6.13a}$$

subject to:

$$\begin{array}{l} \beta_a^{\min} \leq \beta_a \leq \beta_a^{\max} \quad a \in \kappa; \\ \beta_a = 0 \quad a \notin \kappa \end{array} \tag{6.13b}$$

where v is the solution of a lower level TAP (Program L) Eq. 6.9, parameterised in the vector of capacity enhancements for the fixed demand case. We map the travel times to the travel costs by means of Eq. 6.14.

$$c_a(v_a) = \begin{cases} t_a^0(1 + 0.15(\frac{v_a}{Cap_a^0 + \beta_a})^4) & \text{if } a \in \kappa \\ t_a^0(1 + 0.15(\frac{v_a}{Cap_a^0})^4) & \text{if } a \notin \kappa \end{cases} \tag{6.14}$$

6.4.3.2 Previous Work on CNDP

The CNDP has been investigated by many researchers and various solution algorithms have so far been proposed. Meng et al transformed the CNDP into a single level continuously differentiable problem using a marginal function and solved the resulting formulation with the Augmented Lagragian method [75]. Chiou investigated several variants of the descent based Karush-Khun-Tucker (KKT) approaches [24]. Stochastic meta-heuristics have also been used; GAs were applied in [27] and the use of SA has been reported in [43].

6.4.3.3 Example 1: Hypothetical Network

The network for the first example is taken from [27] and reproduced in Fig. 6.3. This network has 6 nodes and 2 OD pairs; the first between nodes 1 and 6 of 10 trips and the second, between nodes 6 and 1 of 20 trips. Please refer to [27] for the link parameter details. Note that $\beta_a^{\min} = 0$ and $\beta_a^{\max} = 20$, $\forall a \in \kappa, \kappa \subseteq A$ as in [27]. We assumed a population size, π, of 20 and allowed a maximum of 150 iterations. Table 6.3 summarises the results that have been reported previously and compares it with the results reported in our paper. UPO refers to the value of (upper level) objective in Eq. 6.13. NFE is the number of function evaluations. Note that the number of lower level programs solved equal to population size multiplied by the maximum number of iterations allowed. SD is the standard deviation over 30 runs. Our results are based on the mean of these 30 runs. Though the SD of the GA method is much lower, [27] also reported using local search method to aid the search process which accounts for the higher NFE as well.

6.4.3.4 Example 2: Sioux Falls Network

The second example is the CNDP for the Sioux Falls (South Dakota,USA) network with 24 nodes, 76 links and 552 OD pairs. The network parameters and OD details are found in [75]. Only 10 links out of the 76 are subject to improvements.

While this network is clearly larger and arguably more realistic, the problem dimension (i.e. leader's variables simultaneously optimised) is smaller than in Example 1, since 10 links are subject to improvement rather than the 16 links in the former. This offers an explanation as to why the number of function evaluations (NFE) reported in all studies compared is less than for the first example. The results are compared in Table 6.4. Our results show the mean of 30 runs with different random seeds.

It can be deduced from Table 6.4 that DEBLP is able to locate the global optimum; again with a lesser number of iterations than the SA method in [43]. More interestingly, DEBLP required less iterations than the deterministic method of [75]. The standard deviation is also very low which suggests that this heuristic is reasonably robust as well.

Table 6.3 Comparison of existing against DEBLP results for CNDP Example 1

Method:	Deterministic		Stochastic		
Source	[24]	[75]	[43]	[27]	DEBLP
UPO	534	532.71	528.49	519.03	522.71
NFE	29	4,000	24,300	10,000	3,000
SD	–Not Reported–			0.403	1.34

Table 6.4 Comparison of existing against DEBLP results for CNDP Example 2

Method:	Deterministic		Stochastic	
Source	[24]	[75]	[43]	DEBLP
UPO	82.57	81.75	80.87	80.74
NFE	10	2,000	3,900	1,600
SD	—-Not Reported—-			0.002

6.5 Applications to Parameter Estimation Problems

In this section we derive the Error-In-Variables model and show that it can be formulated as a BLPP and apply it to 2 examples from [49]. Parameter estimation is an important step in the verification and utilization of mathematical models in many fields of science and engineering [37, 49, 59]. In the classical least-squares approach to parameter estimation, it is implicitly assumed that the set of independent variables is not subject to measurement errors [46]. On the other hand, the error-in-variables (EIV) approach assumes that there are measurement errors in *all* variables [16, 98].

6.5.1 Formulation of EIV Model

We consider models of the implicit form as in Eq. 6.15.

$$f(x,y) = 0 \tag{6.15}$$

In Eq. 6.15, x is the vector of n_1 unknown parameters, y is the vector of n_2 measurement variables and f is the system of algebraic functions. The measured variables are the sum of the true values ζ_m which are unknown and the additive error term ε_m at the data point m as shown in Eq. 6.16.

$$y_m = \zeta_m + \varepsilon_m \tag{6.16}$$

We assume that the error is normally distributed with zero mean and possessing a known covariance matrix. The vector of unknown parameters x can be estimated from the solution of the constrained optimisation problem in Eq. 6.17.

$$\begin{aligned} &\min_{\hat{x},\hat{y}} \sum_{m=1}^{M} (\hat{\mathbf{y}}_m - \mathbf{y}_m)^{\mathsf{T}} \Lambda^{-1} (\hat{\mathbf{y}}_m - \mathbf{y}_m) \\ &\text{subject to} \\ &f(\hat{\mathbf{y}}_m, \hat{x}) = 0,\ m = 1, \ldots, M \end{aligned} \tag{6.17}$$

As mentioned, we do not know the true values of ζ_m. However they are approximated from the optimisation as the fitted variables $\hat{\mathbf{y}}_m$. Assuming that the covariance matrix Λ is the same in each experiment and diagonal, we write Eq. 6.17 as Eq. 6.18.

$$\begin{aligned} &\min_{\hat{x},\hat{y}} \sum_{m=1}^{M} \sum_{i=1}^{n_2} \frac{(\hat{y}_{m,i} - y_{m,i})^2}{\sigma_i^2} \\ &\text{subject to} \\ &f(\hat{\mathbf{y}}_m, \hat{x}) = 0,\ m = 1, \ldots, M \end{aligned} \tag{6.18}$$

In Eq. 6.18, σ_i is the standard deviation of variable i in all the experiments. Following [49], we can write the EIV model as a BLPP of the form of Eq. 6.19.

$$\text{Program } U \left\{ \min_{\hat{x}} \sum_{m=1}^{M} \sum_{i=1}^{n_2} \frac{(\hat{y}_{m,i} - y_{m,i})^2}{\sigma_i^2} \right. \tag{6.19a}$$

where for given $\mathbf{x}$, $\mathbf{y}$ is the solution to the lower level program (Program L):

$$\text{Program } L \left\{ \begin{aligned} &\min_{\hat{x},\hat{y}} \sum_{m=1}^{M} \sum_{i=1}^{n_2} \frac{(\hat{y}_{m,i} - y_{m,i})}{\sigma_i^2} \\ &\text{subject to} \\ &\mathbf{f}(\hat{y}_{m,i}, \hat{x}) = 0,\ m = 1, \ldots, M, i = 1, \ldots, n_2 \end{aligned} \right. \tag{6.19b}$$

For a survey of the alternative optimisation based formulations of the EIV model, the reader is referred to [59].

6.5.2 Examples

We present 2 examples of the EIV model that were solved using deterministic methods in the cited references. Note that we consider only a single common variance term for all variables and we can eliminate it from further consideration. In all our experiments of DEBLP we assumed a population size, π, of 20 and allowed a maximum of 100 iterations.

6.5.2.1 Example 1: "Kowalik Problem"

Consider the model due to Moore et al in [76] known as the "Kowalik Problem" where we estimate the equation of the form in Eq. 6.20.

$$\hat{y}_{m,1} = \frac{x_1 y_{m,2}^2 + x_1 x_2 y_{m,2}}{y_{m,2}^2 + y_{m,2} x_3 + x_4} \tag{6.20}$$

We have 11 data points for this model (see [49] for the data set). It is assumed that $y_{m,1}$ contains errors, and $y_{m,2}$ is error-free. The resulting BLPP is shown in Eq. 6.21. Notice that the lower level equality constraint in Eq. 6.21 is the model formulation hypothesised in Eq. 6.20.

$$\begin{aligned} &\min_{\hat{x}} \sum_{m=1}^{11} (\hat{y}_{m,1} - y_{m,1})^2 \\ &\text{subject to} \\ &\min_{y} \sum_{m=1}^{11} (\hat{y}_{m,1} - y_{m,1})^2 \\ &\hat{y}_{m,1}(y_{m,2}^2 + y_{m,2} x_3 + x_4) - x_1 y_{m,2}^2 - x_1 x_2 y_{m,2} = 0 \end{aligned} \tag{6.21}$$

30 runs of DEBLP were performed for this problem with a maximum of 100 iterations allowed per run and a population size, π of 20. Following [49], the parameter bounds are assumed to be between -0.2892 and 0.2893 for each of the 4 upper level variables x_1,x_2,x_3 and x_4. Table 6.5 shows the results which clearly agrees with that reported in [49]. In this table UPO refers to the objective of the upper level in Eq. 6.21. Note that the standard deviation of the UPO over the 30 independent DEBLP runs conducted was less than 1×10^{-5}.

6.5.2.2 Example 2: "Linear Fit"

The model we intend to estimate is a linear equation of the form in Eq. 6.22. The 10 data points are from [49]. Compared to Example 1, here we assume that measurement errors are present in *both* $y_{m,1}$ and $y_{m,2}$, $m = \{1, \ldots, 10\}$.

$$\hat{y}_{m,2} = x_1 + x_2 \hat{y}_{m,1} \tag{6.22}$$

Assuming a common variance for each data tuple $\{y_{m,1}, y_{m,2}\}$, we can estimate the vector of unknown x parameters via the BLPP in Eq. 6.23.

$$\begin{aligned}
&\min_{\hat{x}} \sum_{m=1}^{10} \sum_{i=1}^{2} (\hat{y}_{m,i} - y_{m,i})^2 \\
&\text{subject to} \\
&\min_{y} \sum_{m=1}^{10} \sum_{i=1}^{2} (\hat{y}_{m,i} - y_{m,i})^2 \\
&\hat{y}_{m,2} - x_1 - x_2 \hat{y}_{m,1} = 0
\end{aligned} \tag{6.23}$$

The results of 30 runs of DEBLP (with a maximum of 100 iterations allowed per run and a population size π of 20) for this problem are shown in Table 6.6. Again the standard deviation over the 30 runs was less than 1×10^{-5}. As with Example 1, the results obtained by DEBLP agrees with those reported in [49].

Table 6.5 Parameter Estimation Example 1 ("Kowalik Problem")

Variable	DEBLP	[49]
x_1	0.1928	0.1928
x_2	0.1909	0.1909
x_3	0.1231	0.1231
x_4	0.1358	0.1358
UPO	0.000307	0.000307

Table 6.6 Parameter Estimation Example 2 ("Linear Fit")

Variable	DEBLP	[49]
x_1	5.7840	5.784
x_2	-0.544556	-0.54556
UPO	0.61857	0.61857

6.6 Handling Upper Level Constraints

The keen reader would notice that up to this point our discussions and our numerical examples have neglected mention of constraints in the upper level problems (cf. Eqn. 6.1a). We have in fact thus far only assumed the presence of bound constraints and described a technique to ensure that the population remains within the search domain which was sufficient for the problem examples investigated. Before proceeding to our next application area for BLPPs, we outline in this section, necessary modifications to DEBLP to enable it to handle them effectively.

6.6.1 Overview of Constraint Handling Techniques with Meta-Heuristics

In their most basic form, meta-heuristics do not have the capability to handle general constraints aside from bound constraints. However since real world problems generally have linear and nonlinear constraints, a large amount of research effort has been expended on the topic of constraint handling with such algorithms. In the past few years many techniques have been proposed. Among others these include penalty methods [121], adaptive techniques [104], techniques based on multiobjective optimisation [25, 65] etc.

The penalty method transforms the constrained problem into an unconstrained one. However one of the drawbacks of this method when applied with meta-heuristics is that the solution quality is sensitive to the penalty parameter used. The penalty parameter itself is problem dependent [99]. This method also encounters difficulties when solutions lie at the boundary of the feasible and infeasible space.

Recall the selection criteria of the DEBLP in Algorithm 10. In the presence of constraints, when we are deciding whether to accept or reject the child, $\mathbf{w}_k$, it is no longer a case of comparing the values of objective U attained. The key consideration is how one would say, decide between a infeasible individual with low U and a feasible individual but higher U.

Intuitively one could conclude that a feasible individual is better than the infeasible individual because the aim is to ultimately seek solutions that minimise the objective function and satisfy all the constraints. This viewpoint however ignores the fact that the meta-heuristics are generally stochastic by design. There exists the possibility that the infeasible individuals could in fact be better than the feasible one at some iterations during the algorithm [124]. The question then is how to strike the right balance between objective and constraints.

6.6.2 *Stochastic Ranking*

Runarsson and Yao [99, 101] proposed an alternative constraint handling method known as stochastic ranking (SR)[7] to aid in answering this question. To use SR, the first step is to obtain a measure of the constraint violation, $v(\mathbf{x}_k)$, of vector $\mathbf{x}_k$ using Eqn. 6.24. The first term on the RHS of Eq. 6.24 sums the maximum of either 0 or the value of the inequality constraint $G_j(\mathbf{x}_k), j \in \{1,\ldots,q_1\}$[8]. The second term sums the absolute value of each of the equality constraints $E_j(\mathbf{x}_k), j \in \{1,\ldots,r_1\}$.

$$v(\mathbf{x}_k) = \sum_{j=1}^{q_1} \max\{0, G_j(\mathbf{x}_k)\} + \sum_{j=1}^{r_1} \left|E_j(\mathbf{x}_k)\right| \tag{6.24}$$

The key operation of SR involves counting how many comparisons of adjacent pairs of solutions are dominated by the objective function and constraint violations. This is accomplished in SR through a stochastic bubble sort like procedure that is used to rank[9] the population. This comparison is illustrated in Algorithm 11 where $rand(0,1)$ is a pseudo random real number between 0 and 1. The method requires a probability factor η which should be less than 0.5 to create a bias against infeasible solutions [99].

[7] The source code of SR is available at `http://notendur.hi.is/tpr/index.php?page=software/sres/sres`, accessed Oct 2011.

[8] In Eq. 6.1a, all the upper level inequality constraints are in the form "≤ 0".

[9] With π population members, ranking results in the best ranked 1 (highest rank) and the worst ranked π (lowest rank).

Suppose we have two individuals $\mathbf{x}_{k1}$ and $\mathbf{x}_{k2}, k1 \neq k2$. If both do not violate constraints *or* if a pseudo random real number is less than or equal to η, we swap their rank order based on the objective function obtained, with the lower one being assigned a higher rank. Otherwise we swap their ranks based on the constraint violations, again with the lower constraint violation being assigned a higher rank. Working our way through the population to be ranked, we continue comparing adjacent members according to Algorithm 11 and swapping ranks. When no change in rank order occurs, SR terminates.

Algorithm 11: Stochastic Ranking

```
if v(x_k1) = 0 and v(x_k2) = 0 or rand(0,1) ≤ η then
    rank based on objective function value only
else
    rank based on constraint violation only
end if
```

6.6.3 *Revised DEBLP with Stochastic Ranking*

DEBLP-SR, as presented in Algorithm 12, is the result of incorporating SR in DEBLP. *Italics* highlight the changes between DEBLP in Algorithm 10 and DEBLP-SR in Algorithm 12. These are summarised as follows:

1. Evaluation of both the upper level objective and constraint violation for each member of the parent and child population.
2. Instead of the one to one selection criteria discussed in Section 7.4, we propose to pool the parent and child population (along with the corresponding objective values and constraint violations) together as an input into SR.
3. Combining parents and children will lead to a population size of 2π. Hence the selection process will only retain the top π ranked individuals output by SR to constitute the population at the next iteration. The remainder are discarded.

In the next section, we apply DEBLP-SR to a examples of BLPPs that are in fact characterised by the presence of upper level constraints. It will be shown that DEBLP-SR continues to be a robust meta-heuristic in such applications.

6.7 Applications to Generalised Nash Equilibrium Problems

Game theory [116] is a branch of social science that provides methodologies to study behaviour when rational agents seek to maximise personal gains in the presence of others symmetrically doing the same simultaneously. The solution concept of such games was devised by Nash in [83, 84]. The game attains a Nash Equilibrium (NE) if no one player can unilaterally improve her payoff given the strategic decisions of

Algorithm 12: DEBLP with Stochastic Ranking (DEBLP-SR)

1. Randomly generate parent population $\mathscr{P}$ of π individuals.
2. Evaluate $\mathscr{P}$ *and obtain constraint violations using Eq. 6.24*
set iteration counter $it = 1$
3. While stopping criterion not met, do:
    For each individual in $\mathscr{P}^{i}t$, do:
        a) Apply Mutation and Crossover to create a single child from individual.
        b) Evaluate child *and obtain constraint violations using Eq. 6.24*
    End For
4. *Combine parents and children violations and objectives.*
5. *Apply stochastic ranking*
6. Selection: *retain the top π ranked individuals to form new population* $\mathscr{P}^{it+1}$
   $it = it + 1$
   End While

all other players. While establishing that an outcome is not a NE (by establishing that a player can profitably deviate) is usually not difficult, locating the NE itself is more challenging. In this section we show how the process of determining NE in some games can be formulated as a BLPP and illustrate the performance of DEBLP on some example problems from the literature.

6.7.1 The Generalised Nash Equilibrium Problem

We are concerned with a specific Nash Game known as the Generalised Nash Equilibrium Problem (GNEP). In the GNEP, the players' payoffs and their strategies are continuous (and subsets of the real line) but most critically the GNEP embodies the distinctive feature that players face constraints depending on the strategies their opponents choose. This distinctive feature is in contrast to a standard Nash Equilibrium Problem (NEP) where the utility/payoff/reward the players obtain depend solely on the decisions they make and their actions are not restricted as a result of the strategies chosen by others. The ensuing constrained action space in GNEPs makes them more difficult to resolve than standard NEPs discussed in monographs such as [116]. As will be demonstrated in this section, the technique here can nevertheless be applied to standard NEPs.

The GNEP under consideration is a single shot[10] game with a set Γ of players indexed by $i \in \{1,2,...,\rho\}$ and each player can play a strategy $x_i \in X_i$ which all players are assumed to announce simultaneously. X is the collective action space for all players. In a standard NEP, $X = \prod_{i=1}^{\rho} X_i$, i.e. X is the Cartesian product.

In contrast, in a GNEP, the feasible strategies for player $i, i \in \Gamma$ depend on the strategies of all other players [4, 39, 53, 112]. We denote the feasible strategy space

[10] It is one-off and not played repeatedly in a dynamic sense.

of each player by the point to set mapping: $\mathscr{C}^i : X^{-i} \rightarrow X^i, i \in \Gamma$ that emphasises the ability of other players to influence the strategies available to player i [39, 51, 112]. The distinction between a conventional Nash game and a GNEP can be viewed as analogous to the distinction between unconstrained and constrained optimisation.

To give stress to the variables chosen by player i, we sometimes write $\mathbf{x} = (x_i, x_{-i})$ where x_{-i} is the combined strategies of all players in the game *excluding* that of player i i.e. $x_{-i} = (x_1, ..., x_{(i-1)}, x_{(i+1)}, ..., x_\rho)$. Note that the notation$(x_i, x_{-i})$ *does not* mean that the components of $\mathbf{x}$ are somehow reordered such that x_v becomes the first block. In addition, let $\phi_i(\mathbf{x})$ be the payoff/reward to player $i, i \in \Gamma$ if $\mathbf{x}$ is played.

Definition 2. *[112] A combined strategy profile $\mathbf{x}^* = (x_1^*, x_2^*, ..., x_\rho^*) \in X$ is a Generalised Nash Equilibrium for the game if:*

$$\begin{aligned} &\phi_i(x_i^*, x_{-i}^*) \geq \phi_i(x_i, x_{-i}^*), \\ &\forall x_i \in \mathscr{C}(x_{-i}^*), i \in \Gamma \end{aligned} \tag{6.25}$$

At a Nash Equilibrium no player can benefit (increase individual payoffs) by unilaterally deviating from her current chosen strategy. Players are also assumed not to cooperate and in this situation each is doing the best she can given what her competitors are doing [45, 62, 116]. For a GNEP, the strategy profile $\mathbf{x}^*$ is a Generalised Nash Equilibrium (GNE) if it is *both* feasible with respect to the mapping $\mathscr{C}^i$ *and* if it is a maximizer of each player's utility over the constrained feasible set [51].

6.7.2 *Nikaido Isoda Function*

The Nikaido Isoda (NI) function in Eq. 6.26 is an important construct much used in the study of Nash Equilibrium problems [39, 52, 53]. Its interpretation is that each summand shows the increase in payoff a player will receive by unilaterally deviating and playing a strategy $y_i \in \mathscr{C}(x_{-i})$ while all other players play according to x.

$$\Psi(\mathbf{x}, \mathbf{y}) = \sum_{1}^{\rho} [\phi_i(y_i, x_{-i}) - \phi_i(x_i, x_{-i})] \tag{6.26}$$

The NI function is always non-negative for any combination of $\mathbf{x}$ and $\mathbf{y}$. Furthermore, this function is everywhere non-positive when either $\mathbf{x}$ or $\mathbf{y}$ is a NE by virtue of Definition 2 since at a NE no one player should be able to increase their payoff by unilaterally deviating. This result is summarised in Definition 3.

Definition 3. *[53] A vector $\mathbf{x}^* \in X$ is called a Generalised Nash Equilibrium (GNE) if $\Psi(\mathbf{x}, \mathbf{y}) = 0$.*

6.7.3 *Solution of the GNEP*

Proposition 6.2 establishes the key result that the GNEP can be formulated as a BLPP.

Proposition 6.2. *The Generalised Nash Equilibrium is the solution to the BLPP in Eq. 6.27.*

$$\min_{(\mathbf{x},\mathbf{y})} \quad f(\mathbf{x},\mathbf{y}) = (\mathbf{y}-\mathbf{x})^T(\mathbf{y}-\mathbf{x}) \tag{6.27a}$$

$$\text{subject to } x^i \in \mathscr{C}^i(x^{-i}), \forall i \in \Gamma. \tag{6.27b}$$

where $\mathbf{y}$ solves

$$\max_{(\mathbf{x},\mathbf{y})}(\phi_1(y^1,x^{-1}) + \ldots + \phi_\rho(y^\rho,x^{-\rho})) = \max_{(\mathbf{x},\mathbf{y})} \sum_{i=1}^{n} [\phi_i(y_i,x_{-i}) - \phi_i(x_i,x_{-i})] \tag{6.28a}$$

$$\text{subject to } y^i \in \mathscr{C}^i(x^{-i}), \forall i \in \Gamma. \tag{6.28b}$$

Proof. For a proof of Proposition 6.2, see [112]. □

The upper level problem (Eq. 6.27a) is a norm minimization problem subject to strategic variable constraints (Eq. 6.27b). The objective function of the lower level problem (Eq. 6.28) is exactly the Nikado Isoda function (Eq. 6.26).

Proposition 6.3. *The optimal value of the upper level objective in Eq. 6.27a, $f(\mathbf{x},\mathbf{y})$, is 0 at the Generalised Nash Equilibrium.*

Proof. For a proof of Proposition 6.3, see [14, 112]. □

Proposition 6.3 serves the critical role of being the termination criteria of the DEBLP. Although DEBLP and DEBLP-SR are heuristic in nature, Proposition 6.3 enables us to detect that we have found the solution to the GNEP.

6.7.4 *Examples*

In this section, we present four numerical examples of GNEPs sourced from the literature. The first case study is in fact a standard NEP and it serves to demonstrate that the BLPP formulation proposed here can also be applied in this situation. We then impose a constraint which transforms the standard NEP into a GNEP which serves as the second example. The third example has origins in pollution abatement modeling while the last example is an internet switching model from [58].

6.7.4.1 Example 1

Example 1 is a non-linear Cournot-Nash Game with 5 players from [81]. As mentioned, this is a standard NEP i.e. where the feasible strategies of each player is unconstrained. The profit function for player $i, i \in \{1,...,5\}$, comprising the difference between revenues and production costs, is given by: $\phi_i(\mathbf{x}) = (5000^{\frac{1}{1.1}} (\sum_{i=1}^{5} x_i)^{-(\frac{1}{1.1})})x_i - \omega_i x_i + (\frac{\alpha_i}{\alpha_i+1})\gamma_i^{\frac{-1}{\alpha_j}} x_i^{\frac{\alpha_i+1}{\alpha_i}}$. The player dependent parameters (ω_i, γ_i and α_i) are found in [81, 87].

The feasible space for this problem is the positive axis since production cannot be negative. The solution of the NEP is $\mathbf{x}^* = [36.9318, 41.8175, 43.7060, 42.6588, 39.1786]^{\mathsf{T}}$ [50, 81].

6.7.4.2 Example 2

Using the same parameters as in Example 1, and introducing a production constraint on total output of all players [11] as in [87], Example 1 is transformed into a GNEP.

The feasible space for the resulting GNEP is defined by [87]:

$$X = \{\mathbf{x} \in \mathbb{R}^5 | x_i \geq 0 \ \forall i \in \{1, ..., 5\}, \sum_{i=1}^{5} x_i \leq 100\}$$

$\mathbf{x}^*$ is $[14.050, 17.798, 20.907, 23.111, 24.133]^{\mathsf{T}}$ [54].

6.7.4.3 Example 3

This problem describes an internet switching model with 10 players originally proposed in [58] and also studied in [54]. The cost function for player $i, i \in \{1, \ldots, 10\}$ is given by $\phi_i(\mathbf{x}) = -(\frac{x_i}{(x_1 + \cdots + x_{10})})(1 - \frac{(x_1 + \ldots + x_{10})}{1})$. The feasible space is $X = \{\mathbf{x} \in \mathbb{R}^{10} | x_i \geq 0.01, i \in \{1, \ldots, 10\}, \sum_{i=1}^{10} x_i \leq 1\}$. The NNE is $x_i^* = 0.09, i = \{1, ..., 10\}$ [53].

6.7.5 *Discussion*

As highlighted earlier, Proposition 6.3 states that when the upper level objective (UPO) (cf. Eqn. 6.27a) , $f(x,y)$, in Program U reaches 0, we have successfully solved the GNEP. Hence this allows us to provide a termination criteria of the DEBLP-SR algorithm. In all other examples, we have always stopped the DEBLP after a user specified number of maximum iterations. In practice, we terminate each run when the UPO attains the value of 1×10^{-8} or less, which we judge to be sufficiently close to 0.

In all these examples, we used DEBLP-SR i.e. Algorithm 12 with a population size, $\pi = 50$ and allowed a maximum of 250 iterations. Following [99, 101], the probability factor, η used in SR was set to 0.45. Table 6.7 reports the mean,median and standard deviations (SD) of the number of function evaluations (NFE) over the 30 independent runs of DEBLP-SR to meet the convergence criteria (i.e. UPO attains the value of at least 1×10^{-8}).

---

[11] One can think of this as simulating a cartel limiting production to keep prices high.

Table 6.7 Summary of Performance of DEBLP-SR on GNEP Examples

Example	1	2	3	4
mean NFE	3993	7378	2825	10098
median NFE	4075	7325	2850	10025
SD	410	2758	273	1659
Constraint Violation	NA	0	0	0

While it is clear that all the examples are easily solved using DEBLP-SR, three observations are pertinent from Table 6.7. Firstly, comparing Problem 3 and 4 for example, we can see that as the dimensions increase, the NFE required to meet the convergence criteria also increase significantly. This is a manifestation of the so called "curse of dimensionality" [10] which plagues optimisation algorithms in general and meta-heuristics in particular. Secondly the mean and median NFE required to solve the GNEP (Example 2) is almost twice that required to solve the NEP (Example 1). This should not come as a surprise because constrained problems are known [121] to be harder to solve than unconstrained ones. Finally, the constraint violation of all examples at termination is 0 as shown in the last row of Table 6.7. Thus we can conclude that the SR method for handling constraints is effective for the examples given.

6.8 Summary and Conclusions

6.8.1 Summary

In this chapter, we have outlined a meta-heuristic algorithm DEBLP to solve bilevel programming problems. These hierarchical optimisation problems are typically characterised by non convexity and non smoothness. DEBLP is designed to synergise the well-documented global search capability of Differential Evolution with the application of robust deterministic optimisation techniques to the lower level problem. Most importantly, DEBLP is fully consistent with the Stackleberg framework upon which the BLPP is founded where the leader takes into account the follower's decision variables when optimizing his objective and where the follower treats the leader's variables as exogenous when solving his problem.

DEBLP was subsequently demonstrated on a number of BLPPs arising from several disciplines. These include control problems in Transportation Systems where we studied the Continuous Optimal Toll Pricing and the Continuous Network Design Problem. In these situations we postulated that the leader/upper level player was the regulatory agency and the followers were users of the highway network. The BLPPs from this field were shown to be MPECs as the lower level problem arises naturally as a Variational Inequality. We also examined examples from Parameter Estimation Problems, a key step in the development of models in science and engineering applications, which could also be formulated as BLPPs. In order to enable DEBLP to solve BLPPs where the upper level problem was also subject to general constraints, we integrated the stochastic ranking algorithm from [99] into DEBLP to produce DEBLP-SR. Stochastic ranking is a constraint handling technique that seeks to balance the dominance of the objective and constraint violations in the search process of meta-heuristic algorithms. We demonstrated the operation of DEBLP-SR on a series of Generalised Nash Equilibrium Problems which could be formulated as BLPPs characterised by upper level constraints. Developments in the literature of GNEPs also enabled us to even specify a specific termination criteria for the proposed BLPP and hence provides additional justification for the use of a meta-heuristic for these problems.

Due to space constraints, we could not illustrate BLPPs where the leader's decision variables and/or the follower's variables were restricted to be discrete or binary. However there exists a large body of literature of DE being used for such problems, albeit single level ones [91, 92]. Thus we conjecture the techniques proposed therein could be integrated into DEBLP to solve such problems as well. Additionally discrete and mixed integer lower level problems can already be solved using established techniques available in the deterministic optimisation literature [9, 70, 85].

6.8.2 *Further Research*

In this chapter, we have demonstrated that DEBLP is an effective meta-heuristic for a variety of BLPPs. Nevertheless there are several topics that still require additional research before robust methodologies can be developed. The study of some of these problems is still in its infancy but we argue that meta-heuristic paradigms such as Differential Evolution can provide a viable alternative solution framework for these.

6.8.2.1 Multiple Optimisation Problems at Lower Level

The BLPP we have formulated assumes the existence of a single optimisation problem at the lower level. Both DEBLP and DEBLP-SR are unable to handle the situation of *multiple followers* i.e. presence of multiple optimisation problems at the lower level. See e.g. [69, 114] for examples of these. However we have neglected consideration of such problems in this paper but should be the subject of further research.

6.8.2.2 Bi-Level Multiobjective Problems

Recall that in our formulation of the BLPP in Eq. 6.1 we assumed the function mappings: $U, L : \mathbb{R}^{n_1} \times \mathbb{R}^{n_2} \to \mathbb{R}^1$. In other words, the objectives in both the upper and lower levels are restricted to be scalar. However there are also problems where the objectives are vectors. Such problems are known as multiobjective (MO) problems i.e. where the decision maker has multiple, usually conflicting, objectives.

In such problems the Pareto Optimality criteria is used to identify optimum solutions [30, 90]. One of the major advantages of using population based meta-heuristic algorithms for MO Problems is that because of their population based structure, they are able to identify multiple Pareto Optimal solutions in a single run [29].

Two categories of these problems have been discussed in the literature. Firstly there is the case where only the upper level objective is vector based or secondly where *both* the upper and lower level objectives are vector based. For problems occurring in the first category, advances in meta-heuristics to solve MO problems (e.g. [30]) could be easily integrated into DEBLP to transform it into an algorithm able of handle MO-BLPPs of the type described in e.g. [44, 110, 123]. Problems of the second category are relatively novel in the literature and have only recently been investigated [31]. Further research should introduce new methodologies to enable DEBLP to solve problems in this latter category.

6.8.2.3 Multiple Leader Follower Games

In Section 7.5 we provided an example of the COTP which models a highway regulatory agency optimising the total travel time on the highway system by levying toll charges. With the trend in recent years towards privatization together with constrained governmental budgets, it is quite possible that instead of a welfare maximizing authority setting the tolls in future, this task could potentially be consigned to private profit maximising entities. The latter obtain concessions to collect tolls from users on these private toll roads [36, 119] in return for providing the capital layout of investments in new road infrastructure. When setting such tolls, these private firms could also be in competition with others doing the same on other roads in the network.

The problem just described is in fact an example of a class of Equilibrium problems with Equilibrium Constraints (EPEC). In EPECs, the decision variables of the private firms are constrained by a variational inequality describing equilibrium in some parametric system [62]. For example in the case of competition between the private toll road operators just highlighted, the equilibrium constraint is just Wardrop's User Equilibrium condition. The study of EPECs has recently been given greater emphasis by researchers in many disciplines [55, 68, 77, 80, 119, 125]. Though it is still in a period of infancy it has emerged as major area of research [22, 35, 109] in applied mathematics.

Formally an EPEC is a mathematical program to find an equilibrium point that simultaneously solves a set of MPECs where each MPEC is parameterised by decision variables of other MPECs [125]. Compared to the MPEC, the focus in the EPEC is shifted away from finding minimum points to finding *equilibrium* points [78, 79]. Figure 6.4 gives a multi-leader generalization of the BLPP that constitutes a Multi-leader-follower game [68] where there are now $\rho, \rho > 1$ leaders instead.

In this multi-leader generalization of the Stackelberg game researchers have conjectured that there could be two possible behaviours of the leaders at the upper level [78, 88]. At one end, leaders could cooperate which results in a multiobjective problem subject to an equilibrium constraint at the lower level [120]. At the other end, the leaders could act non-cooperatively and play a Nash game amongst themselves resulting in a Non Cooperative EPEC (NCEPEC). EPECs are extremely difficult to solve and the current emphasis has been on the use of nonsmooth methods and nondifferentiable optimisation techniques [78, 79]. We believe that meta-heuristic algorithms offer a powerful alternative solution methodology for EPECs in both cases. In the case when leaders are assumed to cooperate, we have pointed out that because they operate with populations, population based meta-heuristics are able to identify multiple Pareto Optimal solutions in a single simulation run. This is key to solving multiobjective problems. For the NCEPECs, a DE based algorithm exploiting a concept from [71] was proposed and demonstrated on a range of EPECs occurring in transportation and electricity markets in [62].

Most importantly, whatever solution algorithms are proposed in future, when searching for an equilibrium amongst the players at the upper level they must continue to take the reaction of the followers at the lower level into account. This serves to ensure that proposals are entirely consistent with the Stackleberg paradigm which remains applicable in EPECs.

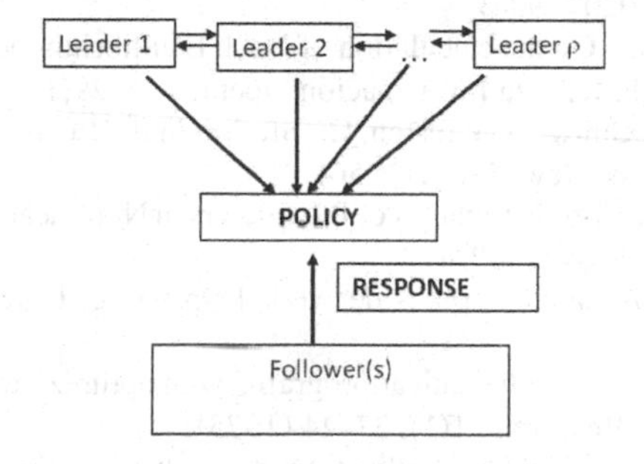

Fig. 6.4 Pictoral Representation of an EPEC

Acknowledgements. The research reported here is funded by the Engineering and Physical Sciences Research Council of the UK under Grant EP/H021345/1 (July 2010 to July 2013).

References

1. Aarts, E., Korst, J.: Simulated Annealing and Boltzmann Machines. John Wiley, Chichester (1988)
2. Aiyoshi, E., Shimizu, K.: Hierarchical decentralized systems and its new solution by a barrier method. IEEE Transactions on Systems, Man, and Cybernetics 11(6), 444–449 (1981)
3. Allsop, R.E.: Some possibilities for using traffic control to influence trip distribution and route choice. In: Buckley, D.J. (ed.) Transportation and Traffic Theory: Proceedings of the Sixth International Symposium on Transportation and Traffic Theory, pp. 345–375. Elsevier, New York (1974)

4. Aussel, D., Dutta, J.: Generalized Nash equilibrium problem, variational inequality and quasiconvexity. Operations Research Letters 36(4), 461–464 (2008)
5. Bard, J.F.: An efficient point algorithm for a linear two-stage optimization problem. Operations Research 31(4), 670–684 (1983)
6. Bard, J.F.: Convex two-level optimization. Mathematical Programming 40(1), 15–27 (1988)
7. Bard, J.F.: Practical Bilevel Optimisation: Algorithms and Applications. Kluwer, Dordrecht (1998)
8. Bard, J.F., Plummer, J., Sourie, J.C.: A bilevel programming approach to determining tax credits for biofuel production. European Journal of Operational Research 120(1), 30–46 (2000)
9. Bazaraa, M.S., Sherali, H.D., Shetty, C.M.: Nonlinear Programming: Theory and Algorithms, 3rd edn. Wiley, Hoboken (2006)
10. Bellman, R.E.: Adaptive control processes: a guided tour. Princeton University Press, Princeton (1961)
11. Ben-Ayed, O., Blair, C.E.: Computational difficulties of bilevel linear programming. Operations Research 38(3), 556–560 (1990)
12. Ben-Ayed, O.: Bilevel linear programming. Computers & Operations Research 20(5), 485–501 (1993)
13. Bialas, W.F., Karwan, M.H.: Two-Level Linear Programming. Management Science 30(8), 1004–1020 (1984)
14. Bouza Allende, G.B.: On the calculation of Nash Equilibrium points with the aid of the smoothing approach. Revista Investigación Operacional 29(1), 71–76 (2008)
15. Beckmann, M., McGuire, C., Winsten, C.: Studies in the Economics of Transportation. Yale University Press, New Haven (1956)
16. Britt, H., Luecke, R.: The Estimation of Parameters in Nonlinear Implicit Models. Technometrics 15(2), 233–247 (1973)
17. Braess, D.: Über ein paradoxon aus der verkehrsplanung. Unternehmenforschung 12, 258–268 (1968)
18. Bracken, J., McGill, J.: Mathematical programs with optimization problems in the constraints. Operations Research 21(1), 37–44 (1973)
19. Bracken, J., McGill, J.: Defense applications of mathematical programs with optimization problems in the constraints. Operations Research 22(5), 1086–1096 (1978)
20. Candler, W., Norton, R.: Multi-level programming. Technical Report DRD-20, World Bank (January 1977)
21. Candler, W., Townsley, R.: A linear two-level programming problem. Computers & Operations Research 9(1), 59–76 (1982)
22. Červinka, M.: Hierarchical structures in equilibrium problems. PhD Thesis, Charles University, Prague, Czech Republic (2008)
23. Chen, Y., Florian, M.: The nonlinear bilevel programming problem: Formulations, regularity and optimality conditions. Optimization 32(3), 193–209 (1995)
24. Chiou, S.: Bilevel programming formulation for the continuous network design problem. Transportation Research 39B(4), 361–383 (2005)
25. Coello-Coello, C.A.: Treating constraints as objectives for single-objective evolutionary optimization. Engineering Optimization 32(13), 275–308 (1999)
26. Colson, B., Marcotte, P., Savard, G.: An overview of bilevel optimization. Annals of Operations Research 153(1), 235–256 (2007)
27. Cree, N.D., Maher, M.J., Paechter, B.: The continuous equilibrium optimal network design problem: a genetic approach. In: Bell, M.G.H. (ed.) Transportation Networks: Recent Methodological Advances, Pergamon, pp. 163–174. Pergamon Press, London (1996)

28. Dafermos, S.C.: Traffic equilibrium and variational inequalities. Transportation Science 14(1), 42–54 (1980)
29. Deb, K.: Multi-objective genetic algorithms: Problem difficulties and construction of test problems. Evolutionary Computation 7(3), 205–230 (1999)
30. Deb, K.: Multi-objective Optimization using Evolutionary Algorithms. John Wiley, Chichester (2001)
31. Deb, K., Sinha, A.: An efficient and accurate solution methodology for bilevel multi-objective programming problems using a hybrid evolutionary-local-search algorithm. Evolutionary Computation 18(3), 403–449 (2010)
32. Dempe, S.: Foundations Of Bilevel Programming. Kluwer, Dordrecht (2002)
33. Dempe, S.: Annotated bibliography on bilevel programming and mathematical programs with equilibrium constraints. Optimization 52(3), 333–359 (2003)
34. Dorigo, M., Stützle, T.: Ant Colony Optimization. MIT Press, Cambridge (2004)
35. Ehrenmann, A.: Equilibrium Problems with Equilibrium Constraints and their Application to Electricity Markets. PhD Thesis, Fitzwilliam College, University of Cambridge, UK (2004)
36. Engel, E., Fischer, R., Galetovic, A.: A new approach to private roads. Regulation 25(3), 18–22 (2002)
37. Esposito, W.R., Floudas, C.A.: Global Optimization in Parameter Estimation of Nonlinear Algebraic Models via the Error-in-Variables Approach. Industrial Engineering and Chemistry Research 37(5), 1841–1858 (1998)
38. Facchinei, F., Jiang, H., Qi, L.: A smoothing method for mathematical programs with equilibrium constraints. Mathematical Programming 85B(1), 107–134 (1999)
39. Facchinei, F., Kanzow, C.: Generalized Nash equilibrium problems. Annals of Operations Research 175(1), 177–211 (2010)
40. Fisk, C.S.: Game theory and transportation systems modelling. Transportation Research 18B(4-5), 301–313 (1984)
41. Fortuny-Amat, J., McCarl, B.: A representation and economic interpretation of a two-Level programming problem. Journal of the Operational Research Society 32(9), 783–792 (1981)
42. Friesz, T.: Harker PT Properties of the iterative optimization-equilibrium algorithm. Civil Engineering Systems 2(3), 142–154 (1985)
43. Friesz, T., Cho, H., Mehta, N., Tobin, R., Anandalingam, G.: A simulated annealing approach to the network design problem with variational inequality Constraints. Transportation Science 26(1), 18–26 (1992)
44. Friesz, T., Anandalingam, G., Mehta, N.J., Nam, K., Shah, S., Tobin, R.: The multiobjective equilibrium network design problem revisited: A simulated annealing approach. European Journal of Operational Research 65(1), 44–57 (1993)
45. Gabay, D., Moulin, H.: On the uniqueness and stability of Nash-equilibria in non cooperative games. In: Bensoussan, A., Dorfer, P.K., Tapiero, C. (eds.) Applied Stochastic Control in Econometrics and Management Science, pp. 271–293. North Holland, Amsterdam (1980)
46. Gau, C.Y., Stadtherr, M.A.: Deterministic global optimization for error-in-variables parameter estimation. AIChE Journal 48(6), 1192–1197 (2002)
47. Goldberg, D.E.: Genetic Algorithms in Search, Optimization, and Machine Learning. Addison-Wesley, Reading (1989)
48. Glover, F., Laguna, M.: Tabu Search. Kluwer, Norwell (1997)
49. Gümüş, Z.H., Floudas, C.A.: Global optimization of nonlinear bilevel programming problems. Journal of Global Optimization 20(1), 1–31 (2001)

50. Harker, P.T.: A variational inequality approach for the determination of oligopolistic market equilibrium. Mathematical Programming 30(1), 105–111 (1984)
51. Harker, P.T.: Generalized Nash games and quasi-variational inequalities. European Journal of Operations Research 54(1), 81–94 (1991)
52. Haurie, A., Krawczyk, J.: Optimal charges on river effluent from lumped and distributed sources. Environment Modelling and Assessment 2(3), 177–189 (1997)
53. von Heusinger, A., Kanzow, C.: Relaxation methods for generalized Nash equilibrium problems with inexact line search. Journal of Optimization Theory and Applications 143(1), 159–183 (2009)
54. von Heusinger, A.: Numerical Methods for the Solution of the Generalized Nash Equilibrium Problem. PhD Thesis, Institute of Mathematics, University of Wüzburg, Würzburg (2009)
55. Hu, X., Ralph, D.: Using EPECs to model bilevel games in restructured electricity markets with locational prices. Operations Research 55(5), 809–827 (2007)
56. Karamardian, S.: Generalized complementarity problems. Journal of Optimization Theory and Applications 8(3), 161–168 (1971)
57. Kennedy, J.: Eberhart RC Particle swarm optimization. In: Proceedings of the IEEE International Conference on Neural Networks, pp. 1942–1948 (1995)
58. Kesselman, A., Leonardi, S., Bonifaci, V.: Game-Theoretic Analysis of Internet Switching with Selfish Users. In: Deng, X., Ye, Y. (eds.) WINE 2005. LNCS, vol. 3828, pp. 236–245. Springer, Heidelberg (2005)
59. Kim, I.W., Liebman, M.J., Edgar, T.F.: Robust nonlinear error-in-variables estimation method using nonlinear programming techniques. AIChE Journal 36(7), 985–993 (1990)
60. Koh, A., Watling, D.: Traffic Assignment Modelling. In: Button, K., Vega, H., Nijkamp, P. (eds.) A Dictionary of Transport Analysis, Edward Elgar,Cheltenham, pp. 418–420 (2010)
61. Koh, A.: Solving transportation bi-level programs with Differential Evolution. In: Proceedings of the IEEE Congress on Evolutionary Computation, pp. 2243–2250 (2007)
62. Koh, A.: An evolutionary algorithm based on Nash dominance for equilibrium problems with equilibrium constraints. Applied Soft Computing (2011) (in press), `http://dx.doi.org/10.1016/j.asoc.2011.08.056`
63. Korilis, Y., Lazar, A., Orda, A.: Avoiding the Braess paradox in non-cooperative networks. Journal of Applied Probability 36(1), 211–222 (1999)
64. Krawczyk, J., Uryasev, S.: Relaxation algorithms to find Nash equilibria with economic applications. Environmental Modelling and Assessment 5(1), 63–73 (2000)
65. Lampinen, J.: A constraint handling approach for the differential evolution algorithm. In: Procedings of the IEEE Congress on Evolutionary Computation, pp. 1468–1473 (2002)
66. Lawphongpanich, S., Hearn, D.W.: An MPEC approach to second-best toll pricing. Mathematical Programming 101B(1), 33–55 (2004)
67. Leyffer, S.: The penalty interior-point method fails to converge. Optimization Methods and Software 20(4-5), 559–568 (2005)
68. Leyffer, S., Munson, T.: Solving multi-leader-common-follower games. Optimization Methods and Software 25(4), 601–623 (2010)
69. Liu, B.: Stackelberg-Nash equilibrium for multilevel programming with multiple followers using genetic algorithms. Computers & Mathematics with Applications 36(7), 79–89 (1998)
70. Luenberger, D.G., Ye, Y.: Linear and Nonlinear Programming. Springer, Berlin (2008)

71. Lung, R.I., Dumitrescu, D.: Computing Nash equilibria by means of evolutionary computation. International Journal of Computers, Commmunications and Control III, 364–368 (2008)
72. Luo, Z.Q., Pang, J.S., Ralph, D.: Mathematical Programs with Equilibrium Constraints. Cambridge University Press, Cambridge (1996)
73. Matthieu, R., Pittard, L., Anandalingam, G.: Genetic algorithm based approach bi-level linear programming. RAIRO-Operations Research 28(1), 1–21 (1994)
74. Marcotte, P., Savard, G.: Bilevel programming: applications. In: Floudas, C.A., Pardalos, P.M. (eds.) Encyclopedia of Optimization, 2nd edn., pp. 241–242. Springer, New York (2009)
75. Meng, Q., Yang, H., Bell, M.G.H.: An equivalent continuously differentiable model and a locally convergent algorithm for the continuous network design problem. Transportation Research 35B(1), 83–105 (2000)
76. Moore, R.E., Hansen, E., Leclerc, A.: Rigorous methods for global optimization. In: Floudas, C.A., Pardalos, P.M. (eds.) Recent Advances in Global Optimization, pp. 321–342. Princeton University Press, Princeton (1992)
77. Mordukhovich, B.S., Outrata, J.V., Červinka, M.: Equilibrium problems with complementarity constraints: Case study with applications to oligopolistic markets. Optimization 56(4), 479–494 (2007)
78. Mordukhovich, B.S.: Optimization and equilibrium problems with equilibrium constraints. Omega 33(5), 379–384 (2005)
79. Mordukhovich, B.S.: Variational Analysis and Generalized Differentiation, I: Basic Theory. Springer, Berlin (2006)
80. Mordukhovich, B.S.: Variational Analysis and Generalized Differentiation, II: Applications. Springer, Berlin (2006)
81. Murphy, F.H., Sherali, H.D., Soyster, A.L.: A mathematical programming approach for determining oligopolistic market equilibrium. Mathematical Programming 24(1), 92–106 (1982)
82. Nagurney, A.: Network Economics: A Variational Inequality Approach. Kluwer, Boston (1999)
83. Nash, J.: Equilibrium points in N-person games. Proceedings of the National Academy of Science of the USA 36(1), 48–49 (1950)
84. Nash, J.: Non-Cooperative games. Annals of Mathematics Second Series 54(2), 286–295 (1951)
85. Nocedal, J., Wright, S.J.: Numerical Optimization. Springer, New York (1999)
86. Oduguwa, V., Roy, R.: Bi-level optimisation using genetic algorithms. In: Proceedings of the IEEE Conference on Artificial Intelligence Systems, pp. 322–327 (2002)
87. Outrata, J.V., Kočvara, M., Zowe, J.: Nonsmooth Approach to Optimization Problems with Equilibrium Constraints. Kluwer, Dordrecht (1998)
88. Outrata, J.V.: A note on a class of equilibrium problems with equilibrium constraints. Kybernetika 40(5), 585–594 (2004)
89. Ortuzar, J.D.: Willumsen L Modelling Transport. Wiley, Chichester (1998)
90. Osyzcka, A.: Multicriterion optimization in engineering with FORTRAN programs. Ellis Horwood, Wiley, Chichester (1998)
91. Pampara, G., Engelbrecht, A.P., Franken, N.: Binary differential evolution. In: Proceedings of the IEEE Congress on Evolutionary Computation, pp. 1873–1879 (2006)
92. Pan, Q., Tasgetiren, M.F., Liang, Y.C.: A discrete differential evolution algorithm for the permutation flowshop scheduling problem. Computers & Industrial Engineering 55(4), 795–816 (2008)

93. Poorzahedy, H., Abulghasemi, F.: Application of ant system to network design problem. Transportation 32(3), 251–273 (2005)
94. Price, K.: An Introduction to Differential Evolution. In: Corne, D., Dorigo, M. (eds.) New Techniques in Optimization, pp. 79–108. McGraw Hill, London (1999)
95. Price, K., Storn, R., Lampinen, J.: Differential evolution: a practical approach to global optimization. Springer, Berlin (2005)
96. Rajesh, J., Gupta, K., Kusumakar, H.S., Jayaraman, V.K., Kulkarni, B.: A tabu search based approach for solving a class of bilevel programming problems in chemical engineering. Journal of Heuristics 9(4), 307–319 (2003)
97. Rosen, J.B.: Existence and uniqueness of equilibrium points for concave N-person games. Econometrica 33(3), 520–534 (1965)
98. Rod, V., Hančil, V.: Iterative estimation of model parameters when measurements of all variables are subject to error. Computers & Chemical Engineering 4(2), 33–38 (1980)
99. Runarsson, T.P., Yao, X.: Stochastic ranking for constrained evolutionary optimization. IEEE Transactions on Evolutionary Computation 4(3), 284–294 (2000)
100. Sahin, K.H., Ciric, A.R.: A dual temperature simulated annealing approach for solving bilevel programming problems. Computers & Chemical Engineering 23(1), 11–25 (1998)
101. Runarsson, T.P., Yao, X.: Search biases in constrained evolutionary optimization. IEEE Transactions on Systems, Man and Cybernetics Part C 35(2), 233–243 (2005)
102. Scheel, H., Scholtes, S.: Mathematical Programs with Complementarity Constraints: Stationarity, Optimality, and Sensitivity. Mathematics of Operations Research 25(1), 1–22 (2000)
103. Sheffi, Y.: Urban Transportation Networks: Equilibrium Analysis with Mathematical Programming Methods. Prentice Hall, Englewood Cliffs (1985)
104. Smith, A., Tate, D.: Genetic optimization using a penalty function. In: Proceedings of the 5th International Conference on Genetic Algorithms, pp. 499–503 (1993)
105. Smith, M.J.: The existence, uniqueness and stability of traffic equilibria. Transportation Research 13B(4), 295–304 (1979)
106. von Stackelberg, H.H.: The theory of the market economy. William Hodge, London (1952)
107. Steenbrink, P.A.: Optimization of Transportation Networks. John Wiley, London (1974)
108. Storn, R., Price, K.: Differential Evolution - a simple and efficient heuristic for global optimization over continuous spaces. Journal of Global Optimization 11(4), 341–359 (1997)
109. Su, C.: Equilibrium problems with equilibrium constraints: stationarities, algorithms and applications. PhD Thesis, Stanford University, California, USA (2005)
110. Sumalee, A., May, A., Shepherd, S.: Road user charging design: dealing with multi-objectives and constraints. Transportation 36(2), 167–186 (2009)
111. Sumalee, A.: Optimal Road Pricing Scheme Design. PhD Thesis, Institute for Transport Studies, University of Leeds, UK (2004)
112. Sun, L.: Equivalent Bilevel Programming Form for the Generalized Nash Equilibrium Problem. Journal of Mathematics Research 2(1), 8–13 (2010)
113. Tjoa, T., Biegler, L.: A reduced successive quadratic programming strategy for errors-in-variables estimation. Computers & Chemical Engineering 16(6), 523–533 (1992)
114. Wang, Y., Jiao, Y., Li, H.: An evolutionary algorithm for solving nonlinear bilevel programming based on a new constraint handling scheme. IEEE Transactions on Systems, Man and Cybernetics Part C 35(2), 221–232 (2005)
115. Wardrop, J.G.: Some theoretical aspects of road traffic research. Proceedings of the Institution of Civil Engineers Part II 1(36), 325–378 (1952)

116. Webb, J.N.: Game theory: Decisions, Interaction and Evolution. Springer, London (2007)
117. Wen, U., Hsu, S.: Linear bi-level programming problems – a review. Journal of the Operational Research Society 42(2), 125–133 (1991)
118. Yang, H., Lam, H.K.: Optimal road tolls under conditions of queuing and congestion. Transportation Research 30A(5), 319–332 (1996)
119. Yang, H., Feng, X., Huang, H.: Private road competition and equilibrium with traffic equilibrium constraints. Journal of Advanced Transportation 43(1), 21–45 (2009)
120. Ye, J.J., Zhu, Q.J.: Multiobjective optimization problem with variational inequality constraints. Mathematical Programming 96A(1), 139–160 (2003)
121. Yeniay, Ö.: Penalty function methods for constrained optimization with genetic algorithms. Mathematical and Computational Applications 10(1), 45–56 (2005)
122. Yin, Y.: Genetic Algorithm based approach for bilevel programming models. ASCE Journal of Transportation Engineering 126(2), 115–120 (2000)
123. Yin, Y.: Multiobjective bilevel optimization for transportation planning and management problems. Journal of Advanced Transportation 36(1), 93–105 (2002)
124. Yuchi, M., Kim, J.H.: Grouping-based evolutionary algorithm: seeking balance between feasible and infeasible individuals of constrained optimization problems. In: Proceedings of the IEEE Congress on Evolutionary Computation, pp. 280–287 (2004)
125. Zhang, X.P.: Overview of electricity market equilibrium problems and market power analysis. In: Zhang, X.P. (ed.) Restructured Electric Power Systems: Analysis of Electricity Markets with Equilibrium Models, pp. 99–137. John Wiley, Hoboken (2010)
126. Zhao, Z., Gu, X.: Particle swarm optimization based algorithm for bilevel programming problems. In: Proceedings of the Sixth IEEE International Conference on Intelligent Systems Design and Applications, pp. 951–956 (2006)

Chapter 7
Matheuristics and Exact Methods for the Discrete $(r|p)$-Centroid Problem

Ekaterina Alekseeva and Yury Kochetov

7.1 Introduction

In the $(r|p)$-centroid problem, there are two decision makers which we refer to as a leader and a follower. They compete to serve customers from a given market by opening a certain number of facilities. The decision makers open facilities in turn. At first, the leader decides where to locate p facilities taking into account the follower's reaction. Later on, the follower opens other r facilities. We assume that the customers' preferences among the opened facilities are based only on the distances to these facilities rather than the quality of service provided by the decision makers. We consider a binary preference (all or none) model where each customer chooses the closest opened facility. In case of ties, the leader's facility is preferred. The binary model is important from a theoretical point of view and useful for certain applications where the product can be considered homogeneous and facilities are assumed to be identical. Each customer has a weight (purchasing power or demand). We assume that the weights are essential, that is goods must be consumed, and each customer visits one facility to get them. The weight of each customer is fixed and does not depend on how far from, or close to a facility, the customer is. The leader and the follower obtain a profit from serving the customer which coincides with the weight of the customer. Each decision maker maximizes his own profit or market share. The problem is to define the p facilities which should be opened by the leader to maximize his market share.

Following Hakimi [22], we call this problem the *$(r|p)$-centroid problem*. In the literature this problem can be found under such names as leader-follower problem, competitive p-median problem, competitive location model with foresight, or preemptive capture problem.

Ekaterina Alekseeva · Yury Kochetov
Sobolev Institute of Mathematics, 4 pr. Akademika Koptuga, Novosibirsk, Russia
e-mail: {ekaterina2, jkochet}@math.nsc.ru

E.-G. Talbi (Ed.): *Metaheuristics for Bi-level Optimization*, SCI 482, pp. 189–219.
DOI: 10.1007/978-3-642-37838-6_7 © Springer-Verlag Berlin Heidelberg 2013

Three types of possible facility locations can be considered:

- at the nodes of a graph (discrete case);
- at the nodes and anywhere on the edges of a graph (absolute case);
- anywhere on a plane (continuous case).

In this chapter we consider the discrete case only when all customers and facilities are located at the nodes of a complete bipartite graph.

We may reagard the discrete $(r|p)$-centroid problem as a game with two players (the leader and the follower). They compete with each other. The players have the same goal which is to catch as large a market share as possible. Each player has exactly one move. The first player (the leader) moves taking into account that once he selects his p facilities, the second player (the follower) will select the best possible r places for his facilities. The payoff to the follower is the loss to the leader. The discrete $(r|p)$-centroid problem can be considered as a Stackelberg game. It is a strategic game in which two players move sequentially. A Stackelberg solution is a pair (X_p^*, Y_r^*) where Y_r^* is the optimal strategy of the follower if the leader has p facilities located at X_p^* and X_p^* is the optimal preemptive strategy of the leader.

The discrete $(r|p)$-centroid problem is a one-round discrete Voronoi game (or Thiessen polygon) [47], when $p = r$. The Voronoi game is a geometric model for the competitive facility location problem with two players where each one must place the same number of facilities on a graph. Each node of a graph is a potential facility or a customer. Each customer is dominated by the player who owns the nearest placed facility generating a profit for the player. Each player aims to obtain the maximum profit. When the follower starts playing after all leader's p facilities are placed, the game is called the one-round Voronoi game. Thus, approaches coming from economics or game theory could be applied. Since the problem can be considered as a game, some questions concerning an equilibrium state arise.

In the voting theory, there is an interesting problem that can be modeled as a discrete $(p|p)$-centroid problem. This problem consists in finding a p-Simpson solution, i. e. a set of p facilities, such that the maximum number of customers closer to another set of p facilities is minimum. For more details, see [8].

In this chapter we present an overview of the recent results for the discrete $(r|p)$-centroid problem. It begins with different formulations of the problem and a brief overview of the related works in Section 7.2. The bi-level mixed integer linear formulation is introduced in Section 7.2.1, the min-max formulation, in Section 7.2.2, and the single-level mixed integer linear formulation, in Section 7.2.3. The complexity status of the problem is discussed in Section 7.3. The heuristic algorithms are described in Section 7.4. We present matheuristics based on the p-median problem, an alternative heuristics, and the hybrid heuristics. We describe the exact approaches in Section 7.5. The branch-and-cut algorithm and the iterative exact method are presented in Sections 7.5.1 and 7.5.2, respectively. Finally, we discuss the comparative computational results in Section 7.6 and conclude with possible further research directions in Section 7.7.

7.2 The Problem Statement

In this section we formulate the discrete $(r|p)$-centroid problem as a bi-level mixed integer linear program, a min-max problem, and a single-level mixed integer linear problem with polynomially many variables and exponentially many constraints.

7.2.1 The Bi-level Mixed Integer Linear Formulation

Let $I = \{1,\ldots,m\}$ be a set of potential facilities locations and $J = \{1,\ldots,n\}$ be a set of customers locations. The elements of matrix (d_{ij}) define the distances between each customer $j \in J$ and each facility $i \in I$. The components of positive vector (w_j) define the weight of each customer $j \in J$. Let us introduce the following decision variables:

$$x_i = \begin{cases} 1 & \text{if facility } i \text{ is opened by the leader,} \\ 0, & \text{otherwise,} \end{cases}$$

$$y_i = \begin{cases} 1 & \text{if facility } i \text{ is opened by the follower,} \\ 0, & \text{otherwise,} \end{cases}$$

$$z_j = \begin{cases} 1 & \text{if customer } j \text{ is serviced by the leader,} \\ 0 & \text{if customer } j \text{ is serviced by the follower.} \end{cases}$$

Denote $x = (x_i)$, $y = (y_i)$ $i \in I$, and $z = (z_j)$ $j \in J$, for short. Now we can define the set of facilities which allows the follower to capture customer j if the leader uses a solution x:

$$I_j(x) = \{i \in I \mid d_{ij} < \min_{l \in I \mid x_l = 1} d_{lj}\}, \quad j \in J.$$

Note that we consider conservative customers. It means that if a customer has the same distances to the closest leader's and the closest follower's facilities, he prefers the leader's facility. So, the follower never opens a facility at a site where the leader has opened a facility. Now the discrete $(r|p)$-centroid problem can be written as a linear 0–1 bi-level programming model:

$$\max_x \sum_{j \in J} w_j z_j^* \tag{7.1}$$

subject to

$$\sum_{i \in I} x_i = p, \tag{7.2}$$

$$x_i \in \{0,1\}, \quad i \in I, \tag{7.3}$$

where z^* is a component of the optimal solution to the follower's problem:

$$\max_{y,\, z} \sum_{j \in J} w_j (1 - z_j) \tag{7.4}$$

subject to

$$\sum_{i \in I} y_i = r, \tag{7.5}$$

$$1 - z_j \leq \sum_{i \in I_j(x)} y_i, \quad j \in J, \tag{7.6}$$

$$x_i + y_i \leq 1, \quad i \in I, \tag{7.7}$$

$$y_i, z_j \in \{0,1\}, \quad i \in I, j \in J. \tag{7.8}$$

The objective function (7.1) defines the market share of the leader. Equation (7.2) guarantees that the leader opens exactly p facilities. The objective function (7.4) defines the market share of the follower. Equation (7.5) guarantees that the follower opens exactly r facilities. Constraints (7.6) determine the values of (z_j) by the decision variables (y_i) of the follower. Constraints (7.7) guarantee that each facility can be opened by at most one decision maker. Actually, these constraints are redundant due to the definition of set $I_j(x)$. It does not make sense for the follower to open facilities at the same places which the leader has already occupied. Nevertheless, we use them to reduce the feasible domain of the follower's problem. Furthermore, this constraint becomes important in case of curious customers, that is when

$$I_j(x) = \{i \in I \,|\, d_{ij} \leq \min_{l \in I | x_l = 1} d_{lj}\}, \quad j \in J.$$

In this situation customers choose the facility opened by the second player whenever there are several equally distant opened facilities. They prefer a newly opened facility. However, this case is of no interest since the follower can occupy the same places where the leader has already opened his own facilities so as to seize the whole market in case $r \geq p$ [22].

We can drop the integrality requirements on variables (z_j), due to the fact that the optimal value for the leader's problem does not change. Thus, formulation (7.1)–(7.8) can be rewritten as a mixed integer linear bi-level program. The upper level (7.1)–(7.8) is called the leader's problem or the $(r|p)$-centroid. The lower level (7.4)–(7.8) is called the follower's problem or the $(r|X_p)$-medianoid, where X_p is a set of facilities opened by the leader.

Following [3], let us introduce some definitions which help us understand the nature of the bi-level problem deeper.

Definition 7.1. The triple (x,y,z) is called a *semi-feasible solution* to the bi-level problem (7.1)–(7.8) if and only if x satisfies the constraints (7.2)–(7.3) and the pair (y,z) satisfies the constraints (7.5)–(7.8).

When we need to consider only x which satisfies constraints (7.2)–(7.3), we will call it a *leader's solution.*

Definition 7.2. The semi-feasible solution (x,y,z) is called a *feasible solution* to the bi-level problem (7.1)–(7.8) if and only if the pair (y,z) is an optimal solution to the follower's problem (7.4)–(7.8).

Note that semi-feasible solutions can be found in a polynomial time. To find feasible solutions, we have to solve the follower's problem, which is NP-hard in the strong sense.

For the feasible solution (x,y,z), let us denote the value of the leader's objective function as $L(x,y,z)$. Before we define the optimal solution to the leader's problem, we should note that the follower's problem may have several optimal solutions for a given x. As a result, the leader's problem turns out to be ill-posed. Thus, we should distinguish two extreme cases:

- cooperative follower's behavior (altruistic follower). In case of multiple optimal solutions, the follower always selects one providing the best objective function value for the leader.
- non-cooperative follower's behavior (selfish follower). In this case, the follower always selects the solution providing the worst objective function value for the leader.

Definition 7.3. The feasible solution (x,y,z) is called a *cooperative solution* if and only if $L(x,y,z) \geq L(x,\bar{y},\ \bar{z})$ for each feasible solution $(x,\bar{y},\ \bar{z})$.

Definition 7.4. The feasible solution (x,y,z) is called a *non-cooperative solution* if and only if $L(x,y,z) \leq L(x,\bar{y},\ \bar{z})$ for each feasible solution $(x,\bar{y},\ \bar{z})$.

Once the follower's behavior is determined, we can give the definition of the optimal solution to the leader's problem.

Definition 7.5. The cooperative solution (x^*,y^*,z^*) is called an *optimal solution under cooperative follower's behavior* (optimistic solution) if and only if $L(x^*,y^*,z^*) \geq L(x,y,z)$ for each cooperative solution (x,y,z).

Definition 7.6. The non-cooperative solution (x^*,y^*,z^*) is called an *optimal solution under non-cooperative follower's behavior* (pessimistic solution) if and only if $L(x^*,y^*,z^*) \geq L(x,y,z)$ for each non-cooperative solution (x,y,z).

If the follower's behavior is unknown and thus cannot be predicted, the bilevel problem is ill-posed. Considering cooperative and non-cooperative behaviors yields upper and lower bounds to the maximal value for the leader, respectively. Note that in our case the sum of leader and follower objective function values equals $\sum_{j\in J} w_j$, i.e. a constant. Thus, each feasible solution (x,y,z) is cooperative and non-cooperative at the same time, and all feasible solutions for x produce the same leader's market share. Hence, we do not need to distinguish between optimistic and pessimistic cases. They coincide here and we can define the optimal solution to the leader as the best feasible solution. In other competitive models, for example, where the sum of objective functions is not a constant, the follower's behavior should be defined properly. Suppose that f_i is a fixed cost for opening the facility i and the follower's objective function is the following: $\max(\sum_{j\in J} w_j(1-z_j) - \sum_{i\in I} f_i y_i)$. In this case, multiple optimal solutions of the follower can provide the leader with different values. Thus, to find the cooperative and non-cooperative solutions, auxiliary optimization problems should be defined, as described in [3, 5].

7.2.2 The Min-Max Formulation

In this section we present the discrete $(r|p)$-centroid problem as a min-max problem. Note that the customer j is serviced by the leader if and only if the follower does not open a facility from the set $I_j(x)$, that is $z_j = \prod_{i \in I_j(x)}(1 - y_i)$. Now we can exclude the variables (z_j) and rewrite formulation (7.1)–(7.8) as follows:

$$\max_x \sum_{j \in J} w_j \prod_{i \in I_j(x)} (1 - y_i^*) \tag{7.9}$$

subject to

$$\sum_{i \in I} x_i = p, \tag{7.10}$$

$$x_i \in \{0,1\}, \quad i \in I, \tag{7.11}$$

where y^* is the optimal solution to the follower's problem:

$$\max_y \sum_{j \in J} w_j (1 - \prod_{i \in I_j(x)} (1 - y_i)) \tag{7.12}$$

subject to

$$\sum_{i \in I} y_i = r, \tag{7.13}$$

$$y_i \in \{0,1\}, \ i \in I. \tag{7.14}$$

Consider the pseudo–Boolean function $P(x,y) = \sum_{j \in J} w_j - \sum_{j \in J} w_j \prod_{i \in I_j(x)} (1 - y_i)$. Then the discrete $(r|p)$-centroid problem can be formulated as the following min-max problem [30]:

$$\min_x \max_y \{P(x,y) \mid \sum_{i \in I} x_i = p, \ \sum_{i \in I} y_i = r, \ x_i, y_i \in \{0,1\}, \ i \in I\}.$$

In this formulation we assume that each customer will be served either by the leader or the follower. In [37] other competitive location models are considered. Suppose that each customer has a patronizing set of facilities. These sets may arise in different ways. For example, it may be the set of facilities lying within a threshold distance from a customer. He could be captured by the leader or by the follower if and only if at least one opened facility belongs to his patronizing set. As a result, some customers in a feasible solution could be unserved. In this case the problem cannot be formalized as a min-max problem. We have to consider the cooperative and non-cooperative solutions, and the definition of an optimal solution should be accurately done.

7.2.3 The Single-Level Mixed Integer Linear Formulation

In Section 7.3 we will discuss the complexity status of the problem. Going ahead, we notice that the discrete $(r|p)$-centroid problem is Σ_2^P-hard. It means that it is more difficult than any NP-complete problem. In spite of its complexity status, the problem admits a single level linear programming formulation with polynomially many variables and exponentially many constraints. Nevertheless, neither a polynomial formulation nor a formulation where all constraints can be separated in a polynomial time is possible unless NP = Σ_2^P [38].

Let us introduce new binary variables:

$$z_{ij} = \begin{cases} 1 & \text{if the facility } i \text{ is the leader's nearest facility to the customer } j, \\ 0, & \text{otherwise,} \end{cases}$$

and a positive variable W which means the leader's market share. Let $\mathscr{F}$ be the set of all possible follower's solutions. Each $y^f \in \mathscr{F}$ defines a set of r facilities opened by the follower. The set $\mathscr{F}$ is composed of $\binom{m}{r}$ solutions. For each y^f we define

$$I_j(y^f) = \{i \in I \mid d_{ij} \leq \min_{l \in I, y_l^f = 1} d_{lj}\}, j \in J$$

which is a set of the facilities which allows the leader to keep the customer j if the follower uses the solution y^f. Following [38], the discrete $(r|p)$-centroid problem can be reformulated as a single–level mixed integer linear problem:

$$\max_{W,\, x,\, z} W \tag{7.15}$$

subject to

$$\sum_{i \in I} x_i = p, \tag{7.16}$$

$$0 \leq z_{ij} \leq x_i, \quad i \in I, j \in J, \tag{7.17}$$

$$\sum_{i \in I} z_{ij} = 1, \quad j \in J, \tag{7.18}$$

$$W \leq \sum_{j \in J} \sum_{i \in I_j(y^f)} w_j z_{ij}, \quad y^f \in \mathscr{F}, \tag{7.19}$$

$$W \geq 0, \quad x_i, z_{ij} \in \{0,1\}, i \in I, j \in J. \tag{7.20}$$

The objective function (7.15) maximizes the total market share of the leader. Constraint (7.16) indicates as before that the leader has to choose precisely p facilities. Constraints (7.17) ensure the consistency between the variables x_i and z_{ij}. The set of constraints (7.18) indicates that exactly one facility of the leader is the nearest facility to each customer. Finally, constraints (7.19) ensure that, the follower chooses the best solution among all feasible ones leaving as small a market share for the leader as possible. Note that we can remove the integrality constraint for variables (z_{ij}).

Reformulation (7.15)–(7.20) contains a polynomial number of variables and an exponential number of constraints, due to the exponential cardinality of the set $\mathscr{F}$. It is not the only single–level formulation of the problem. In [2, 12] other single–level formulations are considered. But these formulations are worse due to an exponentially great number of variables and constraints as far as we know, formulation (7.15)– (7.20) is the best.

7.2.4 The Brief Overview of Related Works

The discrete $(r|p)$-centroid problem belongs to the competitive facility location models. This class of problems has aroused interest since the seventies of the last century due to many private sector applications. In a competitive environment, a location first deemed desirable may become undesirable because competitors locate additional facilities to achieve their own objectives. It is necessary, therefore, to find facility locations that not only improve performance within a short time, but also protect performance from future competitive encroachment [45]. The study of competitive location models is rooted in the Hotelling spatial duopoly model of two vendors on a beach [25] which was the basis of a number of studies on spatial competition. In 1981 Hakimi has proposed the concept of "centroids" [24]. He introduced the terms $(r|p)$-centroid and $(r|X_p)$-medianoid and formalized the leader-follower location problem in networks. Hakimi has considered different models of customers' preference behavior and type of demands. He was the first to obtain the complexity results for the basic $(r|p)$-centroid problem.

The main ingredients of competitive location models are as follows:

- the nature of the space where facilities and customers can be located. *Discrete* and *continuous* cases are considered. In the discrete case potential facility locations are taken from a finite set or of the nodes of a graph. In the continuous case they are on a plane or on the edges of a graph.
- the character of the goods and the demand. *Essential goods* must be consumed and customers visit one or more facilities to get them. Each customer satisfies his purchasing power entirely. *Inessential goods* are dispensable and customers may decide not to visit any facility if they consider the travel distance too long. Essential and inessential goods correspond to *inelastic* and *elastic* demand, respectively.
- the interpretation of matrix elements (d_{ij}). They might be interpreted not only as distances, but as any kind of preferences for customers. For example, a customer may prefer a facility with a minimal traveling and waiting time, or with maximum known brand name rather than with a minimal distance. In [48] facility location models with general customers' preferences are studied. However, we preserve the distances for simplicity.

- the customers' preference behavior. In *binary preference behavior* each customer patronizes one closest facility among all those opened and satisfies his demand entirely. In *partially binary preference behavior* each customer satisfies his demand from several opened facilities. He may select from time to time either the leader's or the follower's facilities, but the percentages of times when he chooses the leader's or the follower's are inversely proportional to some functions of the distances. In the models with *proportional preferences* each customer patronizes all opened facilities with probabilities that are inversely proportional to the function of the distances to $(p+r)$ opened facilities.
- the various objective functions can be considered:
  - each player wants to maximize his own market share;
  - each player wants to minimize the competitors' market share;
  - a player wants to maximize the difference between his market share and the competitors' market share;
  - a player wants to insure that his market share is not less than the competitor's one.

  If the customer's demand is to be totally satisfied by the players, then the total demand is distributed among the competitors and these objective functions are similar. Otherwise, we have got different models [37].

In 1984 Ghosh and Craig [19] considered the competitive location model of retail convenience stores on a certain planning horizon. They assumed that the total consumer expenditure is affected by the distance separating the consumer from that store. That is, demand is elastic with respect to distance. Each consumer allots a typical budget for expenditure on the class of good sold by convenient stores. However, if the nearest convenient store is relatively inaccessible, the consumer may obtain the product at alternative types of stores or completely forgo it on some occasions. A consumer who lives far away from a fast food outlet, for example, may occasionally forgo consumption.

In 1994 Serra and ReVell [42] considered competitive models with uncertainty in the information about weights of customers and the number of facilities opened by the follower. They have started their investigation from the basic Maximum Capture problem. In this problem the leader wants to enter a market and obtain the maximum capture given the locations of the follower. Then they gradually extended this problem relaxing their assumptions about equal weights of customers, the follower's reaction to the leader's entrance, and the fixed number of competitors' locations. They considered binary customers' preference behavior and inelastic demand.

An exhaustive review of recent developments in the field of sequential competitive location problems with a focus on the network problems can be found in [31]. We concentrate on the discrete $(r|p)$-centroid problem with essential goods and binary customers' preference behavior.

7.3 Complexity Status

In this section we discuss the complexity of the discrete $(r|p)$-centroid problem. Hansen and Labbé have developed a polynomial time algorithm for finding the $(1|1)$-centroid of a network. However, multiple competitive location problems are much harder than their counterparts in a single competitive location. The previously obtained results demonstrate that these problems are extremely hard. In [23] Hakimi has proved that to locate r follower facilities optimally under one opened leader's facility is an NP-hard problem. Furthermore, Hakimi has proved that the $(r|X_p)$-medianoid problem is NP-hard on a network. In [15] Davydov et al. have shown that even under Euclidean distances between facilities and customers, the $(r|X_p)$-medianoid problem is NP-hard in the strong sense. It means that finding a feasible solution to the discrete $(r|p)$-centroid problem in this case is a difficult problem. However, there is a polynomial case on a tree network considered in [33].

Various complexity results have been obtained by Spoerhase and Wirth in [40, 43]. They have investigated the complexity status of the problem on general graphs and on special graph structures, such as trees, paths, spider graphs, and pathwidth bounded graphs. They have developed an $O(pm^4)$ polynomial algorithm on a m-node path. However, they have proved that the discrete $(r|p)$-centroid on a spider graph, a tree where only one node has a degree larger than 2 is NP-hard. In terms of approximability, they have showed in [40] that the discrete $(r|X_p)$-medianoid problem is approximable within $\frac{e}{(e-1)}$ but not within $\frac{e}{(e-1)} - \varepsilon$ for any $\varepsilon > 0$ unless P=NP. They studied the approximability of the discrete $(r|p)$-centroid problem. It has turned out that this problem is essentially not approximable at all unless P=NP, that is the problem cannot be approximated within $m^{1-\varepsilon}$ for any $\varepsilon > 0$ unless P = NP. In particular, this holds for any fixed $r \geq 1$. Roughly speaking, there is no approximation algorithm for this problem with a reasonable worst case behavior.

Spoerhase and Wirth have rightly noticed that investigating a complexity status is not only a theoretical aspect of any research. This question has some practical implications. The knowledge that a problem is NP-complete, allows ones to apply various heuristic approaches. Many heuristics are based on the fact that the objective function of the underlying problem is polynomially computable. This concerns, for example, the greedy strategy as well as many metaheuristics such as hill climbing, tabu search, simulated annealing, genetic algorithms and others. For the discrete $(r|p)$-centroid, in order to calculate the leader's objective function, we need to solve the $(r|X_p)$-medianoid problem which is NP-hard in the strong sense. Hence, we could not immediately apply these heuristic approaches. Thus, the bi-level methods which solve heuristically the leader's problem and solve heuristically the follower's problem to evaluate the leader's solutions are worth developing. Unfortunately, such bi-level metaheuristics can find only semi-feasible solutions. They do not solve the follower's problem exactly. But the hybridization of heuristics to the upper level with exact approaches to the lower level allows one to find optimal or near optimal feasible solutions. Below we will describe some of these hybrid methods for the centroid problem.

To investigate the computational complexity of the bilevel problems use is made of a special class Σ_2^P. This class is a part of the polynomial time hierarchy and contains all decision problems decidable in a polynomial time by a non-deterministic Turing machine with access to an oracle for NP. In particular, this class contains decision problems which can be described using a formula of the form $\exists x \forall y \phi$, where ϕ is a quantifier-free formula. It is widely assumed that Σ_2^P is a proper superset of NP. Hence, the problems from this class turn out to be even more complex than the well-known NP-complete decision problems.

It is known [35] that the corresponding decision version of the discrete $(r|p)$-centroid is Σ_2^P-complete. Moreover, this problem remains Σ_2^P-hard even in the case of Euclidean distances between customers and facilities [15]. Thus, it is substantially harder than the $(r|X_p)$-medianoid problem. Table 7.1 summarizes some complexity results concerning the discrete $(r|p)$-centroid problem.

Table 7.1 Complexity of the discrete $(r|p)$-centroid problem. Facilities and customers are located at nodes of a graph, binary choice, essential demand

Problem	Complexity status	Author (year)
$(r\|X_1)$-medianoid	NP-hard on a network	Hakimi (1983)
$(r\|X_p)$-medianoid	NP-hard on a network	Hakimi (1983)
	NP-hard in the strong sense even in the case of Euclidean distances	Davydov et al. (2013)
	$O(m^2 r)$ polynomial algorithm on a tree network	Megiddo et al. (1983)
$(1\|1)$-centroid	polynomial solvable on a network	Hansen and Labbé (1988)
$(1\|p)$-centroid	NP-hard on a general networks	Hakimi (1983)
	$O(m^2(\log m)^2 \log \sum w(j))$ on a tree	Spoerhase and Wirth (2009)
	NP-hard on pathwidth bounded graph	Spoerhase and Wirth (2009)
$(r\|p)$-centroid	polynomial solvable on paths	Spoerhase (2010)
	NP-hard on a spider	Spoerhase (2010)
	Σ_2^P-complete on a graph	Noltemeier et al. (2007)
	Σ_2^P-hard in the case of Euclidean distances	Davydov et al. (2013)

7.4 Heuristics for the Discrete $(r|p)$-Centroid Problem

Due to the combinatorial nature of the discrete $(r|p)$-centroid problems some approaches based on enumeration ideas have been proposed [12, 19]. Nevertheless, due to the laboriousness of such approaches it is worth developing heuristic algorithms. In our case of solving a bi-level problem, pure heuristics could produce only good quality semi-feasible solutions without proving their feasibility. For this reason matheuristics, which integrate heuristics for the upper level and mathematical programming tools for the lower level, are a balanced approach to tackle the bi-level problem. We describe the developed heuristic approaches in this section and the exacts methods in the next one.

In [42] Serra and ReVelle have suggested two heuristic algorithms based on so-called one-opt heuristics for the leader's problem. Algorithms are distinguished by the approaches for solving the follower's problem. It can be solved using integer programming or the one-opt heuristics which is applied to the leader's problem as well. In essence, the one-opt heuristics is a local improvement algorithm under the swap neighborhood. This heuristics is very efficient in terms of running time but it does not guarantee optimality for the follower's problem. Thus, it produces semi-feasible solutions only.

In [4] Benati and Laporte have solved the discrete $(r|p)$-centroid problem by tabu search. They have developed a bi-level tabu search algorithm as it combines tabu search for the $(r|X_p)$-medianoid problem within tabu search for the centroid problem. They have built a greedy procedure in tabu search for $(r|X_p)$-medianoid based on the properties of submodular $(r|X_p)$-medianoid objective function. The algorithm works with semi-feasible solutions; however, it can find optimal solutions for relatively small test instances of the size $m = n = 15$, $p, r \in \{2, 3\}$.

In [11] Campos-Rodríguez et al. have proposed a particle swarm optimization (PSO) procedure with two swarms. It is an evolutive optimization technique based on the social behaviour in Nature. Originally, it was developed for the continuous problems. In [10] an application of PSO with two swarms has been shown for solving the $(r|p)$-centroid problem on the plane. The authors have adapted Jumping Particle Swarm Optimization for the discrete bi-level problem. Semi-feasible solutions are modeled as members of a swarm which moves in the solution space influenced by the inertia effect and the attraction exerted by the best positioned particles in the swarm. The algorithm finds semi-feasible solutions to the instances with $m = n = 25$, $p = 3$, and $r = 2$.

Below we describe the most prominent matheuristics which combine some heuristical strategies for the upper level with mathematical programming tools for the lower level of the problem.

7.4.1 Median Heuristics

The first and simplest approach arising naturally to tackle the discrete $(r|p)$-centroid problem is based on the classical p-median problem [29, 34]. In [41] this idea has already been used to create an initial solution. The approach consists of two steps. At the first step, the classical p-median problem is optimally solved by a branch-and-bound algorithm via, for example, CPLEX software. The leader opens facilities according to the optimal p-median solution. At the second step, the follower's problem is solved by a branch-and-bound algorithm in order to get a feasible solution to the bi-level problem (7.1)–(7.8).

In the p-median problem the leader opens facilities to minimize the total distance between customers and his facilities ignoring the follower. He wishes to service all customers. To formulate the p-median problem, let us introduce the following additional binary variables:

$$x_{ij} = \begin{cases} 1 & \text{if the customer } j \text{ is serviced from the facility } i \\ 0, & \text{otherwise.} \end{cases}$$

Then the well-known p-median problem is the following:

$$\min \sum_{i\in I}\sum_{j\in J} w_j d_{ij} x_{ij} \tag{7.21}$$

subject to

$$\sum_{i\in I} x_{ij} = 1, \quad j \in J, \tag{7.22}$$

$$x_i \geq x_{ij}, \quad i \in I, j \in J, \tag{7.23}$$

$$\sum_{i\in I} x_i = p, \tag{7.24}$$

$$x_i, x_{ij} \in \{0,1\}, \quad i \in I, j \in J. \tag{7.25}$$

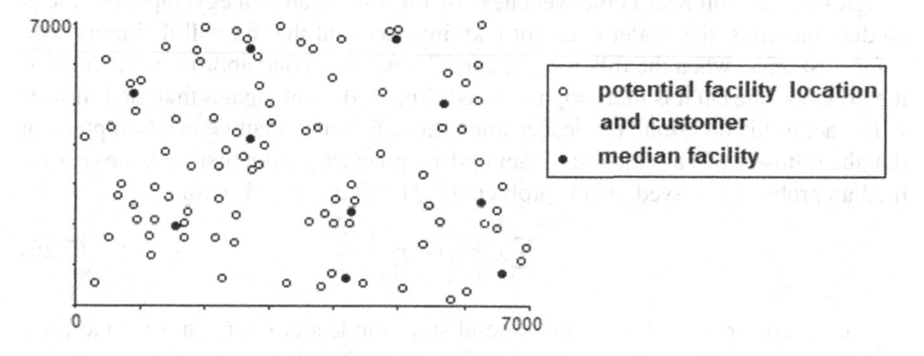

Fig. 7.1 Optimal p-median solution

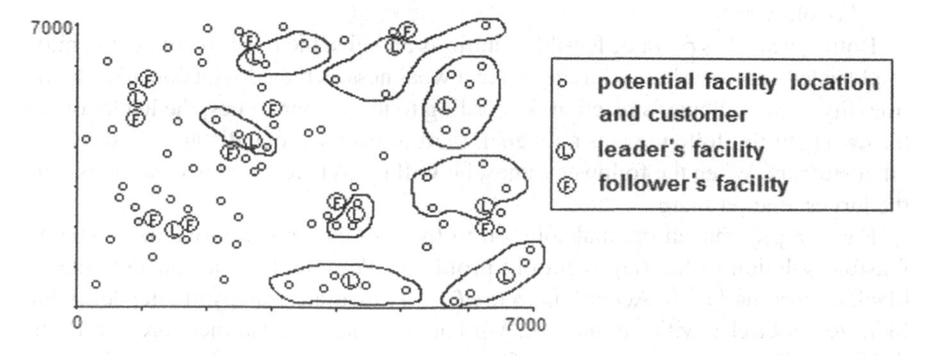

Fig. 7.2 Feasible solution based on the p-median strategy

Fig. 7.1–7.2 show the structure of the feasible solution obtained by the this approach. We have carried out experiments on the instances with 100 potential facilities locations and customers chosen randomly on a 7000×7000 square, $I = J$, $p = r = 10$, d_{ij} is the Euclidean distance between the customer j and the facility location i, and $w_j = 1$ for all $j \in J$.

Fig. 7.1 shows the optimal solution to the p-median problem. The black circles mean the facilities which the leader opens according to the p-median solution. This strategy forces the leader to open his own facilities as closer to customers as possible. Since here the follower is absent, all the customers are served by the leader. Fig. 7.2 shows the corresponding feasible solution to the centroid problem. An interesting observation concerns the follower's reaction, namely, the places where he locates his facilities to catch as many customers as possible. We can notice that in the central areas of the square, where the customers' density is high, the follower opens two facilities in the close vicinity of the leader's facility. He attacks the most profitable leader's facilities from two sides. In the distant areas of the square the follower opens either one facility near the leader or none. We encircle only customers served by the leader, the rest of the customers are served by the follower. Each group of customers corresponds to one facility opened by the leader. The feasible solution based on the p-median strategy delivers 41 customers for the leader in this instance.

Ignoring the follower is the weakness of the p-median strategy. Opening the p median facilities, the leader does not take into account that he will definitely lose some customers when the follower appears. Thus, the second approach is similar to the previous one but it is more sophisticated. The leader anticipates that the follower will react to his decision. The leader finds more facilities than he needs supposing that the follower will use the rest of them. More precisely, at the first step, the $(p+r)$-median problem is solved, that is problem (7.21)–(7.23), (7.25) with

$$\sum_{i \in I} x_i = p + r, \tag{7.26}$$

instead of constrain (7.24). At the second step, the leader opens those p facilities among $(p+r)$ facilities which catch the largest market share. At the third step, the follower's problem is solved exactly and the market is divided between the leader and the follower.

Both approaches produce feasible solutions and give the lower bound to the maximal leader's market share. But they have a weakness as they do not consider the rationality in the follower's reaction. According to these approaches, the leader opens his own facilities following his only goal: to minimize the total distance in servicing all customers. When the follower comes, he will try to take over the customers with the largest market share.

Fig. 7.3 presents an optimal solution to the $(p+r)$-median problem or a semi-feasible solution to the $(r|p)$-centroid problem. We draw the median facilities as black circles marked L. According to the $(p+r)$-median strategy, the leader opens facilities in circles with the mark L. We can see that 65 customers are served by the leader. But in the corresponding feasible solution, the follower does not stay at the black circles. Fig. 7.4 shows the feasible solution obtained by this strategy. We can observe that the follower attacks the leader's facilities and the $(p+r)$-median strategy delivers only 33 customers for the leader.

In Fig. 7.5 we draw the optimal solution to the $(r|p)$-centroid problem. The leader's market share is 50 customers against 33 customers in the previous case. Furthermore, the optimal locations differ from the median solutions significantly. Comparing the feasible solutions obtained by the median strategies in Fig. 7.2 and 7.4, we can notice that the second strategy provides the leader's market share less than the first one.

7.4.2 Alternative Heuristics

In the third approach we try to use the optimal follower's solution to improve the leader's solution. This heuristics is an iterative process. It starts from an arbitrary feasible leader solution. Then, given a set of leader's locations, the follower's problem is solved. Once that is done, and knowing the follower's facilities located at Y_r the leader tries to reoptimize his solution by solving the $(p|Y_r)$-medianoid problem. This process is repeated until some stopping criterion is satisfied. In other words,

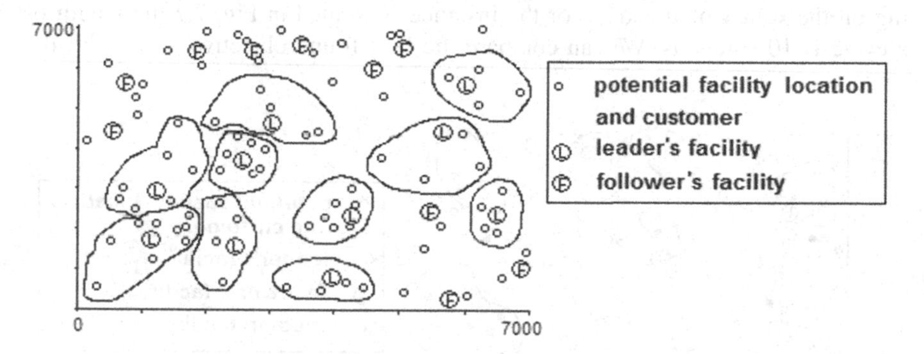

Fig. 7.3 Semi-feasible solution based on the $(p+r)$-median strategy

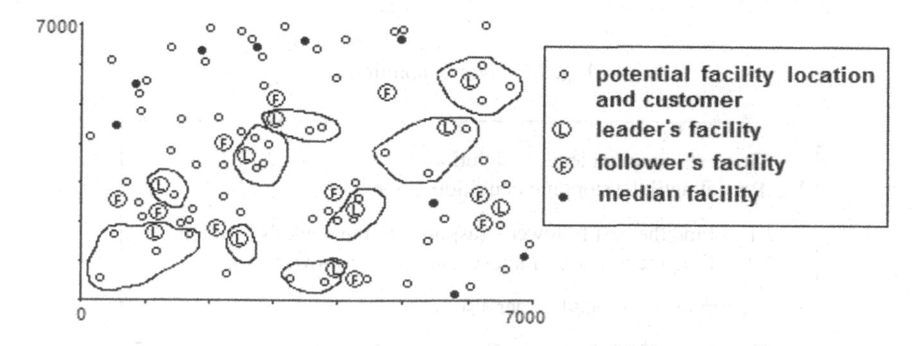

Fig. 7.4 Feasible solution based on the $(p+r)$-median strategy

both players alternately solve the medianoid problems. The best found leader's solution is the result of the heuristics. Originally, it has been studied for the centroid problem in the Euclidean plane [6]. We adopt it for the discrete case. This heuristics is schematically presented in Fig. 7.6.

The idea of this approach comes from the best response strategy for finding the Nash equilibrium. Under the assumption that the players have equal rights, this alternative optimization process may lead to the Nash equilibrium. As we have observed in our computational experiments, Nash equilibrium may not exist. Nevertheless, the alternative heuristics finds a feasible solution to the bi-level problem at each iteration. After some iterations of the alternative heuristics a cycle is revealed. By a cycle we mean a set of leader's solutions obtained since a once found leader's solution appears again. We can stop the alternative heuristics as soon as a cycle is revealed. The cycle existence means that it is useless to increase the total number of iterations. We should restart the alternative heuristics from another point as the cardinality of cycles can be different for the same instance depending on the initial leader's solution.

Fig. 7.7 shows the typical behavior of the alternative heuristics which we observed in previously mentioned test instances. The computational results show that the length of cycles can be quite large. It varies from 10 to 5000 solutions depending on the values of p and r. For the instance presented in Fig. 7.7 the length of a cycle is 10 solutions. We can compare the best found objective value with the

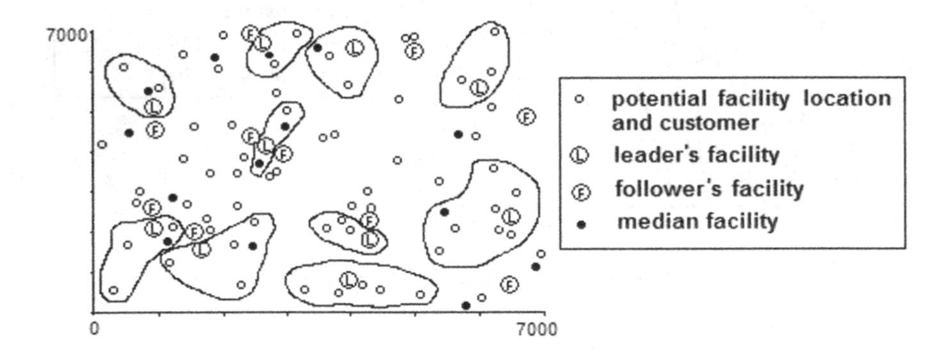

Fig. 7.5 Optimal solution

1 Generate an initial leader's solution.
2 Repeat until the stopping condition is met:
  2.1 Find the best follower's response to the leader's solution.
  2.2 Find the best leader's response to the follower's solution.
3 Return the best found leader's solution.

Fig. 7.6 Alternative heuristics

results obtained by the p-median heuristics (dashed line), and the optimal value (horizontal line). As we can see, the alternative heuristics finds solutions better than the p-median heuristics but it still does not find the optimal solution. Although the alternative heuristics finds good feasible solutions, it is a time consuming procedure due to the necessity of solving the medianoid problem optimally at each step.

7.4.3 Hybrid Heuristics

The fourth strategy, which seems to be the most promising, is to solve problem (7.1)–(7.8) for each element of the randomized neighborhood by hybrid algorithms. Following [32] by hybrid approaches we mean algorithms that combine the use of exact techniques with metaheuristic frameworks. We suggest using a metaheuristic framework based on stochastic tabu search on the upper level and linear programming by the branch-and-bound method on the lower one. The fundamental ideas of tabu search have been proposed by Glover [21] for single–level combinatorial problems. The tabu search based algorithm uses the information on the search history to force local search procedures to overcome local optimality. The local search procedure focuses on the binary leader's variables (x_i). The basic attribute of any heuristics based on local search procedures is a neighborhood. We adopt the well-known swap neighborhood. It contains all the leader solutions which can be obtained from a current solution by closing one leader's facility and opening another one. The size of the neighborhood is also a crucial attribute and should be tuned. In our case the size of the swap neighborhood is $p(m-p)$. As exploring the entire neighborhood might be time-consuming, we use a randomization procedure, which independently includes each element of the swap neighborhood with a fixed probability q in the randomized neighborhood (denoted N_q).

To prevent the local search from coming back to the previously visited solutions and cycling, there is a tabu list. This list contains the components of the leader's solutions, namely the pairs of swapping leader's facilities, which have been changed during the move to the best neighboring solution. They are stored in the list for a certain number of iterations. The number of elements in the tabu list may be fixed

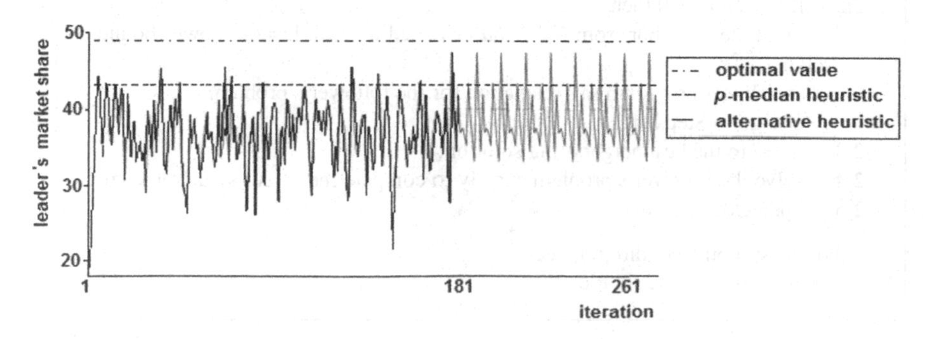

Fig. 7.7 Alternative heuristics's behavior

or changed from time to time at random for the diversification of the search process. All the neighboring leader's solutions with components from the current tabu list are in a *Tabu* set.

Fig. 7.8 shows the general framework of our hybrid algorithm. There are many strategies of creating an initial leader solution at Step 1. Any previously described heuristics can be used for that. Recall that in order to find the best neighbor we have to solve the follower's problem (7.4)–(7.8) and compute the objective function value for each element of the neighborhood. It is a time-consuming procedure. To reduce the running time, in addition to the randomization of the neighborhood, we use the first improvement pivoting rule and the linear programming relaxation to estimate the neighboring solutions. It allows us to get an upper bound for the follower's market share and consequently, a lower bound for the leader's market share. Thus, exploring the neighborhood takes a polynomial time at Step 2.2.

Once the best neighboring solution is found, we solve the follower's problem exactly and calculate the leader's market share. Then we update the tabu list by including a new pair of swapping leader's facilities and removing the oldest one. We use the total number of iterations as a stopping criterion.

Table 7.2 shows the comparative results of the lower bounds obtained by four previously described heuristics. As expected, the hybrid algorithm demonstrates the best results for a series of instances. Moreover, the lower bounds obtained by this algorithm coincide with the optimal values [38].

Note that the hybrid algorithm can be modified easily at some points. To estimate neighboring solutions at Step 2.2 other approaches can be used. In [14] Davydov has suggested a modification of the hybrid algorithm. It differs from Fig. 7.8 in Step 2.2 and concerns with the involvement of subgradient optimization. The Lagrangian relaxation technique has been applied to evaluate the leader solutions. According to reported computational results this algorithm finds optimal solutions with high

1 Generate an initial feasible solution, create an empty tabu list.
2 Repeat

    2.1 generate the randomized neighborhood N_q;
    2.2 if $N_q \setminus Tabu \neq \emptyset$ then
        find the neighbor from $N_q \setminus Tabu$ provided the best leader's lower bound by solving
        the linear programming relaxation for the follower's problem
        else go to Step 2.1;
    2.3 move to the best neighboring solution;
    2.4 solve the follower's problem exactly to compute the leader's market share;
    2.5 update the tabu list

3 until the stopping condition is met.
4 Return the best found solution.

Fig. 7.8 Hybrid algorithm: stochastic tabu search with linear programming

Table 7.2 The lower bounds. $m = n = 100$, $p = r = 10$, $w_j = 1, j \in J$

Instance	p-median	$(p+r)$-median	Alternative heuristics	Hybrid algorithm
1	41	31	49	50
2	41	36	45	49
3	46	41	44	48
4	41	39	45	49
5	48	40	47	48
6	42	37	46	47
7	49	37	48	51
8	42	37	44	48
9	47	35	46	49
10	46	33	47	49

frequency. It has been shown that the algorithm generates a Markov chain on a finite set of outcomes. This chain is irreducible and non-periodic under the proper restrictions on the length of the tabu list. These properties guarantee that the algorithm can reach an optimal solution from an arbitrary starting point. Furthermore, other states of the art of heuristics, such as variable neighborhood search [27], genetic algorithm [16], artificial immune systems algorithm [7], particle swarm optimization [11] etc. can be adopted instead of the stochastic tabu search for the upper level and / or instead of the linear programming for the lower level. The choice of an appropriate heuristics is not clear. Every time it constitutes an interesting line of research. In [2] a memetic matheuristics based on an exact solution of the follower's problem is suggested.

7.5 Exact Methods

Some exact methods have been proposed for the discrete $(r|p)$-centroid problem. Table 7.3 summarizes them. In [19] Ghosh and Craig have explored a competitive location model with elastic demand, with respect to distance, and given the length of the planning horizon. As their model is not a constant sum game, they have designed an enumerative search algorithm with simple heuristic rules to improve efficiency and reduce the computational burden. They have also used myopic solution ideas, based on the p-median problem to create an initial leader's solution. This exact approach could be used for our model, but as it does not take into account the specificity of the $(r|p)$-centroid, it is not efficient in our case.

In [22] Hakimi has noticed that the $(1,1)$-centroid on a tree is equivalent to the 1-median which can be determined in linear time [18] and on a general graph $(1,1)$-centroid can be found in polynomial time [9, 26]. In [43] Spoerhase and Wirth have solved the discrete $(r|p)$-centroid on a path and the discrete $(1|p)$-centroid on a tree in a polynomial time.

Table 7.3 Exact approaches to the discrete $(r|p)$-centroid problem. Facilities and customers are located at nodes of a graph, demand is essential

Authors (year)	Algorithms	The largest size of solved instances
Ghosh and Craig (1984)	Enumerative search method	$m = 21, n = 48, p \leq 4, r \leq 3$
Campos-Rodríguez and Moreno-Pérez (2010)	Partially enumerative method	$m = 50, n = 100, p = r \in \{2,4\}$
Alekseeva et al. (2010)	Iterative exact method	$m = n = 100, p = r \leq 5$
Roboredo and Pessoa (2012)	The branch-and-cut method	$m = n = 100, p = r \leq 15$

In [12] Campos-Rodríguez et al. have developed a partially enumerative method for general graphs. They use the single–level integer linear programming reformulation for the discrete $(r|p)$-centroid problem with an exponential number constraints and variables. At each iteration of this approach, a small subfamily of good follower's solutions is created. The method stops when all leader's solutions have been examined. The authors avoid full enumeration by organizing a special elimination process. They have solved the instances where customers and facilities are randomly distributed on a (50×50) grid graph. The authors report that the average computational time for solving the instances with 20 potential facilities, 30 customers, $p = 4$, $r = 2$ was about 50 minutes, while larger instances consumed more than 2 hours on the average.

To the best of our knowledge, two the most successful and exact approaches have been developed in [2] and [38]. These approaches have some similarities, but they use different ideas. Both of them use the single-level reformulation. Due to the exponential size of that reformulation, these methods operate with a small subfamily F instead of a full family $\mathscr{F}$. The iterative exact method from [2] uses matheuristics to find an optimal subfamily by enlarging the subfamily at each iteration, whereas, the branch-and-cut method from [38] uses the upper bound obtained by linear programming relaxation for the leader's problem with a subfamily F. To enlarge F, a special separation problem is built in [38] and solved by a greedy procedure or IP optimization solver depending on the size of the initial problem.

Originally, a single-level reformulation with an exponential number of constraints and variables was presented in [8]. At that time it was the first single–level reformulation for the discrete $(r|p)$-centroid problem. In [2] another reformulation with an exponential number of constraints and variables was presented. Later on, a new, improved reformulation (7.15)–(7.20) with an exponential number of constraints and polynomially many variables is proposed in [38]. An ability for improving the iterative exact method has occurred. Also, an original contribution of [38] is strengthened inequalities, which makes it possible to improve both exact approaches. Below we describe in some details these exact methods taking account of the improvements completed for the iterative exact method from [2]. In Section 7.6 we present comparative computational results for these methods.

7.5.1 The Branch-and-Cut Method

In [38] Roboredo and Pessoa have recently developed a branch-and-cut method. To set forth its key aspects, let us go back to reformulation (7.15)–(7.20). This program has an exponential number of constraints due to the exponential cardinality of $\mathscr{F}$, which contains all possible follower's solutions, and polynomially many variables. Since the number of variables x_i is smaller than that of z_{ij}, the branch is performed over the x_i variables. To produce the upper bound, the linear programming relaxation of the single-level reformulation is solved. But due to the exponential number of constraints it is necessary to solve a separation problem associated with (7.19) in order to include only the necessary constraints into the reformulation. To improve the formulation, a family of strengthened valid inequalities has been suggested.

The main idea of these inequalities is based on considering other follower's solutions, in addition to y^f when we compute the upper bound for W in the cases where the leader locates one facility opened in y^f. To this end, we define a function $H : y^f \to I$ that gives an alternative place for each facility opened by the follower to be used if the original place has already been used by the leader. Then, in addition to the follower's solution y^f, we consider the solutions that replace some facilities i such that $y_i^f = 1$, by $H(i)$. Now the new family of inequalities is as follows:

$$W \leq \sum_{j\in J} \sum_{i\in I_j(y^f)\cup \widetilde{I}_j(y^f)} w_j z_{ij}, \quad y^f \in F, \tag{7.27}$$

where

$$I_j(y^f) = \{i \in I \mid y_i^f = 0 \text{ and } d_{ij} \leq \min_{k\in I}(d_{kj}|y_k^f = 1)\},$$

$$\widetilde{I}_j(y^f) = \{i \in I \mid y_i^f = 1 \text{ and } d_{ij} \leq \min\{\min_{k\in I}(d_{kj}|y_k^f = 1), d_{H(i)j}\}\}, j \in J.$$

These sets contain the leader's facilities which allow the leader to keep customer j when the follower uses the solution y^f and wants to open some facilities at the same places where the leader has already opened his own facilities.

For finding some violated cuts, Roboredo and Pessoa have defined the separation problem. Given a fractional solution $(\bar{W}, \bar{x}, \bar{z})$ that satisfies (7.16)–(7.18), and some of the constraints (7.27), the separation problem is to find the best follower's solution $y^f \in \mathscr{F}$ and a corresponding function $H : y^f \to I$ that minimize the number of customers caught by the leader, that is they minimize [38]:

$$\sum_{j\in J} w_j \sum_{i\in I|d_{ij}\leq d_{k_j^\star j}} \bar{z}_{ij} - \sum_{j\in J|d_{H(k_j^\star)j}<d_{k_j^\star j}} w_j \bar{z}_{k_j^\star j}, \tag{7.28}$$

where $k_j^\star$ is the number of the closest follower's facility in solution y^f for the customer j, that is $k_j^\star = argmin_{\{k\in I|y_k^f=1\}} d_{kj}$.

The separation problem can be written as IP formulation. Let us introduce the binary variables:

$$s_k = \begin{cases} 1 & \text{if } y_k = 1 \text{ and } y \in F, \\ 0, & \text{otherwise,} \end{cases}$$

$$t_{jk} = \begin{cases} 1 & \text{if } y_k = 1, y \in F \text{ and } k \text{ is the closest facility to the customer } j, \\ 0, & \text{otherwise,} \end{cases}$$

$$t'_{jk} = \begin{cases} 1 & \text{if the closest facility } k \text{ to the customer } j \text{ is situated farther than } H(k), \\ 0, & \text{otherwise,} \end{cases}$$

and

$$h_{kl} = \begin{cases} 1 & \text{if } H(k) = l, \\ 0, & \text{otherwise.} \end{cases}$$

The separation problem is as follows:

$$\min \sum_{j\in J}\sum_{k\in I}(w_j \sum_{i\in I|d_{ij}\le d_{kj}} \bar{z}_{ij})t_{jk} - \sum_{j\in J}\sum_{k\in I}(w_j\bar{z}_{kj})t'_{jk}, \tag{7.29}$$

subject to

$$\sum_{k\in I} s_k = r, \tag{7.30}$$

$$t_{jk} \le s_k, \quad j\in J, k\in I, \tag{7.31}$$

$$\sum_{k\in I} t_{jk} = 1, \quad j \in J, \tag{7.32}$$

$$t'_{jk} \le t_{jk}, \quad j\in J, k\in I, \tag{7.33}$$

$$t'_{jk} \le \sum_{l\in I|d_{kj}>d_{lj}} h_{kl}, \quad j\in J, k\in I, \tag{7.34}$$

$$\sum_{l\in I} h_{kl} = s_k, \quad k\in I, \tag{7.35}$$

$$s_k, t_{jk}, t'_{jk}, h_{kl} \in \{0,1\}, \quad k\in I, j\in J, l\in I. \tag{7.36}$$

The value of the objective function (7.29) is equivalent to the sum (7.28). Constraint (7.30) ensures that the follower has exactly r facilities. Constraints (7.31) ensure the consistency between the variables t_{jk} and s_k. Constraints (7.32) ensure that, for each customer j, there is only one facility k, such that $y_k = 1$, closest to this customer. Constraints (7.33) and (7.34) ensure the consistency between the variables t'_{jk} and t_{jk} and between the variables t'_{jk} and h_{kl}, respectively. Finally, constraints (7.35) ensure that for each customer j, if $y_k = 1$ then there is only one facility l such that $H(k) = l$.

This problem can be solved exactly by the IP optimization solver. To avoid integer programming optimization, the authors have proposed a greedy heuristics. The heuristics first greedily constructs the solution y^f by choosing r facilities one at a time as follows. At each iteration, we choose a facility that causes a minimum increase in the value of the positive terms of (7.28). Next, for each facility i such that $y_i^f = 1$, we choose $H(i)$ to maximize the value of the negative terms of (7.28).

The entire exact method is the branch-and-cut, where the authors apply the separation procedures described above to get the upper bound and use the lower bounds obtained in [2] and [14]. This method solves small instances on $G_{50\times 50}$ significantly faster than the other previously developed approaches. It also allows one to solve the large instances with $m = n = 100$ and $p = r \in \{5, 10, 15\}$ from [1].

7.5.2 An Iterative Exact Method

Here we describe another exact approach based on the same single-level reformulation presented in Section 7.2.3. We present an improved version of the iterative exact method developed in [2]. The improvement results from using a formulation with a polynomial number of variables and strengthened inequalities introduced by Roboredo and Pessoa.

Let us go back again to reformulation (7.15)–(7.20). Note, that if we take any subset F of $\mathscr{F}$ and solve problem (7.15)–(7.20) with F instead of $\mathscr{F}$, we get an upper bound $W(F)$ to the leader's market share. This problem is NP-hard even when the subset F contains only two follower's solutions. It can be shown by reduction of the well-known set partitioning problem [20]. We do not give the proof of this statement here so as not to distract the reader's attention from the exact method.

The arbitrary feasible solution obtained by the matheuristics described above produces a lower bound. The main idea of the iterative exact method is to find a subfamily F so that the corresponded upper bound coincides with the lower bound. It would mean that the optimal solution is found. Fig. 7.9 presents the general framework of the method.

1 Choose an initial subfamily F.
2 Solve problem (7.15)–(7.20) with F instead of $\mathscr{F}$ exactly and find the leader's solution $x(F)$ and the upper bound $W(F)$.
3 Solve the follower's problem exactly, find the follower's solution $y(F)$, and calculate $LB(F)$.
4 If $W(F) = LB(F)$ then STOP.
5 Include the solution $y(F)$ into the subfamily F and go to Step 2.

Fig. 7.9 An iterative exact method

It is easy to make sure that the iterative exact method presented in 7.9 is exact and finite. Indeed, assume that we solve the follower's problem at Step 3 and find $y(F)$, but $y(F)$ has already belonged to F. From (7.19) we have

$$LB(F) = \sum_{j \in J} w_j z_j^*(F) \geq W(F).$$

Thus, $LB(F) = W(F)$ and $x(F)$ is the optimal solution to the bi-level problem. The method is finite because $|F| \leq \binom{m}{r}$.

Since the cardinality of F is increased as the number of iterations grows, Step 2 becomes the most time-consuming. We have to solve exactly a large scale optimization problem. If we use an IP optimization solver, we get $W(F)$ and $x(F)$, but it spends a lot of time to prove its optimality. Actually, we need a solution only. Therefore, we may reduce the running time if we replace problem (7.15)–(7.20) by a feasibility problem.

Denote the optimum for the bi-level problem (7.1)–(7.8) as W^* and consider the following feasibility problem:

$$W \geq W^*, \tag{7.37}$$

$$W \leq \sum_{j \in J} \sum_{i \in I_j(y^f) \cup \widetilde{I}_j(y^f)} w_j z_{ij}, \quad y^f \in F, \tag{7.38}$$

$$\sum_{i \in I} x_i = p, \tag{7.39}$$

$$z_{ij} \leq x_i, \quad i \in I, j \in J, \tag{7.40}$$

$$\sum_{i \in I} z_{ij} = 1, \quad j \in J, \tag{7.41}$$

$$W \geq 0, x_i, z_{ij} \in \{0,1\}, i \in I, j \in J. \tag{7.42}$$

If we have a feasible solution $x(F)$ to this system, we include $y(F)$ into the subset F and repeat the calculations. Otherwise, we can stop the search with the appropriate subfamily F. The feasibility problem is easier than the optimization one. We do not need to prove the optimality. We can apply IP solvers with a convenient objective function or metaheuristics. To cut down the size of F, we use the strengthened inequalities (7.38) suggested by Roboredo and Pessoa with the closest distance function H. We do not know the optimal value of W^*. Thus, we use the best value W' found by metaheuristics and update it during the search. Fig. 7.10 presents the framework of the modified exact method.

The laboriousness of the modified exact method strongly depends on the size of the subfamily. Ideally, if we can find the optimal subfamily F by a metaheuristics at Step 1, then we check the feasibility of the system only once. Otherwise, we have to solve the feasibility system by a solver many times. In the branch-and-cut method described above the authors suggest the use of the greedy procedure to solve the separation problem and thereby avoid IP optimization. Here we apply the modified stochastic tabu search as in Fig.7.8 for solving the feasibility system. We try to find the leader's solution with a maximal market share against the subfamily F. At Step 2.2 instead of solving the linear programming relaxation of the follower's problem we calculate the market share of the leader for each element of the randomized neighborhood directly. To accelerate the computations, we have adapted

1	Apply metaheuristics to create the subfamily F and find W'.
2	Find a feasible solution $x(F)$ to system (7.37)–(7.42). If it is infeasible then return the best found solution and stop.
3	Solve the follower's problem exactly, find optimal solution $y(F)$ and calculate $LB(F)$.
4	If $W' < LB(F)$ then $W' := LB(F)$.
5	Include $y(F)$ into the subfamily F and go to Step 2.

Fig. 7.10 The modified exact method

the procedure of Resende and Werneck [39] developed for the p-median problem. This procedure finds the most prominent pair of open-close facilities for a current leader's solution and result in the best neighboring solution. Due to special data structures to store the partial results, it is significantly faster than the standard evaluation for each possible pair. If we cannot find a feasible solution of the system by metaheuristics, we have to apply the branch-and-bound method from, say, CPLEX software to check infeasibility.

This modified exact method can be slightly changed to find solutions with a given gap of the optimal solution. To this end, we introduce a positive parameter ε and replace inequality (7.37) in the feasibility problem by the following:

$$W \geq (1+\varepsilon)W^*.$$

In this case, the method allows us to find approximate solutions with at most an ε relative gap of the optimum. Our computational experiments presented in the next section demonstrate that we are able to find optimal and approximate solutions within a reasonable time.

7.6 Computational Experiments

The exact methods have been tested and compared on a group of the instances from the benchmark library *Discrete Location Problems* [1]. This electronic library contains the test instances for the facility location problems including the competitive ones. For the discrete $(r|p)$-centroid problem we have 10 instances which are differentiated by (d_{ij}) matrix. For all instances, customers and facilities are in the same sites, that is $m = n$. The sites are chosen at random uniformly in a square 7000×7000. The elements of matrix (d_{ij}) are the Euclidean distances between the sites i, j. There are two cases for each instance with respect to the customer's weight. In the first case, all customers are identical and $w_j = 1$, for all $j \in J$. In the second case, all customers are different and w_j is chosen from (0,200) interval uniformly. The size of the instances is $m = n = 100$, $p = r \in \{5, 10, 15, 20\}$.

Our experiments have been carried out in a PC Intel Xeon X5675, 3 GHz, RAM 96 Gb, running under the Windows Server 2008 operating system. We have used the CPLEX 12.3. as an optimization solver. Below we present comparative computational results which have been obtained on the computers with other specifications. Namely, our earlier iterative exact method suggested in [2] have been realized on a PC Pentium Intel Core 2, 1.87 GHz, RAM 2 Gb, running under the Windows XP Professional operating system. The branch-and-cut method by Roboredo and Pessoa [38] has been realized on a PC Pentium Intel Core 2 duo, 2.13 GHz, RAM 2 Gb, and the authors used the CPLEX 12.1.

The presented exact approaches use the metaheuristics which we have discussed before to find the solution provided the lower bound of a good quality. Each exact method tries to improve it or, in case it has already been the optimal solution, to prove it. Actually, the chosen test instances from [1] with $m = n = 100$, $p = r \in \{5, 10, 15, 20\}$ are not difficult for the metaheuristics. They can find the optimal solutions [2, 14]. However, proving its optimality is a burdensome problem for exact methods. Here, we focus on the comparative results concerning the computational time for the exact methods. We demonstrate their capacities with respect to the parameters p and r.

Tables 7.4 – 7.6 present computational results. The column *Instance* indicates the code name of the test instance. The column *Opt* contains the leader's market share for the optimal solution, the columns *IEM*, *MEM* and *BC* show the total CPU time in minutes consumed by the iterative exact method suggested in [2], their modified exact method presented in Section 7.5.2 and the branch-and-cut method by Roboredo and Pessoa from Section 7.5.1, respectively.

Table 7.4 shows that the modified exact method (the column *MEM*) consumes significantly less time than its earlier version (the column *IEM*) even taking into account the differences in the machine specifications. The columns *MEM* and *BC* in Tables 7.4, and 7.5 show that the modified exact method is better for the relatively

Table 7.4 $m = n = 100$, $p = r = 5$

	$w_j = 1, j \in J$				$w_j \in (0, 200), j \in J$			
		Time (in min)				Time (in min)		
Instance	*Opt*	*IEM*	*MEM*	*BC*	*Opt*	*IEM*	*MEM*	*BC*
111	47	120	6	44.2	4139	65	1	27.4
211	48	60	1	58.7	4822	37	4	159.9
311	45	3600	26	202.8	4215	5460	38	313.7
411	47	150	2	52.0	4678	900	69	74.1
511	47	120	1	39.4	4594	720	16	469.0
611	47	90	3	51.0	4483	660	2	25.7
711	47	180	4	53.4	5153	2550	5	130.9
811	48	42	1	35.7	4404	720	2	195.5
911	47	160	2	44.8	4700	2520	13	290.3
1011	47	165	2	66.2	4923	30	1	18.3

small test instances with $m = n = 100$, $p = r = 5$ and 10 than the branch-and-cut method in comparison with the computational time.

Table 7.6 shows that the *MEM* and *BC* methods take a lot of computational efforts for the case $p = r = 15$. It means that these instances become difficult for both methods. Discussed in [2] have been the reasons why this case requires such a long running time. It deals with the growth of the subfamily F as the values of p and r increase. It has been noticed that under the fixed values of n and m the problem becomes more difficult in the instances with the values of p and r equal to about a one-third of m. The computational time for the *BC* algorithm increases as both the number of branch-and-bound nodes and the number of cuts generated expand [38]. Thus, for more difficult instances we have to develop other exact approaches or do with approximate solutions.

Table 7.7 shows the comparative results concerning the approximate solutions. The column $5\% \cdot Opt$ contains a leader market share with a gap of at most 5% of the optimal value for the *MEM*. The column *GapBC* shows a root gap between the best known lower bound and the best upper bound obtained by the branch-and-cut algorithm [38]. We can see that even for the problems where the gap has been about 2%, the running time has reached 10 hours. It is unlikely that *MEM* would be able to find solutions with a lesser gap for admissible computational efforts but to find solutions within 5% of the optimum by the *MEM* takes a reasonable time on a powerful computer. We can conclude that the existing exact methods are good enough since they are able to tackle previously open instances with up to 100 customers, 100 potential facilities and $p = r = 15$ for a reasonable time. Nevertheless, they leave room for further improvements.

Table 7.5 $m = n = 100$, $p = r = 10$

	$w_j = 1, j \in J$			$w_j \in (0,200), j \in J$		
		Time (in min)			Time (in min)	
Instance	*Opt*	*MEM*	*BC*	*Opt*	*MEM*	*BC*
111	50	13	38,1	4361	60	170,3
211	49	20	78,5	5310	42	158,1
311	48	195	222,8	4483	146	317,9
411	49	135	188,6	4994	33	229,1
511	48	270	315,4	4906	399	1340,2
611	47	900	381,8	4595	143	859,7
711	51	12	42,2	5586	73	339,2
811	48	145	259,9	4609	152	446,8
911	49	102	187,3	5302	6	39,6
1011	49	180	237,1	5005	97	562,7

Table 7.6 $m = n = 100$, $p = r = 15$

	$w_j \in (0, 200)$, $j \in J$		
		Time (in min)	
Instance	Opt	MEM	BC
111	4596	72	162,53
211	5373	3845	1349,27
311	4800	395,00	461,78
411	5064	1223	1402,33
511	5131	2120	1318,32
611	4881	2293	472,37
711	5827	1320	810,00
811	4675	4570	1919,73
911	5158	>600	>600
1011	5195	>600	1200,57

Table 7.7 $m = n = 100$, $p = r = 20$

		$w_j \in (0, 200)$, $j \in J$		
Instance	$5\% \cdot Opt$ MEM	Time MEM (in min)	Gap (%) BC	Time BC (in min)
111	4737,6	1	3,83	>600
211	5703,6	185	3,35	>600
311	5137,65	248	3,05	>600
411	5677,81	5	1,99	>600
511	5600,7	110	1,61	>600
611	5199,6	190	1,90	>600
711	6187,65	97	7,28	>600
811	5100,9	570	2,46	>600
911	5731,95	165	2,41	>600
1011	5668,95	130	1,64	>600

7.7 Conclusions

Since the nineties of the last century, the competitive facility location models have been increasingly asked-for and have been created an active field of research. In this chapter we have discussed the classical model in this field, a so-called discrete $(r|p)$-centroid problem. With respect to other competitive models, it is the basic model. There are many extensions of this challenging problem concerning the customers' behavior, type of demands, arising additional costs etc. [31]. A new line of studies in this field is the competitive facility location and design models. The players aim at finding the location and attractiveness of each facility to be opened so as to maximize their own market shares or profits. Some insights in this area can be found in [28, 36, 44].

We have reviewed the heuristics and exact methods for the $(r|p)$-centroid problem. The stochastic tabu search shows excellent results. To be sure, it is a time-consuming approach but we have obtained a global optimum in a relatively small number of steps. We believe that the matheuristics are useful for bi-level optimization. We can apply these methods to solve the discrete bi-level problems but have to make a lot of efforts for computing objective function values. We have discussed two exact approaches. They realize different ideas but they are based on the single level reformulation. It is easy to suggest the single level reformulation for the min-max problems. The cooperative and noncooperative cases coincide here. Is it possible to adopt these methods for other competitive location models? For example, we have not touched the continuous location models where the players can open their own facilities anywhere in the Euclidean plane [13]. It is a min-max problem with a finite number of customers. The sum of the leader and follower objective functions values is here a constant too. Nevertheless, we have neither any upper bounds nor any single level reformulations. We believe that exact methods and matheuristics for this case will be great of interest for further research.

Acknowledgements. This research was partially supported by RFBR grants 11-07-00474, 12-01-31090 and 12-01-00077. The work of the first author was partially carried out during the tenure of an ERCIM "Alain Bensoussan" Fellowship Programme. The research leading to these results has received funding from the European Union Seventh Framework Programme (FP7/2007-2013) under grant agreement N246016.

References

1. Alekseeva, E., Beresnev V., Kochetov, Y. et al.: Benchmark library: Discrete Location Problems, `http://math.nsc.ru/AP/benchmarks/Competitive/p_me_comp_eng.html`
2. Alekseeva, E., Kochetova, N., Kochetov, Y., Plyasunov, A.: Heuristic and exact methods for the discrete $(r|p)$-centroid problem. In: Cowling, P., Merz, P. (eds.) EvoCOP 2010. LNCS, vol. 6022, pp. 11–22. Springer, Heidelberg (2010)
3. Ben–Ayed, O.: Bilevel linear programming. Comput. Oper. Res. 20(5), 485–501 (1993)
4. Benati, S., Laporte, G.: Tabu search algorithms for the $(r|X_p)$–medianoid and $(r|p)$–centroid problems. Location Science 2, 193–204 (1994)
5. Beresnev, V., Melnikov, A.: Approximate algorithms for the competitive facility location problem. J. Appl. Indust. Math. 5(2), 180–190 (2011)
6. Bhadury, J., Eiselt, H., Jaramillo, J.: An alternating heuristic for medianoid and centroid problems in the plane. Comp. & Oper. Res. 30, 553–565 (2003)
7. Burke, E.K., Kendall, G.: Introductory Tutorials in Optimization and Decision Support Techniques. Springer, US (2005)
8. Campos-Rodríguez, C.M., Moreno Pérez, J.A.: Multiple voting location problems. European J. Oper. Res. 191, 437–453 (2008)
9. Campos-Rodríguez, C., Moreno-Pérez, J.A.: Relaxation of the condorcet and simpson conditions in voting location. European J. Oper. Res. 145(3), 673–683 (2003)

10. Campos-Rodríguez, C., Moreno-Pérez, J.A., Notelmeier, H., Santos-Peñate, D.R.: Two-swarm PSO for competitive location. In: Krasnogor, N., Melián-Batista, M.B., Pérez, J.A.M., et al. (eds.) NICSO 2008. SCI, vol. 236, pp. 115–126. Springer, Heidelberg (2009)
11. Campos-Rodríguez, C., Moreno-Pérez, J.A., Santos-Peñate, D.R.: Particle swarm optimization with two swarms for the discrete $(r|p)$-centroid problem. In: Moreno-Díaz, R., Pichler, F., Quesada-Arencibia, A. (eds.) EUROCAST 2011, Part I. LNCS, vol. 6927, pp. 432–439. Springer, Heidelberg (2012)
12. Campos-Rodríguez, C.M., Moreno-Pérez, J.A., Santos-Peñate, D.: An exact procedure and LP formulations for the leader–follower location problem. Business and Economics TOP 18(1), 97–121 (2010)
13. Carrizosa, E., Davydov, I., Kochetov, Y.: A new alternating heuristic for the $(r|p)$–centroid problem on the plane. In: Operations Research Proceedings 2011, pp. 275–280. Springer (2012)
14. Davydov, I.: Tabu search for the discrete $(r|p)$–centroid problem. J. Appl. Ind. Math. (in press)
15. Davydov, I., Kochetov, Y., Plyasunov, A.: On the complexity of the $(r|p)$–centroid problem on the plane. TOP (in press)
16. Dréo, J., Pétrowski, A., Siarry, P., Taillard, E.: Metaheuristics for Hard Optimization. In: Chatterjee, A., Siarry, P. (eds.) Methods and Case Studies. Springer, Heidelberg (2006)
17. Friesz, T.L., Miller, T., Tobin, R.L.: Competitive network facility location models: a survey. Regional Science 65, 47–57 (1988)
18. Goldman, A.: Optimal center location in simple networks. Transportation Science 5, 212–221 (1971)
19. Ghosh, A., Craig, C.: A Location allocation model for facility planning in a competitive environment. Geographical Analysis 16, 39–51 (1984)
20. Garey, M.R., Johnson, D.S.: Computers and intractability (a guide to the theory of NP-completeness). W.H. Freeman and Company, New York (1979)
21. Glover, F., Laguna, M.: Tabu Search. Kluwer Acad. Publ., Boston (1997)
22. Hakimi, S.L.: Locations with spatial interactions: competitive locations and games. In: Mirchandani, P.B., Francis, R.L. (eds.) Discrete Location Theory, pp. 439–478. Wiley & Sons (1990)
23. Hakimi, S.L.: On locating new facilities in a competitive environment. European J. Oper. Res. 12, 29–35 (1983)
24. Hakimi, S.L.: On locating new facilities in a competitive environment. In: Annual ORSA-TIMS Meeting, Houston (1981)
25. Hotelling, H.: Stability in competition. Economic J. 39, 41–57 (1929)
26. Hansen, P., Labbé, M.: Algorithms for voting and competitive location on a network. Transportation Science 22(4), 278–288 (1988)
27. Hansen, P., Mladenović, N.: Variable neighborhood search: principles and applications. European J. Oper. Res. 130, 449–467 (2001)
28. Küçükaudin, H., Aras, N., Altinel, I.K.: Competitive facility location problem with attractiveness adjustment of the follower. A bilevel programming model and its solution. European J. Oper. Res. 208, 206–220 (2011)
29. Kariv, O., Hakimi, S.: An algoritmic approach to network location problems. The p-medians. SIAM J. Appl. Math. 37, 539–560 (1979)
30. Kochetov, Y.A.: Facility location: discrete models and local search methods. In: Chvatal., V. (ed.) Combinatorial Optimization. Methods and Applications, pp. 97–134. IOS Press, Amsterdam (2011)

31. Kress, D., Pesch, E.: Sequential competitive location on networks. European J. Oper. Res. 217(3), 483–499 (2012)
32. Maniezzo, V., Stützle, T., Voß, S. (eds.): Matheuristics. Hybridizing Metaheuristics and Mathematical Programming. Annals of Information Systems, vol. 10 (2010)
33. Megiddo, N., Zemel, E., Hakimi, S.: The maximum coverage location problem. SIAM J. Algebraic and Discrete Methods 4, 253–261 (1983)
34. Mladenovic, N., Brimberg, J., Hansen, P., Moreno-Pérez, J.A.: The p-median problem: a survey of metaheuristic approaches. European J. Oper. Res. 179, 927–939 (2007)
35. Noltemeier, H., Spoerhase, J., Wirth, H.: Multiple voting location and single voting location on trees. European J. Oper. Res. 181, 654–667 (2007)
36. Plastia, F., Carrizosa, E.: Optimal location and design of a competitive facility. Math. Program., Ser A. 100, 247–265 (2004)
37. Plastria, F., Vanhaverbeke, L.: Discrete models for competitive location with foresight. Comp. & Oper. Res. 35(3), 683–700 (2008)
38. Roboredo, M.C., Pessoa, A.A.: A branch-and-cut algorithm for the discrete $(r|p)$-centroid problem. European J. Oper. Res. (in press)
39. Resende, M., Werneck, R.: On the implementation of a swap-based local search procedure for the p-median problem. In: Ladner, R.E. (ed.) Proceedings of the Fifth Workshop on Algorithm Engineering and Experiments (ALENEX 2003), pp. 119–127. SIAM, Philadelphia (2003)
40. Spoerhase, J.: Competitive and Voting Location. In: Dissertation. Julius Maximilian University of Würzburg (2010)
41. Serra, D., ReVelle, C.: Competitive location in discrete space. In: Drezner, Z. (ed.) Facility Location - A Survey of Applications and Methods, pp. 367–386. Springer, New York (1995)
42. Serra, D., ReVelle, C.: Market capture by two competitors: the pre-emptive capture problem. J. Reg. Sci. 34(4), 549–561 (1994)
43. Spoerhase, J., Wirth, H. (r, p)-centroid problems on paths and trees. J. Theor. Comp. Sci. Archive. 410(47–49), 5128–5137 (2009)
44. Saidani, N., Chu, F., Chen, H.: Competitive facility location and design with reactions of competitors already in the market. European J. Oper. Res. 219, 9–17 (2012)
45. Smith, J.C., Lim, C., Alptekinoglu, A.: Optimal mixed-integer programming and heuristic methods for a bilevel Stackelberg product introduction game. Nav. Res. Logist. 56(8), 714–729 (2009)
46. Santos-Peñate, D.R., Suárez-Vega, R., Dorta-González, P.: The leader-follower location model. Networks and Spatial Economics 7, 45–61 (2007)
47. Teramoto, S., Demaine, E., Uehara, R.: Voronoi game on graphs and its complexity. In: IEEE Symposium on Computational Intelligence and Games, pp. 265–271 (2006)
48. Vasilev, I.L., Klimentova, K.B., Kochetov, Y.A.: New lower bounds for the facility location problem with clients preferences. Comp. Math. Math. Phys. 49(6), 1055–1066 (2009)

Chapter 8
Exact Solution Methodologies for Linear and (Mixed) Integer Bilevel Programming

Georgios K.D. Saharidis, Antonio J. Conejo, and George Kozanidis

Abstract. Bilevel programming is a special branch of mathematical programming that deals with optimization problems which involve two decision makers who make their decisions hierarchically. The problem's decision variables are partitioned into two sets, with the first decision maker (referred to as the leader) controlling the first of these sets and attempting to solve an optimization problem which includes in its constraint set a second optimization problem solved by the second decision maker (referred to as the follower), who controls the second set of decision variables. The leader goes first and selects the values of the decision variables that he controls. With the leader's decisions known, the follower solves a typical optimization problem in his self-controlled decision variables. The overall problem exhibits a highly combinatorial nature, due to the fact that the leader, anticipating the follower's reaction, must choose the values of his decision variables in such a way that after the problem controlled by the follower is solved, his own objective function will be optimized. Bilevel optimization models exhibit wide applicability in various interdisciplinary research areas, such as biology, economics, engineering, physics, etc. In this work, we review the exact solution algorithms that have been developed both for the case of linear bilevel programming (both the leader's and the follower's problems are linear and continuous), as well as for the case of mixed integer bilevel programming (discrete decision variables are included in at least one of these two problems). We also document numerous applications of bilevel programming models from various different contexts. Although several reviews dealing with bilevel programming

Georgios K.D. Saharidis
Department of Mechanical Engineering, University of Thessaly, Volos,
Greece Kathikas Institute of Research and Technology Paphos, Cyprus

Antonio J. Conejo
Department of Electrical Engineering Univ. Castilla - La Mancha, Spain
e-mail: antonio.conejo@uclm.es

George Kozanidis
Systems Optimization Laboratory Department of Mechanical Engineering,
University of Thessaly, Volos, Greece
e-mail: gkoz@mie.uth.gr

E.-G. Talbi (Ed.): *Metaheuristics for Bi-level Optimization*, SCI 482, pp. 221–245.
DOI: 10.1007/978-3-642-37838-6_8 © Springer-Verlag Berlin Heidelberg 2013

have previously appeared in the related literature, the significant contribution of the present work lies in that a) it is meant to be complete and up to date, b) it puts together various related works that have been revised/corrected in follow-up works, and reports in sequence the works that have provided these corrections, c) it identifies the special conditions and requirements needed for the application of each solution algorithm, and d) it points out the limitations of each associated methodology. The present collection of exact solution methodologies for bilevel optimization models can be proven extremely useful, since generic solution methodologies that solve such problems to global or local optimality do not exist.

8.1 Introduction

Hierarchical optimization deals with mathematical programming problems whose feasible set is implicitly determined by a sequence of nested optimization problems. The most studied case is the case of bilevel programming and especially the linear one. A bilevel program is a problem in which a subset of the variables is required to be an optimal solution of a second mathematical program. This problem can be considered as a two-person game where one of the players, the leader, knows the cost function mapping of the second player, the follower, who may or may not know the cost function of the leader. The follower knows, however, the strategy chosen by the leader and takes it into account when computing his own strategy. The leader can foresee the reactions of the follower and can therefore optimize his strategy choice.

Several surveys on bilevel programming have appeared in the related literature. In the linear bilevel programming survey of Wen and Hsu [88], the authors review the basic models and the characterizations of the problem, the areas of application, the existing solution approaches, and the related models and areas for further research. In the survey on the features of linear bilevel programming by BenAyed [16], the author reviews complexity properties, algorithms, applications, as well as the relationship of the problem with other optimization problems. Vicente and Calamai [86] discuss the main properties, summarize the main solution algorithms, review key applications, and present an extensive bibliography of bilevel and multi-level programming. Dempe [42] [43] discusses alternative formulations, relationships with other optimization problems, theoretical and complexity results, optimality conditions, solution algorithms and applications of bilevel optimization models.

Colson et al. [36] consider various cases (linear, linear-quadratic, nonlinear) of bilevel optimization problems, describe their main properties, and give an overview of solution approaches. An updated version of the same survey was presented by Colson et al. [37]. Chinchuluun et al. [33] discuss algorithmic and theoretical results on multilevel programming, including complexity issues, optimality conditions, and algorithmic methods for solving multilevel programming problems. Finally, the books of Bard [10] and Dempe [41] provide an excellent introduction to the subject.

8.2 Description of the Bilevel Problem

Bilevel programming involves two optimization problems, the first of which (upper) is controlled by the leader and the second of which (lower) is controlled by the follower. The feasible region of the upper optimization problem is determined by its own constraints plus the lower optimization problem. In general, the problem is non-convex, and finding its global optimum is an arduous task.

The decision variables of a bilevel program are partitioned into two sets, with the leader controlling the first subset (x), and the follower controlling the second one (y). If the leader chooses $x = x'$ then the follower responds with $y = y'$ as is shown in figure 8.1. For a given x, the follower solves the lower problem, optimizing F_2. The leader examines the reactions of the follower for each feasible choice of x (the dashed line in the $x-$axis). The set of all feasible solutions of the bilevel problem (the black line in figure8.1) is called inducible region (IR), and is generally non-convex [30]. The optimal solution of the bilevel problem is the point on the inducible region for which the upper level objective function, F_1, takes its optimal value (point A in figure8.1). In the linear case, this is an extreme point of the inducible region. In general, a bilevel programming model may have continuous and/or integer decision variables and constraints in the upper and/or the lower level. In order to simplify the presentation of the main characteristics of the bilevel problem, we consider first the case where only continuous variables (x, which is an $n-$dimensional vector controlled by the leader and y, which is an $m-$dimensional vector controlled by the follower) are present. This leads to the Bilevel Linear Problem (BLP) defined as follows:

$$x \in X \subseteq R^n, y \in Y \subseteq R^m$$

For

$$F_1 : X \times Y \longrightarrow R^1, F_2 : X \times Y \longrightarrow R^1$$

$Max_{x \in X} F_1(x,y) = c_1 x + d_1 y$
s.t. $g_1(x,y) = A_1 x + B_1 y \leq b_1$

$Max_{y \in Y} F_2(x,y) = c_2 x + d_2 y$
s.t. $g_2(x,y) = A_2 x + B_2 y \leq b_2$,

where $c_1, c_2 \in R^n, d_1, d_2 \in R^m, b_1 \in R^p, b_2 \in R^q$,

$$A_1 \in R^{p \times n}, B_1 \in R^{p \times m}, A_2 \in R^{q \times n}, B_2 \in R^{q \times m}$$

Sets X and Y impose additional restrictions on the decision variables, such as upper and/or lower bounds. From the leader's perspective, this model can be viewed as a mathematical program with an implicitly defined non-convex constraint region given by the follower's sub-problem. In general, the following regions and sets (cf. figure 8.1) are defined in linear bilevel optimization problems:

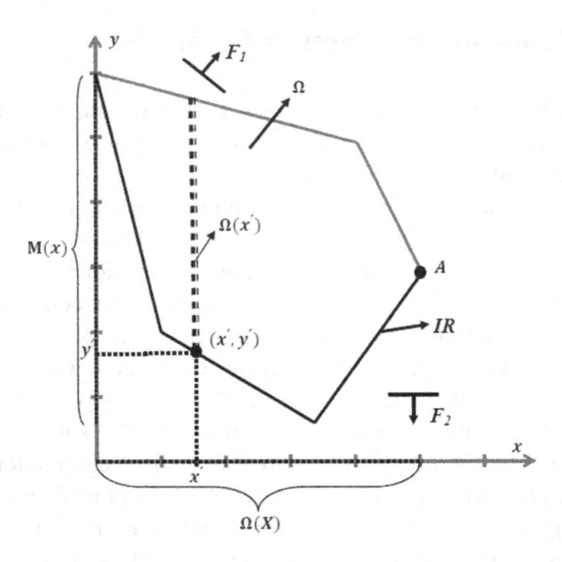

Fig. 8.1 Bilevel linear problem

- The BLP constraint region is defined as: $\Omega = \{(x,y) : x \in X, y \in Y, g_1(x,y) \leq b_1, g_2(x,y) \leq b_2\}$
- The projection of Ω onto the leader's decision space is $\Omega(X) = \{x \in X : \exists y, (x,y) \in \Omega\}$
- The follower's feasible region for $x \in X$ fixed is $\Omega(x) = \{y \in Y : g_2(x,y) \leq b_2\}$
- The follower's rational reaction set is $M(x) = \{y \in Y : y \in argmax(F_2(x,\overline{y}) : \overline{y} \in \Omega(x))\}$
- The Inducible Region (IR) which corresponds to the solution space of the bilevel problem is $IR = \{(x,y) = x \in \Omega(X), y \in M(x)\}$

In order to ensure that the above bilevel problem is well posed, we make the additional assumption that Ω is nonempty and compact and that for each decision taken by the leader, the follower has some room to respond $(\Omega(x) \neq 0)$. The rational reaction set, $M(x)$, defines these responses, while the IR represents the set over which the leader may optimize. Thus, in terms of the above notation, the BLP can be written as:

$$Max(F_1(x,y) : (x,y) \in IR)$$

A bilevel feasible solution is a pair $(\overline{x},\overline{y})$ with $\overline{y} \in M(\overline{x})$for the specific $\overline{x}$. An optimal solution for the bilevel problem is a point (x^*,y^*) if this point is bilevel feasible and for all bilevel feasible points $(\overline{x},\overline{y}) \in IR, F_1(x^*,y^*) \geq F_1(\overline{x},\overline{y})$.

Besides the lower problem's primal decision variables, the upper level optimization problem may also depend on the lower problem's dual decision variables. For example, such is very often the case in the context of an electricity market that adopts a clearance payment scheme which compensates each participating energy producer by a uniform market clearing price for each unit of energy that he provides. In the bilevel programming model that the individual producer must solve in order to maximize his profit, the actual value of this profit depends on the energy quantity that he will inject to the system (a lower level primal decision variable), as well as on the value of this uniform market price, which is defined as the dual variable of the lower level constraint that ensures satisfaction of the demand for energy.

Bialas and Karwan [20] propose an incentive scheme to overcome the problem of multiple optima in the lower level problem. This problem arises when for a particular choice of the upper level variables, the follower's problem has multiple optima, which do not all result in the same objective value for the leader. It is clear that the bilevel program may not have an optimal solution in this case, since the leader has no way of forcing the follower to choose a particular lower level optimal solution. Thus, if one of these alternative optimal solutions is the one that maximizes the objective of the leader, there is no guarantee that it will be selected. The method of Bialas and Karwan [20] perturbs the lower level problem, replacing the original objective by $F_2' = F_2(x,y) + \varepsilon F_1(x,y)$,where the value of $\varepsilon > 0$ is suitably small. As the authors show, this would require a "kick-back" of a small portion of the upper level's earnings to encourage the follower to choose a desirable solution. Concluding, the authors note that, in general, such a perturbation method may still not determine a unique solution, since the leader may have the same objective function value for a number of distinct level two optimal solutions. In that case, however, any of these solutions would be satisfactory for level one.

Another approach that has been proposed to deal with the issue of multiple optimal lower level solutions is the optimistic (pessimistic) approach (see for example [61]). According to this approach, whenever the follower has multiple optimal solutions, he is forced to select the one which is the most (least) favorable to the leader. This implies that in the case of multiple lower level optimal responses to a particular choice, x, of the leader, the follower would have to solve a second optimization problem that would determine his final selected solution as the one that maximizes (minimizes) the leader's objective value among all these alternative solutions.

If the problem involves integer and continuous variables which are separable and appear in linear relations, then the bilevel problem corresponds to a Mixed Integer Bilevel Linear Problem (MIBLP). Let x be an n-dimensional vector of continuous variables controlled by the leader, y be an m-dimensional vector of continuous variables controlled by the follower, z be a t-dimensional vector of integer variables controlled by the leader, and w be a u-dimensional vector of integer variables controlled by the follower. Then, the MIBLP is formulated as follows in its general form:

$$x \in X \subseteq R^n, y \in Y \subseteq R^m, z \in Z_t, w \in Z_u$$

For

$$F_1 : X \times Y \times Z_t \times Z_u \longrightarrow R^1, F_2 : X \times Y \times Z_t \times Z_u \longrightarrow R^1$$

$Max_{x \in X, z \in Z_t} F_1(x,y,z,w) = c_1 x + d_1 y + r_1 z + g_1 w$
s.t. $A_1 x + B_1 y + C_1 z + Q_1 w \leq b_1$

$Max_{y \in Y, w \in Z_u} F_2(x,y,z,w) = c_2x + d_2y + r_2z + g_2w$
s.t. $A_2x + B_2y + C_2z + Q_2w \leq b_2$,

where $c_1, c_2 \in R^n, d_1, d_2 \in R^m, r_1, r_2 \in R^t, g_1, g_2 \in R^u, b_1 \in R^p, b_2 \in R^q$,

$A_1 \in R^{p\times n}, B_1 \in R^{p\times m}, C_1 \in R^{p\times t}, Q_1 \in R^{p\times u}, A_2 \in R^{q\times n}, B_2 \in R^{q\times m}, C_2 \in R^{q\times t}, Q_2 \in R^{q\times u}$

Three general cases can appear: a) only the leader controls integer variables, b) only the follower controls integer variables, and c) both players control integer variables. Figure 8.2 below presents the IR of each of these three cases.

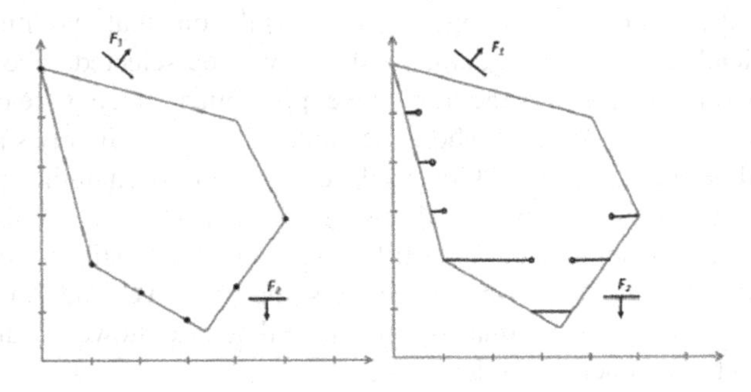

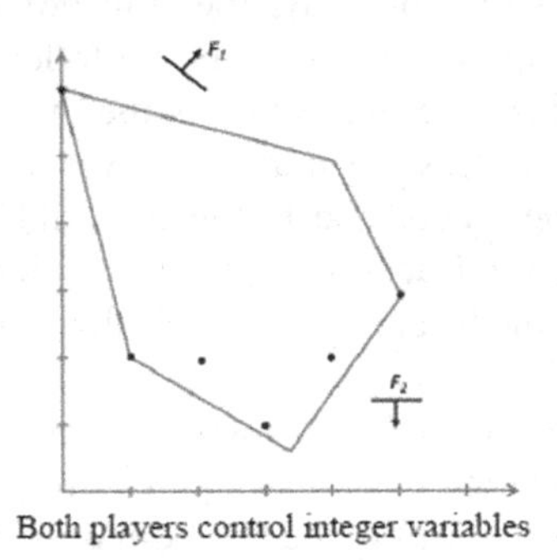

Fig. 8.2 IR of mixed integer bilevel linear problems

8.3 Solution Approaches for the Linear Bilevel Problem

Solution approaches for the BLP can be categorized into 4 main categories. The first one includes reformulation techniques which transform the bilevel program into a single level problem using mainly the lower problem's Karush Kuhn Tucker (KKT) optimality conditions. The second group includes vertex enumeration approaches , which develop a modified version of the simplex method, exploiting the fact that the problem's global optimal solution occurs at an extreme point of IR, which is also an extreme point of Ω [21]. The third group includes algorithms which extract gradient information from the lower problem and use it to compute directional derivatives for the leader's objective function. The fourth group includes penalty algorithms , which introduce a term that incorporates a penalty associated with the violation of certain optimality conditions, leading this way the search towards the optimal solution.

Fortuny-Amat and McCarl [47] propose a solution procedure for the quadratic bilevel problem . In the case of linear bilevel programming with constraints at both levels, this procedure leads to the transformation of the original problem into a single level problem using the lower problem's first order optimality conditions. This problem is then transformed into a mixed integer linear problem through the introduction of one auxiliary binary decision variable for each complementary slackness condition. Each of these variables takes the value 0 or 1, depending on whether the corresponding constraint is binding or the associated multiplier is equal to 0. The problem that results after this transformation is then solved through a suitable enumerative (branch and bound) procedure that checks every possible combination for optimality. The authors apply the proposed procedure on a numerical study, but do not present generic results regarding its computational performance.

Bard and Falk [11] consider the general linear bilevel problem with functional constraints at both levels, and transform it into a single level optimization problem by substituting the (convex) lower level problem with its KKT optimality conditions. Due to the introduction of the complementarity conditions, the problem obtained through this transformation is non-convex. In order to handle this difficulty, the authors reformulate the general form of the complementarity conditions,

$$\sum_i u_i g_i = 0,$$

xhere u_i is the multiplier that corresponds to constraint i as follows:

$$\sum_i u_i g_i = 0 \Rightarrow \sum_i min(u_i, g_i) = 0 \Rightarrow \sum_i \{min(0, g_i - u_i) + u_i\} = 0 \Rightarrow \sum_i \{min(0, w_i) + u_i\} = 0,$$

where

$$w_i = g_i - u_i.$$

Although this reformulation does not eliminate the non-convexity, the resulting problem is separable and can be solved with existing solution methodologies for separable programming [46]. The authors solve this problem with a branch and bound such methodology that partitions the problem's feasible space. They illustrate the proposed methodology on various examples, but they do not present computational results demonstrating its requirements on random problems.

The method proposed by Candler and Townsley [30] considers a formulation with constraints at the lower level only, and assumes that for any given values of the leader's decision variables, there will be a unique solution for the follower. An implicit search is employed, which uses necessary conditions for a better solution to limit the extent of the search. The developed algorithm involves an implicit search of all bases in the leader's decision vectors. The number of bases to be searched explicitly is restricted to those which satisfy (global) necessary conditions for a better basis. Three types of necessary conditions are developed and tested.

The algorithm presented by Bard [8] solves a BLP with the lower level problem acting as the only constraint for the leader's problem. The algorithm uses sensitivity analysis to solve the BLP, developing a parametric model which is equivalent to the initial BLP. The main concept of the underlying theory rests on a set of first order optimality conditions that parallel the KKT conditions associated with a one dimensional parametric linear program. The solution to the original problem is uncovered by systematically varying the parameter over a unit interval and solving the corresponding linear program. For the solution of the resulting parametric model, a one-dimensional grid search algorithm is presented. For the validity and the finite convergence of the proposed algorithm, the author assumes that the solution to the follower's problem is non-degenerate. The paper also discusses other solution techniques including branch and bound and vertex enumeration, and gives an example highlighting their computational and storage requirements.

Utilizing the relationship between biobjective and bilevel programming, Unlu [85] proposes an algorithm for the linear bilevel problem without constraints at the upper level, which modifies the parametric approach that has been proposed by Bard [8]. The algorithm is an iterative procedure that searches among the efficient solutions of the associated biobjective problem (the one that results considering the leader's and the follower's objectives) for bilevel feasibility. The author presents computational results demonstrating that the performance of the algorithm is superior to that of the algorithm presented by Bard [8].

Using counterexamples, Candler [28] and Benayed and Blair [18] independently showed that the grid search algorithm developed by Bard [8] does not always produce the optimal solution. In turn, this contradicted the validity of the algorithm proposed by Unlu [85]. The shortcomings of these two methodologies were also pointed out by Wen and Hsu [87]. Haurie et al. [52] also shows that the efficient point algorithm proposed by Bard [8] does not always converge to the desired solution. A counterexample is provided in which the proposed algorithm fails to converge to the optimal solution. The authors explain that the main reasons for this lack of convergence are: a) the fact that in the case of semi-infinite programming the correct reformulation is obtained when the Fritz-John necessary conditions are used instead of the KKT conditions even if the problem is linear, and b) the fact that the inequalities produced in Bard's method are only valid if the follower's constraints do not depend on the leader's decision variables. The authors conclude by pointing out that even though the algorithm presented by Bard [8] does not find the optimal solution, it can be used as a solution concept in real-life bilevel systems in order to find satisfactory sub-optimal solutions.

Based on the parametric algorithm proposed by Bard [8], White [90] presents another interesting approach for the solution of BLP. The main result of this paper resides in a theorem which reduces the solution of BLP to the solution of a maximin linear problem. The author points out that the general methodology proposed by Bard [8] is not always correct, and develops, for this reason, the suggested methodology. The main idea of this methodology is that if the solution obtained by the parametric maximin linear problem with respect to the decision variables controlled by the follower has the same value with the optimal solution obtained when the follower's problem is optimized over $\Omega(X)$, then this is the BLP optimal solution; otherwise, the selected parameter is increased until the optimality criterion is satisfied.

Three interesting algorithms are presented by Bialas and Karwan [22] for the solution of BLP. The first one finds local optima and can rarely identify a global optimal solution, whereas the other two find the global optimum. The first algorithm assumes that the follower has a unique (non-degenerate) solution and utilizes most of the standard tools of the simplex method for bounded decision variables. At each iteration, the leader's objective function is optimized first and the obtained solution is used in order to optimize the follower's objective function in the same bounded solution space. If the optimal solution of the follower and the leader are the same, then the algorithm stops; otherwise, the algorithm continues using the current basic solution. In the last step of the algorithm, a modified version of the simplex method is employed, which solves the follower's problem to optimality, calculating the reduced cost of the variables which are candidate to enter the basis based on the leader's solution space. The algorithm obtains an extreme point on the rational reactions over Ω; it then moves among the extreme points of the rational reactions over Ω, never allowing the leader's objective function to decrease (in the case of maximization). However, only a local optimal solution is obtained this way. More specifically, the algorithm terminates with an extreme point solution in the leader's solution space, which has the property that all the adjacent extreme points either lead to a decrease in the leader's objective function (in the case of maximization) or do not belong to the leader's solution space.

The second algorithm proposed by Bialas and Karwan [22] is a vertex enumeration procedure called the "k-th best" algorithm. In this approach, the leader solves his problem in both the leader's and the follower's decision vectors (suppressing the follower's objective function), and orders the basic feasible solutions in non-increasing value of his objective (in the case of a maximization problem). Given the optimal values of the decision variables controlled by the leader, the follower solves, at any iteration of the algorithm, his own problem. If the optimal solution and the optimal values of the decision variables obtained this way are the same, then the algorithm stops. Otherwise, the algorithm uses the next optimal solution of the leader and solves again the follower's problem, until both problems find the same optimal solution for the follower's decision variables. Finally, the third algorithm proposed for the solution of BLP by Bialas and Karwan [22] uses the KKT optimality conditions in order to reformulate the original problem. The main idea of this approach is to replace the follower's problem with its KKT optimality conditions and append the resultant system of constraints to the leader's problem.

Júdice and Faustino [54] show that the sequential linear complementarity problem (SLCP) presented by Bialas and Karwan [22] is not able to solve the BLP in all cases. For this reason, they present in a later work [55] a modified version of the SLCP that achieves this goal, introducing at the same time some modifications that can improve the convergence of the algorithm. The improved SLCP algorithm is shown to be competitive with the branch and bound methods that have been proposed for BLPs of small dimension, and becomes much more efficient when the dimension of the BLP increases. The novelty of the proposed method lies in the approach adopted for the solution of the linear complementarity problems arising in the SLCP algorithm. This approach is a hybrid enumerative method with three modifications introduced in the enumerative method to make the SLCP algorithm more robust. The main steps of the hybrid enumerative method incorporate Al-Khayyal's algorithm [1], some new procedures for the generation of nodes, and heuristic rules for the choice of nodes and complementary pairs of variables.

Dempe [40] proposes a solution algorithm for finding local optima to the linear bilevel problem with constraints at the lower level only. The algorithm employs a modified simplex procedure with specialized rules for updating the basis. The author proposes a technique for generating starting points which can be utilized by this algorithm in an attempt to reach global optima. He also proposes the implementation of a branching search procedure exploring the different basic feasible solutions of the problem, which is guaranteed to reach the global optimum. He does not provide computational results, but he illustrates the algorithm on a small example.

The solution methodology suggested by Bard and Moore [12] is valid for BLPs with constraints at both levels. The authors begin by converting the original BLP into a standard mathematical program. This is achieved by replacing the follower's problem with its KKT conditions, and giving control of all the variables to the leader. As was suggested firstly by Fortuny-Amat and McCarl [47], the basic idea of the proposed algorithm is to suppress the complementarity condition term and solve the resulting linear program. At each iteration, a check is carried out to verify if the complementarity condition term is satisfied. If it does, the corresponding point is in the inducible region; hence, it is a potential solution to the BLP. The novelty of the proposed algorithm lies in the course of action followed in the case that the complementarity condition term is not satisfied. Fortuny-Amat and McCarl [47] take the more direct approach of replacing the complementarity condition term with a set of inequalities using the big-M approach. They then solve the resulting problem with a standard zero-one mixed integer code. The novelty of Bard and Moore [12] lies in that they suggest using a branch and bound scheme to implicitly examine all combinations of complementary slackness. The authors present a comparison of the Parametric Complementary Pivot algorithm of Bialas and Karwan [22], their branch and bound scheme, and the separable approach of Bard and Falk [11]. As they demonstrate, the latter is roughly equivalent to the zero-one formulation of Fortuny-Amat and McCarl, in that both approaches lead to problems of nearly identical size and structure. The presented results show that the suggested method outperforms the separable approach and is on equal footing with the Parametric Complementary Pivot method with respect to CPU solution time.

Anandalingam and White [4] and White and Anandalingam [91] propose a penalty function approach for solving the linear bilevel problem without constraints at the upper level. They notice that at the optimal solution to the follower's problem, the duality gap between the objective value of its primal and its dual formulation is equal to 0. Thus, they transform the original problem into a single level optimization problem in the original decision variables of the leader and the primal and dual variables of the follower, which includes the primal and dual constraints of the follower. The objective of this problem consists of the original upper level objective minus a penalty which is proportional to the difference between the dual and the primal objective values of the follower's problem. The algorithm employs a search that updates the current solution until the global optimum is obtained. The authors present computational results demonstrating that their approach outperforms the k-th best algorithm proposed by Bialas and Karwan [22], but requires significantly higher computational effort than the branch and bound algorithm proposed by Bard and Moore [12]. Campelo et al. [26] contradicted the validity of the assumptions of the algorithm proposed by White and Anandalingam [91]. They also provided alternative versions of these assumptions and modified some of the steps of the algorithm, re-establishing this way its validity.

Onal [69] presents a modified simplex approach for solving bilevel linear problems with constraints at the lower level only. At the first step, this approach uses the KKT optimality conditions to reformulate the initial BLP into an equivalent non-linear single level problem (ENLSLP), whereas in the second step, the complementary slackness constraints are incorporated into the leader's objective function as a penalty term with a large coefficient, M, resulting in a parametric quadratic program (PQP). The modified simplex approach is applied into this PQP. This solution approach solves the ENLSLP by searching local solutions of the PQP. The author develops a method for finding a stable local solution of PQP as a feasible solution that remains the same under any arbitrary choice of $M > M_0$ for some large number M_0. If this is valid, then the penalty term vanishes. The difficulty of this methodology is the selection of M. Small values may not eliminate the penalty term, whereas large values may cause computational problems. Campelo and Scheimberg [27] do a more rigorous analysis of the possible cases at a local solution of the PQP. They show that some additional cases might occur that have not been taken into consideration by Onal [69]. Another important point is that Onal's method is valid only if the PQP is not unbounded for sufficiently large M. Campelo and Scheimberg [26] illustrate that this algorithm may not find the global optimal solution in some cases, and explain why this may happen.

Tuy et al. [84] reformulate the linear bilevel problem with constraints at both levels as a single level reverse convex programming problem, by replacing the follower's (minimization) objective with a constraint stating that the value of the follower's objective function cannot exceed the value that it will have at the optimal solution of the problem. Of course, this is a trivial optimality condition, since these two quantities will be equal at optimality. The authors develop a technique to reduce the size of the problem and then solve it through a vertex enumeration solution procedure. They do not present generic computational results, but illustrate the proposed algorithm on a small numerical example.

Liu and Spencer [60] propose a solution algorithm for the linear bilevel problem without constraints at the upper level, which runs in polynomial time if the number of decision variables controlled by the follower is fixed. The algorithm solves a series of linear programs, each of which results from optimizing the follower's objective subject to a different set of constraints determined by nonsingular coefficient matrices of the original bilevel formulation that pertain to the decision variables of the follower. Of course, these are single level programs involving only the decision variables of the follower. Those linear programs that do not have a bounded optimal solution are discarded. Each of the programs that have a bounded solution is used to formulate a new single level linear program that only involves the leader's decision variables. Each of the optimal solutions of these linear programs in the leader's decision variables combined with the optimal values of the follower's decision variables found in the associated problem of the previous stage provides a solution which is candidate for optimality for the original bilevel problem formulation. Of course, out of these solutions, the one which provides the largest value for the objective value of the leader is the optimal solution of the original bilevel problem. The authors do not present generic computational results, but illustrate the proposed algorithm on a small numerical example.

Shi et al. [79] study how to solve a BLP when the upper-level constraint functions are of arbitrary linear form. A new definition for the solution of the BLP is given attempting to take into consideration some cases that the classical definition for the solution of the BLP could not handle properly. The main difference between the two definitions is that, in the new one, the follower's feasible set is defined from the sets of constraints of both levels, and not only from the lower level. The authors motivate their new definition by giving an instance of the BLP for which no solution can be found even though Ω is not empty. They introduce a new type of optimal solution referred to as Pareto optimal solution when the BLP does not have a complete optimal solution. They claim that the fact that a Pareto optimal solution exists but cannot be found using the common definitions is a deficiency of the theory. In order to obtain this Pareto optimal solution, they propose a new definition for the solution of the BLP. The authors apply this definition on two numerical examples, demonstrating how and why the classical approach fails to find a solution and how the new approach finds the Pareto optimal solution. They do not clarify why the obtained solution is a Pareto solution and if this Pareto solution corresponds to the equivalent biobjective Pareto optimal solution.

Based on the results presented by Shi et al. [79] [80] [82], Shi et al. [81] develop 3 extended algorithms which can be considered as extensions of existing algorithms through the incorporation of the new definition presented by Shi et al. [79]. The first algorithm is an extended version of the Kuhn-Tucker approach proposed by Bialas and Karwan [20], Bard and Falk [11], and Hansen et al. [51]; the second algorithm is an extended version of the k-th best approach proposed by Candler and Townsley [30], and Bialas and Karwan [22]; the third algorithm is an extended version of the branch and bound algorithm proposed by Bard and Moore [12]. All these extensions make use of the new definition for the BLP and revise the original algorithms similarly, incorporating the upper level constraints involving lower level decision variables into the follower's constraint region. The authors refer to the theorem given in Shi et al. [79] on the condition under which a BLP has an optimal solution, but the term Pareto optimal solution used and defined in Shi et al. [79] is replaced by the term optimal solution in this theorem, without making clear if this solution is the global optimal solution of the BLP. Through two numerical examples, they demonstrate how the classical approaches fail to reach an optimal solution in a case where Ω is not empty, and show that the new definition in which the leader's problem does not have constraints involving the follower's decision variables gives exactly the same optimal solution.

Audet et al. [6] show that the new definition of BLP presented by Shi et al. [79] is equivalent to transferring the constraints of the leader that involve decision variables of the follower into the follower's problem, resulting in a special case of BLP in which there are no upper level constraints that involve lower level decision variables. The authors point out that the difference between the two definitions is that the definition presented by Shi et al. [79] implies that the lower level is now responsible for enforcing the upper level constraints that involve the follower's decision variables. They also prove that the new definition relaxes the feasible region of the BLP, allowing for infeasible points to be considered as feasible. The aim of the note paper is to show that the deficiency pointed out by Shi et al. [79] is not an actual one, and that if the upper level constraints that involve lower level decision variables are moved into the lower level, the nature of the original BLP changes, since the new problem obtained through this transformation is not equivalent to the original. The authors note that the solution of the original BLP, as it is strictly defined, does not have to be Pareto optimal [85], and explain that this is due to the intrinsic non-cooperative nature of the model. This implies that the fact that no Pareto optimal solution could be found for the example presented in Shi et al. [79] is not really a deficiency, but is due to the fact that no such solution exists, and that this may happen even if Ω is not empty. In the presented note, the authors show that the first numerical example presented in Shi et al. [80] [81] [82] does not have an optimal solution even though Ω is not empty, and that the second numerical example has the same optimal solution just because the leader's formulation does not have constraints that involve the follower's decision variables, resulting in an application of the new BLP definition which is coincidentally the same.

Audet et al. [7] propose a two-phase branch and cut algorithm for the bilevel linear problem with constraints at both levels. The algorithm exploits the relationship between bilevel linear programming and mixed integer programming [5]. The first phase of the algorithm consists of an iterative procedure that generates and adds a valid inequality (cut) to a suitable relaxation of the problem. The authors consider three different types of cuts, i.e., Gomory cuts, simple cuts and extended cuts. The second phase engages a branch and bound enumerative procedure that converges to the global optimum of the problem. The authors test the algorithm on problems with the same size as the problems tested by Hansen et al. [51]. Through these results, they compare different design rules, and assess the impact that they have on total computational performance.

Glackin et al. [49] propose an algorithm for bilevel linear programming with constraints at both levels, which exploits the relationship between multiple objective linear programming and bilevel linear programming, as well as the results for minimizing a linear objective over the efficient set of a multiple objective problem. The algorithm works by searching among the efficient solutions of the associated biobjective linear program through simplex pivots on an expanded tableau. The authors report computational results on problem instances generated similarly as in the procedure adopted by Bard and Moore [13].

8.4 Solution Approaches for the Mixed Integer Bilevel Problem

The differentiation of algorithms for solving bilevel linear problems is based on the presence or not of constraints at the upper level. Some algorithms can only solve BLPs with constraints at the lower level only, while some other algorithms can solve the more general case of BLP where constraints are present at both levels. In the case of mixed integer bilevel linear programming, the situation is more complex and many more different cases exist. The upper (lower) level problem may include integer/continuous variables controlled by the leader (follower), and/or integer/continuous variables controlled by the follower (leader). The discrete variables can be restricted to general integer and/or binary values.

Solution approaches for MIBLP can be classified into 3 main categories. The first one includes reformulation approaches, where, for example, mathematical decomposition techniques are used in order to decompose the initial MIBLP into two single level problems. The second category includes branch and bound/branch and cut techniques, and the third one includes parametric programming approaches.

Exact and heuristic solution procedures based on the branch and bound technique for solving the MIBLP are presented by Wen and Yang [89]. These algorithms assume that the optimal solution of MIBLP is non-degenerate. The authors define some trivial bound information for the optimal solution of MIBLP in the case of maximization problems as follows: a) a trivial lower bound on the optimal value of the leader's objective function is computed as the greatest feasible such value that has been found so far, and b) a trivial upper bound on the optimal solution of MIBLP is the optimal objective value of the problem that results when the lower level objective function is neglected by MIBLP and all the decision variables are controlled by the leader. The exact algorithm that is proposed involves typical steps of any branch and bound algorithm: 1) initialization, 2) branching procedure, 3) calculation of bounds, 4) fathoming, 5) backtracking, 6) calculation of feasible solution, and 7) termination.

The authors point out that if the number of upper level zero-one decision variables grows linearly, then the computational time grows exponentially. For this reason, they propose a heuristic solution procedure, which provides satisfactory near optimal solutions in reasonably short computational time. The heuristic algorithm utilizes a judgment index which is calculated based on the weighted estimated optimal solution for the values of the decision variables controlled by the leader that are obtained by neglecting the follower's objective function, and on the weighted estimated optimal solution for the values of the decision variables controlled by the leader that are obtained by neglecting the leader's objective function. The weights depend on the estimated optimal solution obtained and the number of variables controlled by each of them; the higher the judgment index value of the upper level decision variable, the higher priority is assigned for checking that variable. The numerical results show that the bounding function becomes more effective as the percentage of variables controlled by the leader increases. The heuristic algorithm provides a good approximation to find an optimal or near optimal solution. The proposed methodology can also solve the case where the leader only controls integer variables and constraints are only present at the lower level problem. Another limitation of the algorithm is that the integer decision variables cannot appear in the follower's objective function.

Bard and Moore [13] propose a solution algorithm for the pure binary bilevel problem with constraints at the lower level, binary decision variables at both levels and integer coefficients in both objectives and all constraints . The algorithm performs an enumerative branch and bound search procedure on the decision variables of the leader. More specifically, the authors replace the objective of the leader in the original problem formulation with a constraint that sets the value of this objective greater or equal to α, where α is a parameter, originally set equal to $-\infty$. Optimizing the follower's objective and incrementing the value of α in successive iterations, the algorithm finds a series of bilevel feasible solutions that provide a monotonic improvement on the objective value of the leader. The algorithm can be modified to accommodate the case where the lower level decision variables are allowed to take integer values. The authors present computational results on randomly generated problem instances.

The (heuristic) algorithm of Moore and Bard [66] was later extended by DeNegre and Ralphs [44] through the incorporation of cutting plane techniques for improving the bounds on the optimal objective function value. The formulation addressed by DeNegre and Ralphs [44] includes constraints and pure general integer formulations at both levels. The proposed algorithm employs a branch and cut tree, at each node of which a relaxation of the original problem is solved. If the solution obtained is bilevel feasible, then this solution is the desired in the associated subtree. If not, a suitable cut is added, which excludes this solution without excluding any bilevel feasible solution. The authors claim that the advantage of this procedure over the one of Moore and Bard [66] is that it relies solely on the solution of standard integer linear programs, preserving the typical rules for fathoming and branching. The authors present computational results on random problems with two different branching strategies.

Saharidis and Ierapetritou [75] propose a new algorithm for the solution of MIBLPs. Their algorithm is based on the decomposition of the initial problem into two problems, the restricted master problem (RMP) and a series of problems named slave problems (SPs). The proposed approach is based on the Benders decomposition method where, at each iteration, the set of variables controlled by the leader is fixed, generating the SP. The RMP is a relaxation of the MIBLP composed by all the constraints including only integer decision variables controlled by the leader. The RMP interacts at each iteration with the current SP through the addition of three type of cuts produced using Lagrangean information from the current SP. These cuts are the classical Benders cuts (optimality Benders cut and feasibility Benders cut) and a third cut referred to as exclusion cut which is used if the RMP is not restricted by the last generated Benders cut. The lower and upper bound provided (in the case of minimization) from the RMP and the (best found so far) SP are updated in each iteration, respectively. The algorithm converges when the difference between the upper and lower bound is within a small difference ε. In the case of MIBLP, the KKT optimality conditions cannot be used directly for the inner problem in order to transform the bilevel problem into a single level problem. The proposed decomposition technique, however, allows the use of these conditions and transforms the MIBLP into two single level problems. The algorithm is illustrated through a modified numerical example from the literature. Additional examples from the literature are presented to highlight the algorithm convergence properties, which are comparable with those of other approaches (the same optimal solution is found in small CPU time). The proposed methodology can solve MIBLPs in which the leader controls discrete (binary or general integer) decision variables and these decision variables can appear in any constraint or objective function.

For a bilevel problem with only integer decision variables at the lower level and constraints at both levels, Köppe et al. [57] consider both the case in which the leader's decision variables are continuous and the case in which they are integer. The solution algorithm that the authors develop is based on the theory of parametric integer programming and runs in polynomial time when the number of decision variables of the follower is fixed. If the infimum of the problem is not attained, the algorithm is able, under the same assumption, to find an ε-optimal solution whose objective value approximates the sought infimum in polynomial time, too. The authors do not report computational results demonstrating the performance of the algorithm.

Before closing this section, it is worthwhile discussing a special solution methodology that can be applied whenever the lower problem is linear and continuous, independently of the exact nature of the upper problem. This methodology consists of replacing the lower level problem with corresponding optimality conditions, which set the objective function values of the lower level primal and dual problems equal, while also ensuring primal and dual feasibility. In a sense, this approach seems equivalent to the KKT based approaches; yet it exhibits a significant advantage, which stems from the fact that it avoids the introduction of the (nonlinear) complementarity conditions. Thus, the resulting problem can be readily solved through solution methodologies that can handle problems such as the original upper level problem. Garcés et al. [48] and Baringo and Conejo [15] have utilized this solution methodology in the context of power market optimization.

8.5 Bilevel Programming Applications

Bilevel programming formulations are encountered in the context of several interdisciplinary areas, such as agricultural planning, government policy making, economic planning, financial management, warfare optimization, transportation planning, optimal pricing, ecological programming, chemical design, production planning, optimal resource allocation, etc. One of the main domains of bilevel programming applications is decision-making in electricity markets. In that context, market agents (typically energy producers) aim to maximize the profit they will realize from participating in an energy market. With the decisions of the market agents known, on the other hand, the market operator aims to clear the market in the lowest possible cost (i.e., the highest possible benefit for the energy consumers). Since a separate optimization problem needs to be solved for the clearance of the market, the overall problem, on the individual agent's view, is an optimization problem (agent decision making) constrained by another optimization problem (market clearing), i.e., a bilevel programming model.

Ruiz and Conejo [72] consider supply function offering by a strategic electricity producer using a bilevel model whose upper-level problem represents the strategic behavior of the producer who seeks to maximize his profit, and whose lower-level problem represents the clearing of the market under many demand constraints. Garcés et al. [48] address the electricity transmission expansion planning problem considering a bilevel programming approach, whose upper-level problem represents transmission investment decisions (in power lines), and whose lower level problem describes the clearance of the market. Kazempour et al. [56] analyze generation capacity investments by a strategic electricity producer using a bilevel approach, whose upper-level problem represents the producer investment decisions and a collection of lower-level problems describes market clearing under many operational conditions. Baringo and Conejo [14] address investment problems pertaining to non-dispatchable electricity producers (wind and solar power facilities) via a bilevel model whose upper-level problem represents investment in non-dispatchable production units and a number of lower-level problems describe market clearing under many scenarios of wind/solar energy production. Baringo and Conejo [15] extend this work by simultaneously considering investment decisions in network reinforcements and in renewable production facilities. Finally, Pandzic et al. [71] address the preventive maintenance scheduling of power transmission lines within a yearly time framework using a bilevel approach: the upper-level problem represents line maintenance decisions aiming to maximize security in the system, whereas a collection of lower-level problems model the clearance of the market under different demand conditions.

Hobbs and Nelson [53] introduce another application of bilevel programming in the electric utility industry. The leader is seen as the electric utility who seeks to determine the rates so as to optimize a cost/benefit related objective, whereas the follower represents the customers, who aim to decide their energy consumption, so as to maximize their individual benefit. In another related application, Motto et al. [67] model the vulnerability analysis of an electric grid under terrorist threat using a mixed-integer nonlinear bilevel program. The objectives of the leader and the follower are in direct conflict, with the leader trying to maximize the damage caused on the network, and the follower trying to minimize it. Salmerón et al. (2004), Brown et al. (2006) and Salmerón et al. (2009) consider similar problem types, which involve an informed attacker seeking to maximize system damage, and an informed system operator who is reacting in order to minimize such damage.

Several transportation/traffic problems have been addressed using bilevel optimization methodologies, as also documented by Migdalas [64]. LeBlanc and Boyce [59] formulate the transportation network design problem as a linear bilevel problem. At the upper level of the model, the leader minimizes the cost associated with the implementation of a set of network improvements aimed mainly at increasing its flow capacities, whereas at the lower level, the network users minimize their flow cost under these improvements. The authors propose a modification of the algorithm that has been proposed by Bard [8] in order to solve this problem. A formulation of the network design problem as a bilevel linear program is also presented by BenAyed et al. [18].

Suh and Kim [83] consider two bilevel programming formulations for the network design problem, in which the leader minimizes the cost associated with the increase in capacity of several network links, whereas the follower solves a user-equilibrium route choice problem. The authors develop a descent-type heuristic algorithm for the solution of the problem and compare its performance against that of other existing ones. In a related work, BenAyed et al. [19] formulate the highway network design problem as a linear bilevel optimization model and solve it through a specialized algorithm that decomposes the lower level problem into separate problems.

Constantin and Florian [38] develop a bilevel programming model for the problem of optimizing frequencies of transit lines in a transportation network. In the associated formulation, both the leader and the follower have the same objective function, although their decision variables differ. The authors solve this problem with a projected subgradient algorithm and illustrate its application on different case studies. Labbe et al. [58] formulate a bilevel model, in which the leader wants to maximize revenues from a taxation scheme, whereas the follower reacts to these regulations in order to minimize the cost payments he is entitled to. The authors illustrate the application of the model on optimal highway toll pricing through a numerical example. Maher et al. [62] formulate the trip matrix estimation and the traffic signal optimization problems as bilevel programs. They develop a neighborhood search solution algorithm that can be applied to both problems with minor modifications and illustrate its behavior on simple examples.

Yin [92] considers a multiobjective bilevel optimization model for the problem of setting tolls in a highway network. The objectives optimized by the leader are related to the travel cost and the total revenue, whereas the lower level problem represents a network equilibrium that describes users' route choice behavior. The author applies a genetic algorithm for the solution of the problem, and illustrates its application on a numerical example. Cote et al. [39] introduce a bilevel model for solving the pricing/fare optimization problem in the airline industry. Acting as the airline operator, the leader maximizes revenues, whereas the follower minimizes an aggregate function expressing the disutility of the passengers. The authors illustrate the applicability of the model on a small numerical example. Marinakis et al. [63] introduce a bilevel formulation for the vehicle routing problem. At the upper level of the model, the leader assigns vehicles to customers, whereas at the lower level, the follower finds the minimum cost route for each vehicle. The authors propose a genetic algorithm for the solution of the problem. Erkut and Gzara [45] develop a bilevel programming model for the network design problem for hazardous material transportation, in which the government acts as the leader making decisions that concern the structure of the network, whereas the carriers (follower) optimize their routes, given these decisions. The authors solve the problem with a heuristic solution method that always finds a stable solution.

Saharidis et al. [75] study the berth scheduling problem with customer differentiation. A new methodological approach based on bilevel optimization is used in order to model the different objectives that are present in a container terminal. Such objectives are often non-commensurable, and gaining an improvement on one of them often causes degrading performance on the others. An iterative algorithm based on the k-th best algorithm is proposed in order to solve the resulting problem. Finally, Golias et al. [50] deal with the problem of scheduling inbound trucks at a cross-docking facility with two conflicting objectives: minimize the total service time for all the inbound trucks, and minimize the delayed completion of service for a subset of the inbound trucks, which are considered as preferential customers. The problem is formulated both as a biobjective and as a mixed integer bilevel problem.

In the context of production and process optimization, Clark and Westerberg [34] provide a bilevel formulation for chemical process design optimization, and solve it through two specialized approaches that find local optima. The first approach is based on an active set strategy for the inner problem and the second on a relaxation of the complementarity conditions. Clark and Westerberg [35] extended this work by refining these two approaches and comparing their performance on a process design problem. Nicholls [68] develops a nonlinear bilevel optimization model for aluminum production, and solves it with a vertex enumeration algorithm that is based on the grid search algorithm proposed by Bard [8]. Mitsos et al. [65] consider a bilevel model for parameter estimation in phase equilibrium problems. The model includes multiple lower level problems, each of which is associated with a separate experiment. The leader minimizes the errors between predicted and measured values, whereas each follower minimizes the Gibbs free energy. The authors illustrate the model on several binary mixture case studies.

Ryu et al. [73] address bilevel decision-making problems under uncertainty in the context of enterprise-wide supply chain optimization, with the upper level corresponding to a plant planning problem, and the lower level corresponding to a distribution network problem. The authors solve the problem through a parametric based solution methodology and illustrate its application on a numerical example. Cao and Chen [31] introduce a bilevel optimization model to address the capacitated plant selection problem. At the upper level, a central decision maker selects which plants to establish, so that the related cost is minimized. At the lower level, the aim is to minimize the operational cost of the selected plants. The authors transform the problem into a single level program using the KKT optimality conditions, linearize the nonlinear terms through an equivalent reformulation, and solve the resulting problem with standard commercial optimization software.

Amouzegar and Moshirvaziri [2] introduce a bilevel optimization model for waste capacity planning and facility location. At the upper level of the model, the government, acting as the leader, maximizes social welfare via taxation, whereas at the lower level, the firms optimize their collective location/allocation problem after observing the leader's decisions. The authors solve the problem through a modification of an existing penalty solution methodology for bilevel programming. Dempe et al. [43] develop a discrete bilevel program to model the problem of minimizing the cash-out penalties of a natural gas shipper. The authors reformulate the lower level problem as a generalized transportation problem and solve the resulting bilevel program with the algorithm of White and Anandalingam [91].

In the context of warfare optimization, Bracken and McGill [23] formulate several defense related problems as bilevel optimization models. They study their general characteristics and key properties, and discuss their main similarities and differences. They also point to previous works that can be utilized for their solution. Anandalingam and Apprey [3] examine conflict resolution problems using multi-level (including bilevel) programming and illustrate their methodology on an application in a water conflict problem between two countries.

A review on multi-level approaches for firm-related organizational issues has been presented by Burton and Obel [25]. Cassidy et al. [32] present a bilevel model for optimal resource allocation of government funds to state projects. Acting as the leader, the government tries to allocate the available budget to the states as evenly as possibly in order to minimize their total dissatisfaction resulting from the percentage of the total budget they are entitled to. Acting as the follower, on the other hand, each state maximizes the return attained from the utilization of the budget it gets allocated. The authors solve the problem through a parametric programming enumeration procedure and illustrate its application on small numerical examples.

Onal et al. [70] develop a bilevel optimization model for agricultural credit distribution. The government plays the role of the leader in this model, maximizing the total agricultural output value. The follower's objective is to maximize the total surplus of all consumers and producers. The authors solve the problem through a heuristic procedure which is a modification of the algorithm proposed by Bard [9]. They also illustrate the application of the proposed methodology on a case study.

In one of the earliest applications of bilevel programming, Candler and Norton [29] address a policy development problem, in which the leader (government/manager, etc.) wants to select a policy (tax rates, etc.) at the upper level, so as to maximize some appropriate objective function, whereas the follower (firms/consumers, etc.) optimizes some objective function related to its behavior/operation given that policy. The authors demonstrate the proposed methodology on an agricultural case study.

The final bilevel programming application concerns a procedure that aims to build a model that will be able to provide a description of the behavior of a system under consideration, as well as a prediction for the future. Saharidis et al. [76] present a novel hierarchical bilevel implementation of the cross validation method. In this bilevel scheme, the leader optimization problem builds (trains) the model, whereas the follower checks (tests) the developed model. The problem of synthesis and analysis of regulatory networks is used to compare the classical cross validation method to the proposed methodology referred to as bilevel cross validation.

8.6 Summary

Bilevel programming is a very important branch of mathematical programming that has stimulated the interest of operations researchers for many years. In the present work, we have attempted to review in rational sequence the exact solution methodologies that have been developed for the solution of linear bilevel and mixed integer bilevel problems, as well as to provide the most recent developments on the subject. We have also tried to provide an extensive list of relevant applications that have been treated through the development of bilevel optimization models.

Over the last decades, many significant results have been reported in the bilevel programming related literature, some of which have been revised or restated in later works, in order to be directly applicable and/or mathematically correct. As a consequence, although initially introduced as exact, some of the algorithms described above have turned out to exhibit a rather heuristic behavior. In order to assist researchers who are active in the area, the present survey also documents and categorizes these works. We believe that the present work can be proven particularly useful as a study and reference guide to the numerous researchers who currently engage themselves with bilevel programming problems, as well as to those that will do so in the future.

References

1. Al-Khayyal, F.A.: An implicit enumeration procedure for the general linear complementarity problem. Mathematical Programming Studies 31, 1–20 (1987)
2. Amouzegar, M.A., Moshirvaziri, K.: Determining optimal pollution control policies: An application of bilevel programming. European Journal of Operational Research 119, 100–120 (1999)
3. Anandalingam, G., Apprey, V.: Multi-level programming and conflict resolution. European Journal of Operational Research 51, 233–247 (1991)
4. Anandalingam, G., White, D.J.: A solution method for the linear static stackelberg problem using penalty functions. IEEE Transactions on Automatic Control 35(10), 1170–1173 (1990)
5. Audet, C., Hansen, P., Jaumard, B., Savard, G.: Links between linear bilevel and mixed 0-1 programming problems. Journal of Optimization Theory and Applications 93(2), 273–300 (1997)
6. Audet, C., Haddad, J., Savard, G.: A note on the definition of a linear bilevel programming solution. Applied Mathematics and Computation 181, 351–355 (2006)
7. Audet, C., Savard, G., Zghal, W.: New branch and cut algorithm for bilevel linear programming. Journal of Optimization Theory and Applications 134(2), 353–370 (2007)
8. Bard, J.F.: An efficient point algorithm for a linear two-stage optimization problem. Operations Research 31(4), 670–684 (1983)

9. Bard, J.F.: An investigation of the linear three level programming problem. IEEE Transactions on Systems, Man, and Cybernetics 14(5), 711–717 (1984)
10. Bard, J.F.: Practical bilevel optimization. Kluwer Academic Publishers, Dordrecht (1998)
11. Bard, J.F., Falk, J.E.: An explicit solution to the multilevel programming problem. Computers and Operations Research 9(1), 77–100 (1982)
12. Bard, J.F., Moore, J.T.: A branch and bound algorithm for the bilevel programming problem. SIAM Journal on Scientific and Statistical Computing 11, 281–292 (1990)
13. Bard, J.F., Moore, J.T.: An algorithm for the discrete bilevel programming problem. Naval Research Logistics 39(3), 419–435 (1992)
14. Baringo, L., Conejo, A.J.: Wind power investment within a market environment. Applied Energy 88(9), 3239–3247 (2011a)
15. Baringo, L., Conejo, A.J.: Transmission and wind power investment. IEEE Transactions on Power Systems (2011b) (accepted)
16. BenAyed, O.: Bilevel linear programming. Computers and Operations Research 20(5), 485–501 (1993)
17. BenAyed, O., Blair, C.E.: Computational difficulties of bilevel linear programming. Operations Research 38(3), 556–560 (1990)
18. BenAyed, O., Boyce, D.E., Blair, C.E.: A general bilevel linear programming formulation of the network design problem. Transportation Research - Part B 22B(4), 311–318 (1988)
19. BenAyed, O., Blair, C.E., Boyce, D.E., LeBlanc, L.J.: Construction of a real-world bilevel linear programming model of the highway network design problem. Annals of Operations Research 34, 219–254 (1992)
20. Bialas, W.F., Karwan, M.H.: Multilevel linear programming. Research Report No. 78-1, Operation Research Program, Department of Industrial Engineering, State University of New York at Buffalo (1978)
21. Bialas, W.F., Karwan, M.H.: On two-level optimization. IEEE Transactions on Automatic Control 27(1), 211–214 (1982)
22. Bialas, W.F., Karwan, M.H.: Two-level linear programming. Management Science 30(8), 1004–1020 (1984)
23. Bracken, J., McGill, J.T.: Defense applications of mathematical programs with optimization problems in the constraints. Operations Research 22(5), 1086–1096 (1974)
24. Brown, G., Carlyle, M., Salmerón, J., Wood, K.: Defending critical infrastructure. Interfaces 36, 530–544 (2006)
25. Burton, R.M., Obel, B.: The multilevel approach to organizational issues of the firm - A critical review. OMEGA The International Journal of Management Science 5(4), 395–414 (1977)
26. Campelo, M., Scheimberg, S.: A note on a modified simplex approach for solving bilevel linear programming problems. European Journal of Operations Research 126(2), 454–458 (2000)
27. Campelo, M., Dantas, S., Scheimberg, S.: A note on a penalty function approach for solving bilevel linear programs. Journal of Global Optimization 16, 245–255 (2000)
28. Candler, W.: A linear bilevel programming algorithm: A comment. Computers and Operations Research 15(3), 297–298 (1988)
29. Candler, W., Norton, R.: Multi-level programming and development policy. World Bank, Bank Staff Working Paper No. 258 (1977)
30. Candler, W., Townsley, R.: A linear two-level programming problem. Computers and Operations Research 9, 59–76 (1982)
31. Cao, D., Chen, M.: Capacitated plant selection in a decentralized manufacturing environment: A bilevel optimization approach. European Journal of Operational Research 169, 97–110 (2006)

32. Cassidy, R.G., Kirby, M.J.L., Raike, W.M.: Efficient distribution of resources through three levels of government. Management Science 17(8), 462–473 (1971)
33. Chinchuluun, A., Pardalos, P.M., Huang, H.-X.: Multilevel (hierarchical) optimization: Complexity issues, optimality conditions, algorithms. In: Advances in Applied Mathematics and Global Optimization -Advances in Mechanics and Mathematics, vol. 17, ch. 6, pp. 197–221 (2009)
34. Clark, P.A., Westerberg, A.W.: Optimization for design problems having more than one objective. Computers and Chemical Engineering 7(4), 259–278 (1983)
35. Clark, P.A., Westerberg, A.W.: Bilevel programming for steady state chemical process design - I. Fundamentals and algorithms. Computers and Chemical Engineering 14(1), 87–97 (1990)
36. Colson, B., Marcotte, P., Savard, G.: Bilevel programming: A survey. 4OR 3, 87–107 (2005)
37. Colson, B., Marcotte, P., Savard, G.: An overview of bilevel optimization. Annals of Operations Research 153, 235–256 (2007)
38. Constantin, I., Florian, M.: Optimizing frequencies in a transit network: A non-linear bilevel programming approach. International Transactions in Operational Research 2(2), 149–164 (1995)
39. Cote, J.-P., Marcotte, P., Savard, G.: A bilevel modelling approach to pricing and fare optimisation in the airline industry. Journal of Revenue and Pricing Management 2(1), 23–36 (2003)
40. Dempe, S.: A simple algorithm for the linear bilevel programming problem. Optimization 18(3), 373–385 (1987)
41. Dempe, S.: Foundations of bilevel programming. Kluwer Academic Publishers, Dordrecht (2002)
42. Dempe, S.: Annotated bibliography on bilevel programming and mathematical programs with equilibrium constraints. Optimization 52(3), 333–359 (2003)
43. Dempe, S.: Bilevel programming. In: Audet, C., Hansen, P., Savard, G. (eds.) Essays and Surveys in Global Optimization, ch. 6, pp. 165–194 (2005)
44. DeNegre, S.T., Ralphs, T.K.: A branch and cut algorithm for integer bilevel linear programs. Operations research and cyber-infrastructure, Operations Research/Computer Science Interfaces Series 47(2), Part 1, 65–78 (2009)
45. Erkut, E., Gzara, F.: Solving the hazmat transport network design problem. Computers and Operations Research 35, 2234–2247 (2008)
46. Falk, J.E., Soland, R.M.: An algorithm for separable nonconvex programming problems. Management Science 15, 550–569 (1969)
47. Fortuny-Amat, J., McCarl, B.: A representation and economic interpretation of a two-level programming problem. Journal of the Operational Research Society 32(9), 783–792 (1981)
48. Garcés, L.P., Conejo, A.J., García-Bertrand, R., Romero, R.: A bi-level approach to transmission expansion planning within a market environment. IEEE Transactions on Power Systems 24(3), 1513–1522 (2009)
49. Glackin, J., Ecker, J.G., Kupferschmid, M.: Solving bilevel linear programs using multiple objective linear programming. Journal of Optimization Theory and Applications 140(2), 197–212 (2009)
50. Golias, M.M., Saharidis, G.K.D., Boile, M., Theofanis, S.: Scheduling of inbound trucks at a cross-docking facility: Bi-objective vs bi-level modeling approaches. International Journal of Information Systems and Supply Chain Management 5(1) (2012) (in press)
51. Hansen, P., Jaumard, B., Savard, G.: New branch and bound rules for linear bilevel programming. SIAM Journal on Scientific and Statistical Computing 13(5), 1194–1217 (1992)

52. Haurie, A., Savard, G., White, D.J.: A note on: An efficient point algorithm for a linear two-stage optimization problem. Operations Research 38(3), 553–555 (1990)
53. Hobbs, B.F., Nelson, S.K.: A nonlinear bilevel model for analysis of electric utility demand-side planning issues. Annals of Operations Research 34, 255–274 (1992)
54. Júdice, J.J., Faustino, A.M.: The solution of the linear bilevel programming problem by using the linear complementarity problem. Investigavio Operacional 8, 77–95 (1988)
55. Júdice, J.J., Faustino, A.M.: A sequential LCP method for bilevel linear programming. Annals of Operations Research 34(1), 89–106 (1992)
56. Kazempour, J., Conejo, A.J., Ruiz, C.: Strategic generation investment using a complementarity approach. IEEE Transactions on Power Systems 26(2), 940–948 (2011)
57. Koppe, M., Queyranne, M., Ryan, C.T.: Parametric integer programming algorithm for bilevel mixed integer programs. Journal of Optimization Theory and Applications 146(1), 137–150 (2010)
58. Labbe, M., Marcotte, P., Savard, G.: A bilevel model of taxation and its application to optimal highway pricing. Management Science 44(12), 1608–1622 (1998)
59. LeBlanc, L.J., Boyce, D.E.: A bilevel programming algorithm for exact solution of the network design problem with user-optimal flows. Transportation Research - Part B 20B(3), 259–265 (1986)
60. Liu, Y.H., Spencer, T.H.: Solving a bilevel linear program when the inner decision maker controls few variables. European Journal of Operational Research 81(3), 644–651 (1995)
61. Loridan, P., Morgan, J.: Weak via strong stackelberg problem: New results. Journal of Global Optimization 8, 263–287 (1996)
62. Maher, M.J., Zhang, X., Vliet, D.V.: A bi-level programming approach for trip matrix estimation and traffic control problems with stochastic user equilibrium link flows. Transportation Research - Part B 35, 23–40 (2001)
63. Marinakis, Y., Migdalas, A., Pardalos, P.M.: A new bilevel formulation for the vehicle routing problem and a solution method using a genetic algorithm. Journal of Global Optimization 38, 555–580 (2007)
64. Migdalas, A.: Bilevel programming in traffic planning: Models, methods and challenge. Journal of Global Optimization 7, 381–405 (1995)
65. Mitsos, A., Bollas, G.M., Barton, P.I.: Bilevel optimization formulation for parameter estimation in liquid-liquid phase equilibrium problems. Chemical Engineering Science 64, 548–559 (2009)
66. Moore, J.T., Bard, J.F.: The mixed integer linear bilevel programming problem. Operations Research 38(5), 911–921 (1990)
67. Motto, A.L., Arroyo, J.M., Galiana, F.D.: A mixed-integer LP procedure for the analysis of electric grid security under disruptive threat. IEEE Transactions on Power Systems 20(3), 1357–1365 (2005)
68. Nicholls, M.G.: Aluminum production modeling - A nonlinear bilevel programming approach. Operations Research 43(2), 208–218 (1995)
69. Onal, H.: A modified simplex approach for solving bilevel linear programming problems. European Journal of Operations Research 67(1), 126–135 (1993)
70. Onal, H., Darmawan, D.H., Johnson, S.H.: A multilevel analysis of agricultural credit distribution in East Java, Indonesia. Computers and Operations Research 22(2), 227–236 (1995)
71. Pandzic, H., Conejo, A.J., Kuzle, I., Caro, E.: Yearly maintenance scheduling of transmission lines within a market environment. IEEE Transactions on Power Systems (2011) (in press)
72. Ruiz, C., Conejo, A.J.: Pool strategy of a producer with endogenous formation of locational marginal prices. IEEE Transactions on Power Systems 24(4), 1855–1866 (2009)

73. Ryu, J.-H., Dua, V., Pistikopoulos, E.N.: A bilevel programming framework for enterprise-wide process networks under uncertainty. Computers and Chemical Engineering 28, 1121–1129 (2004)
74. Saharidis, G.K.D., Ierapetritou, M.G.: Resolution method for mixed integer bilevel linear problems based on decomposition technique. Journal of Global Optimization 44(1), 29–51 (2009)
75. Saharidis, G.K.D., Golias, M.M., Boile, M., Theofanis, S., Ierapetritou, M.G.: The berth scheduling problem with customer differentiation: A new methodological approach based on hierarchical optimization. International Journal of Advanced Manufacturing Technology 46, 377–393 (2010)
76. Saharidis, G.K.D., Androulakis, I.P., Ierapetritou, M.G.: Model building using bi-level optimization. Journal of Global Optimization 49, 49–67 (2011)
77. Salmerón, J., Wood, K., Baldick, R.: Analysis of electric grid security under terrorist threat. IEEE Transactions on Power Systems 19(2), 905–912 (2004)
78. Salmerón, J., Wood, K., Baldick, R.: Worst-case interdiction analysis of large-scale electric power grids. IEEE Transactions on Power Systems 24, 96–104 (2009)
79. Shi, C., Zhang, G., Lu, J.: On the definition of linear bilevel programming solution. Applied Mathematics and Computation 160(1), 169–176 (2005a)
80. Shi, C., Lu, J., Zhang, G.: An extended Kuhn-Tucker approach for linear bilevel programming. Applied Mathematics and Computation 162, 51–63 (2005b)
81. Shi, C., Lu, J., Zhang, G.: An extended Kth best approach for linear bilevel programming. Applied Mathematics and Computation 164, 843–855 (2005c)
82. Shi, C., Lu, J., Zhang, G., Zhou, H.: An extended branch and bound algorithm for linear bilevel programming. Applied Mathematics and Computation 180, 529–537 (2006)
83. Suh, S., Kim, T.J.: Solving nonlinear bilevel programming models of the equilibrium network design problem: A comparative review. Annals of Operations Research 34, 203–218 (1992)
84. Tuy, H., Migdalas, A., Varbrand, P.: A global optimization approach for the linear two-level program. Journal of Global Optimization 3, 1–23 (1993)
85. Unlu, G.: A linear bilevel programming algorithm based on bicriteria programming. Computers and Operations Research 14(2), 173–179 (1987)
86. Vicente, L.N., Calamai, P.H.: Bilevel and multilevel programming: A bibliography review. Journal of Global Optimization 5(3), 291–306 (1994)
87. Wen, U.P., Hsu, S.T.: A note on a linear bilevel programming algorithm based on bicriteria programming. Computers and Operations Research 16(1), 79–83 (1989)
88. Wen, U.P., Hsu, S.T.: Linear bi-level programming problems - A review. Journal of the Operational Research Society 42(2), 125–133 (1991)
89. Wen, U.P., Yang, Y.H.: Algorithms for solving the mixed integer two-level linear programming problem. Computers and Operations Research 17(2), 133–142 (1990)
90. White, D.J.: Solving bi-level linear programmes. Journal of Mathematical Analysis and Applications 200(1), 254–258 (1996)
91. White, D.J., Anandalingam, G.: A penalty function approach for solving bilevel linear programs. Journal of Global Optimization 3(4), 397–419 (1993)
92. Yin, Y.: Multiobjective bilevel optimization for transportation planning and management problems. Journal of Advanced Transportation 36(1), 93–105 (2002)

Chapter 9
Bilevel Multi-Objective Optimization and Decision Making

Ankur Sinha and Kalyanmoy Deb

Abstract. Bilevel optimization problems are special kind of optimization problems which require every feasible upper-level solution to satisfy the optimality conditions of a lower-level optimization problem. Due to complications associated in solving such problems, they are often treated as single-level optimization problems, and approximation principles are employed to handle them. These problems are commonly found in many practical problem solving tasks which include optimal control, process optimization, game-playing strategy development, transportation problems, coordination of multi-divisional firms, and others. The chapter addresses certain intricate issues related to solving multi-objective bilevel programming problems, and describes recent methodologies to tackle such problems. The first methodology is a hybrid evolutionary-cum-local-search based algorithm to generate the entire Pareto-frontier of multi-objective bilevel problems. The second methodology is a decision maker oriented approach, where preferences from the upper level decision maker are incorporated in the intermediate steps of the algorithm, leading to reduced computational expense. Both these methodologies are tested on a set of recently proposed test problems. The test problems involve various intricacies which could be encountered in multi-objective bilevel problem solving, and the algorithms have been shown to successfully handle these problems. The study opens up a variety of issues related to multi-objective bilevel programming, and shows that evolutionary methods are effective in solving such problems.

Ankur Sinha
Aalto University School of Economics, P.O. Box 1210, FIN-101, Helsinki, Finland
e-mail: ankur.sinha@aalto.fi

Kalyanmoy Deb
Indian Institute of Technology Kanpur, PIN 208016, India, and Aalto University School of Economics, P.O. Box 1210, FIN-101, Helsinki, Finland
e-mail: deb@iitk.ac.in

E.-G. Talbi (Ed.): *Metaheuristics for Bi-level Optimization*, SCI 482, pp. 247–284.
DOI: 10.1007/978-3-642-37838-6_9 © Springer-Verlag Berlin Heidelberg 2013

9.1 Introduction

Optimization is an inter-disciplinary mathematical science which refers to the process of finding the optimal solutions for a given number of objectives, subject to a set of constraints [Reklaitis *et al.*(1983), Rao(1984)]. Constraint satisfaction provides feasible members for the optimization problem, with one or more feasible members being the optimal solutions. The constraints for optimization problems are commonly in the form of equalities or inequalities, and such problems are single level optimization tasks. However, the practice always seems to offer more, and there can be problems with another optimization task within the constraints, giving rise to two-levels of optimization. The original problem is referred as an upper level optimization problem and the one within the constraints is referred as a lower level optimization problem. Such problems are bilevel optimization problems, where the two levels of optimization tasks are inter-twined in a way that solving the lower level optimization problem along with satisfying the other constraints provide a feasible member for the upper level problem. Bilevel programming problems can be found in many practical optimization problems [Bard(1998)]; for example, a solution is considered feasible for the upper level optimization problem only if it satisfies certain equilibrium, stability or conservation principles. To ensure that the requisite condition or a principle is satisfied, a lower level optimization task is required to be solved.

Lack of efficient methodologies and the computational expense required to solve bilevel problems to optima, have deterred practitioners to solve the problem in its original form. In practice [Bianco *et al.*(2009), Dempe(2002), Pakala(1993)], these problems are not usually treated as bilevel programming problems, instead some approximate methodologies are used to replace the lower level problem. The approximate methodologies are inept to find true optimal solutions for bilevel problems, which provides a motivation to explore better solution methodologies to handle such problems. Bilevel programming problems with single objectives at both level have received some attention from theory [Dempe *et al.*(2006)], algorithm development and application [Alexandrov and Dennis(1994), Vicente and Calamai(2004)], and even using evolutionary algorithms [Yin(2000), Wang *et al.*(2008)]. However, apart from a few recent studies [Eichfelder(2007), Eichfelder(2008), Halter and Mostaghim(2006), Shi and Xia(2001)] and our recent evolutionary multi-objective optimization (EMO) studies [Deb and Sinha(2009a), Deb and Sinha(2009b), Sinha and Deb(2009), Deb and Sinha(2010)], multi-objective bilevel programming studies are scarce in both classical and evolutionary optimization fields. The lack of interests for handling bilevel multi-objective problems is not due to lack of practical problems, but more due to the added complexities offered by multiple objectives, in addition to the complex interactions of the two levels.

In this chapter, we briefly outline a generic multi-objective bilevel optimization problem and then provide an overview of existing studies both on single and multi-objective bilevel programming. The chapter takes a closer look at the intricacies of bilevel multi-objective programming problems and describes a procedure to construct bilevel multi-objective test problems. It describes solution methodologies to handle multiple objectives in bilevel context. The first solution methodology is a hybrid and self-adaptive bilevel evolutionary multi-objective optimization (H-BLEMO) algorithm, which aims at generating the entire Pareto-optimal front for the multi-objective bilevel problem. The second methodology extends the first methodology by incorporating decision maker's preferences at the upper level, and produces the most preferred solution on the Pareto-front. Both the methodologies have been evaluated on the recently proposed set of test problems and a scalability study has been performed by considering different problem sizes ranging from 10 to 40 decision variables.

9.2 Multi-objective Bilevel Optimization Problems

A multi-objective bilevel optimization problem has an upper and lower level of multi-objective optimization task. The lower level optimization problem belongs to the constraint of the upper level optimization problem, such that, a member can be feasible at the upper level only if it is Pareto-optimal for the lower level optimization problem. A general multi-objective bilevel optimization problem can be described as follows:

$$\begin{aligned} \text{Minimize}_{(\mathbf{x}_u,\mathbf{x}_l)} \ & \mathbf{F}(\mathbf{x}) = (F_1(\mathbf{x}),\ldots,F_M(\mathbf{x})), \\ \text{subject to } & \mathbf{x}_l \in \text{argmin}_{(\mathbf{x}_l)} \big\{\mathbf{f}(\mathbf{x}) = (f_1(\mathbf{x}),\ldots,f_m(\mathbf{x})) \big| \\ & \qquad \mathbf{g}(\mathbf{x}) \geq \mathbf{0}, \mathbf{h}(\mathbf{x}) = \mathbf{0}\big\}, \\ & \mathbf{G}(\mathbf{x}) \geq \mathbf{0}, \mathbf{H}(\mathbf{x}) = \mathbf{0}, \\ & x_i^{(L)} \leq x_i \leq x_i^{(U)}, \quad i = 1,\ldots,n. \end{aligned} \tag{9.1}$$

In the above description of a generic multi-objective bilevel problem, $F_1(\mathbf{x}),\ldots,$ $F_M(\mathbf{x})$ are upper level objectives and $f_1(\mathbf{x}),\ldots,f_m(\mathbf{x})$ are lower level objectives. The functions $\mathbf{g}(\mathbf{x})$ and $\mathbf{h}(\mathbf{x})$ determine the feasible space for the lower level problem. The decision vector $\mathbf{x}$ is formed by two smaller vectors $\mathbf{x}_u$ and $\mathbf{x}_l$, such that $\mathbf{x} = (\mathbf{x}_u, \mathbf{x}_l)$. At the lower level, optimization is performed only with respect to the variables $\mathbf{x}_l$, therefore, the solution set of the lower level problem can be represented as a function of $\mathbf{x}_u$, or as $\mathbf{x}_l^*(\mathbf{x}_u)$. This means that the upper level variables $\mathbf{x}_u$, act as a parameter to the lower level problem and hence the lower level optimal solutions $\mathbf{x}_l^*$ are a function of the upper level vector $\mathbf{x}_u$. The functions $\mathbf{G}(\mathbf{x})$ and $\mathbf{H}(\mathbf{x})$ along with the Pareto-optimality to the lower level problem determine the feasible space for the upper level optimization problem. Both sets $\mathbf{x}_l$ and $\mathbf{x}_u$ are decision variables for the upper level problem.

Figure 9.1 shows feasible regions for a multi-objective bilevel problem. The problem has two objectives both at the upper and the lower level. The graphs on the corners of the figure correspond to four different upper level variable vectors $\mathbf{x}_u^{(1)}, \mathbf{x}_u^{(2)}, \mathbf{x}_u^{(3)}$ and $\mathbf{x}_u^{(4)}$. In these graphs, we show with arrows, the possible relationships between members in the lower level space and their positions in the upper level space. For the lower level problem corresponding to $\mathbf{x}_u^{(1)}$, we have shown a few

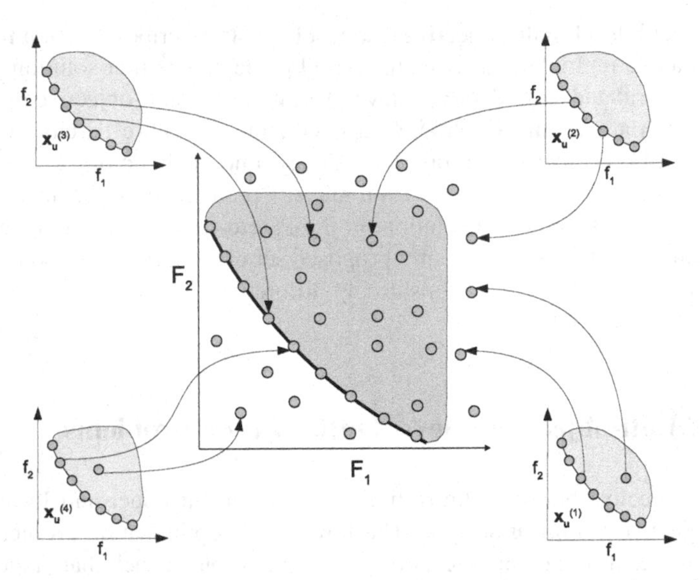

Fig. 9.1 Two objective bilevel minimization problem. The graph in the centre shows the upper level feasible region along with members which are feasible, infeasible or optimal. The corner graphs show the lower level feasible region for different $\mathbf{x}_u$ along with members which are lower level feasible or lower level optimal.

members in the lower level space. A lower level feasible but non-optimal member always corresponds to an infeasible position at the upper level. As shown in the figure, a lower level optimal member may also be infeasible at the upper level because of some equality or inequality upper level constraint. For the lower level problem corresponding to $\mathbf{x}_u^{(2)}$, we observe that one of the lower level optimal members is feasible at the upper level, and another lower level optimal member is infeasible at the upper level because of some upper level constraint. For the lower level problem corresponding to $\mathbf{x}_u^{(3)}$, we observe that one of the lower level optimal members is feasible at the upper level, and the other lower level optimal member is also upper level optimal. Finally, for the lower level problem corresponding to $\mathbf{x}_u^{(4)}$, we observe that a lower level feasible but non-optimal member is infeasible at the upper level, and the lower level optimal member is feasible at the upper level. Therefore, the graphs present the possible scenarios in a bilevel multi-objective problem. It is noteworthy here, if the lower level is not optimized properly we might end up with a non-optimal lower level solution, and hence an infeasible solution at the upper level. This infeasible solution might lie ahead of the upper level frontier, as is the case with $\mathbf{x}_u^{(4)}$, which would dominate and eliminate the true Pareto-optimal solutions at the upper level.

9.2.1 *Real World Problems*

Bilevel optimization problems are commonly found in practice. These problems arise from hierarchical problems where the strategy for solving the overall system depends on optimal strategies of solving a number of subsystems. In this subsection, we illustrate a few examples, where the first two have single objectives at all levels, and the later two are multi-objective bilevel problems.

9.2.1.1 Engineering Problems

Many engineering design problems in practice involve an upper level optimization problem requiring that a feasible solution to the problem must satisfy certain physical conditions, such as satisfying a network flow balance or satisfying stability conditions or satisfying some equilibrium conditions. If simplified mathematical equations for such conditions are easily available, often they are directly used as constraints and the lower level optimization task is avoided. But in many problems establishing whether a solution is stable or in equilibrium can be established by ensuring that the solution is an optimal solution to a derived optimization problem. Such a derived optimization problem can then be formulated as a lower level problem in a bilevel optimization problem. A common source of bilevel problems is in chemical process optimization problems, in which the upper level problem optimizes the overall cost and quality of product, whereas the lower level optimization problem optimizes error measures indicating how closely the process adheres to different theoretical process conditions, such as mass balance equations, cracking, or distillation principles [Dempe(2002)].

9.2.1.2 Environmental Economics

Multi-level solution methodologies are useful in handling a number of problems in environmental economics, we state one such example here [Dempe(2002)]. It is a well known theory in economics that market failures occur when the market is unable to allocate scarce resources to generate the maximum social welfare. Keeping this statement in mind consider an environmental setting, where increase in profit for one firm leads to increase in environmental pollution, and increase in environmental pollution leads to a decrease in profit for another firm. Under such circumstances, the government tries to protect the second firm by levying taxes on the first firm. The tax determination problem for maximizing social welfare in this case is a tri-level problem. The upper level problem is a social welfare function which is to be maximized, and the two lower level problems are profit maximization problem for the two firms. The first firm maximizes its profit subject to the taxes, and the second firm maximizes its profit subject to the production of first firm. A multi-objective social welfare function could also be considered such that the problem becomes multi-objective at the upper level.

9.2.1.3 Transportation Problems

It is easy to find bilevel problems in transportation. Here, we state an example [Wang and Ehrgott(2011)] of a tolled road network, where the objectives at the upper level are to minimize system travel time and total vehicle emissions. The upper level is an optimization task necessary for the government or a planning body. The lower level is a network equilibrium problem of traffic flow which results from the tax proposed by the planning body. The lower level decision makers are users, who have two objectives in their mind, namely, minimize travel time and minimize toll cost. The users are rational in the sense that they will choose one of the efficient paths. Based on this bilevel formulation, a set of efficient toll charges can be obtained.

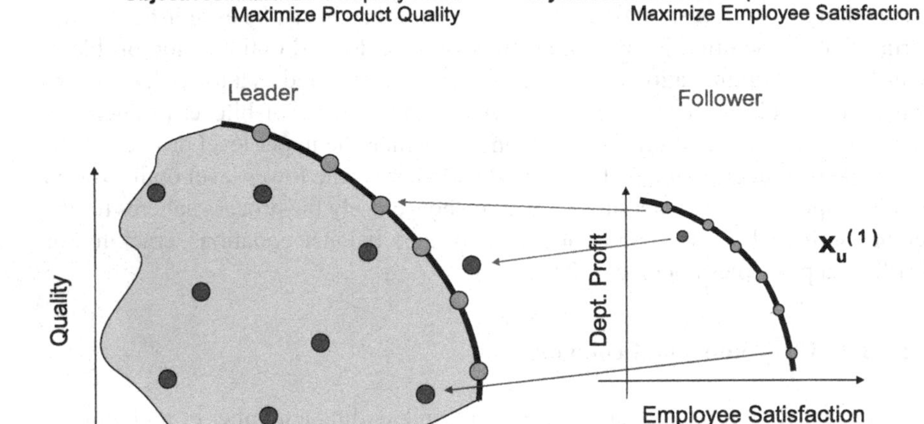

Fig. 9.2 A company scenario with upper and lower level decision makers

9.2.1.4 A Company Scenario

The bilevel problems are also similar in principle to the Stackelberg games [Fudenberg and Tirole(1993), Wang and Periaux(2001)] in which a leader makes the first move and a follower then maximizes its move considering the leader's move. The leader has an advantage that it can control the game by making its move in a way so as to maximize its own gain knowing that the follower will always maximize its own gain. For an example [Zhang *et al.*(2007)], consider a company scenario shown in Figure 9.2, where a CEO (leader) may be interested in maximizing net profits of the company and quality of products, and head of branch (follower) may maximize his own net profit and worker satisfaction. The CEO knows that for each of his strategy, the head of branch will optimize his own objectives. The CEO must then adjust his own decision variables so that CEO's own objectives are maximized. Stackelberg's game model and its solutions are used in many different problem domains, including engineering design [Pakala(1993)], security applications [Paruchuri *et al.*(2008)], and others.

9.3 Existing Classical and Evolutionary Methodologies

The importance of solving bilevel optimization problems, particularly problems having a single objective in each level, has been recognized amply in the optimization literature. The research has been focused in both theoretical and algorithmic aspects. However, there has been a lukewarm interest in handling bilevel problems having multiple conflicting objectives in any or both levels. Here we provide a brief description of the main research outcomes so far in both single and multi-objective bilevel optimization areas.

9.3.1 Theoretical Developments

Several studies exist in determining the optimality conditions for an upper level solution. The difficulty arises due to the existence of another optimization problem as a hard constraint to the upper level problem. Usually the Karush-Kuhn-Tucker (KKT) conditions of the lower level optimization problems are first written and used as constraints in formulating the KKT conditions of the upper level problem, involving second derivatives of the lower level objectives and constraints as the necessary conditions of the upper level problem. However, as discussed in [Dempe *et al.*(2006)], although KKT optimality conditions can be written mathematically, the presence of many lower level Lagrange multipliers and an abstract term involving coderivatives makes the procedure difficult to be applied in practice.

[Fliege and Vicente(2006)] suggested a mapping concept in which a bilevel single-objective optimization problem (one objective each in upper and lower level problems) can be converted to an equivalent four-objective optimization problem with a special cone dominance concept. Although the idea may apparently be extended for bilevel multi-objective optimization problems, no such suggestion with an exact mathematical formulation is made yet. Moreover, derivatives of original objectives are involved in the problem formulation, thereby making the approach limited to only differentiable problems.

9.3.2 Algorithmic Developments

One simple algorithm for solving bilevel optimization problems using a point-by-point approach would be to directly treat the lower level problem as a hard constraint. Every solution ($\mathbf{x} = (\mathbf{x}_u, \mathbf{x}_l)$) must be sent to the lower level problem as an initial point and an optimization algorithm can then be employed to find the optimal solution $\mathbf{x}_l^*$ of the lower level optimization problem. Then, the original solution $\mathbf{x}$ of the upper level problem must be repaired as $(\mathbf{x}_u, \mathbf{x}_l^*)$. The employment of a lower level optimizer within the upper level optimizer for every upper level solution makes the overall search a *nested* optimization procedure, which may be computationally an expensive task. Moreover, if this idea is to be extended for multiple conflicting objectives in the lower level, for every upper level solution, multiple Pareto-optimal solutions for the lower level problem need to be found and stored by a suitable multi-objective optimizer.

Another idea [Herskovits *et al.*(2000), Bianco *et al.*(2009)] of handling the lower level optimization problem having differentiable objectives and constraints is to include the explicit KKT conditions of the lower level optimization problem directly as constraints to the upper level problem. This will then involve Lagrange multipliers of the lower level optimization problem as additional variables to the upper level problem. As KKT points need not always be optimum points, further conditions must have to be included to ensure the optimality of lower level problem. For multi-objective bilevel problems, corresponding multi-objective KKT formulations need to be used, thereby involving further Lagrange multipliers and optimality conditions as constraints to the upper level problem. Despite these apparent difficulties, there exist some useful studies, including reviews on bilevel programming [Colson *et al.*(2007), Vicente and Calamai(2004)], test problem generators [Calamai and Vicente(1994)], nested bilevel linear programming [Gaur and Arora(2008)], and applications [Fampa *et al.*(2008), Abass(2005), Koh(2007)], mostly in the realm of single-objective bilevel optimization.

Recent studies by [Eichfelder(2007), Eichfelder(2008)] concentrated on handling multi-objective bilevel problems using classical methods. While the lower level problem uses a numerical optimization technique, the upper level problem is handled using an adaptive exhaustive search method, thereby making the overall procedure computationally expensive for large-scale problems. This method uses the nested optimization strategy to find and store multiple Pareto-optimal solutions for each of finitely-many upper level variable vectors.

Another study by [Shi and Xia(2001)] transformed a multi-objective bilevel programming problem into a bilevel ε-constraint approach in both levels by keeping one of the objective functions and converting remaining objectives to constraints. The ε values for constraints were supplied by the decision-maker as different levels of 'satisfactoriness'. Further, the lower-level single-objective constrained optimization problem was replaced by equivalent KKT conditions and a variable metric optimization method was used to solve the resulting problem.

Certainly, more efforts are needed to devise effective classical methods for multi-objective bilevel optimization, particularly to handle the upper level optimization task in a more coordinated way with the lower level optimization task.

9.3.3 Evolutionary Methods

Several researchers have proposed evolutionary algorithm based approaches in solving single-objective bilevel optimization problems. As early as in 1994, [Mathieu *et al.*(1994)] proposed a GA-based approach for solving bilevel linear programming problems having a single objective in each level. The lower level problem was solved using a standard linear programming method, whereas the upper level was solved using a GA. Thus, this early GA study used a nested optimization strategy, which may be computationally too expensive to extend for nonlinear and large-scale problems. [Yin(2000)] proposed another GA based nested approach in which the lower level problem was solved using the Frank-Wolfe gradient based linearized optimization method and claimed to solve non-convex bilevel optimization problems better than an existing classical method. [Oduguwa and Roy(2002)] suggested a coevolutionary GA approach in which two different populations are used to handle variable vectors $\mathbf{x}_u$ and $\mathbf{x}_l$ independently. Thereafter, a linking procedure is used to cross-talk between the populations. For single-objective bilevel optimization problems, the final outcome is usually a single optimal solution in each level. The proposed coevolutionary approach is viable for finding corresponding single solution in $\mathbf{x}_u$ and $\mathbf{x}_l$ spaces. But in handling multi-objective bilevel programming problems, multiple solutions corresponding to each upper level solution must be found and maintained during the coevolutionary process. It is not clear how such a coevolutionary algorithm can be designed effectively for handling multi-objective bilevel optimization problems. We do not address this issue in this chapter, but recognize that Oduguwa and Roy's study ([Oduguwa and Roy(2002)]) was the first to suggest a coevolutionary procedure for single-objective bilevel optimization problems. Since 2005, a surge in research in this area can be found in algorithm development mostly using the nested approach and the explicit KKT conditions of the lower level problem, and in various application areas [Hecheng and Wang(2007), Li and Wang(2007), Dimitriou *et al.*(2008), Yin(2000), Mathieu *et al.*(1994)], [Sun *et al.*(2006), Wang *et al.*(2007), Koh(2007), Wang *et al.*(2005)], [Wang *et al.*(2008)].

[Li *et al.*(2006)] proposed particle swarm optimization (PSO) based procedures for both lower and upper levels, but instead of using a nested approach, they proposed a serial application of upper and lower levels iteratively. This idea is applicable in solving single-objective problems in each level due to the sole target of finding a single optimal solution. As discussed above, in the presence of multiple conflicting objectives in each level, multiple solutions need to be found and preserved for each upper level solution and then a serial application of upper and lower level optimization does not make sense for multi-objective bilevel optimization. [Halter and Mostaghim(2006)] also used PSO on both levels, but since the lower level problem in their application problem was linear, they used a specialized linear multi-objective PSO algorithm and used an overall nested optimization strategy at the upper level.

Recently, we proposed a number of EMO algorithms through conference publications [Deb and Sinha(2009a), Deb and Sinha(2009b), Sinha and Deb(2009)] using NSGA-II to solve both level problems in a synchronous manner. First, our methodologies were generic so that they can be used to linear/nonlinear, convex/non-convex, differentiable/non-differentiable and single/multi-objective problems at both levels. Second, our methodologies did not use the nested approach, nor did they use a serial approach, but employed a structured intertwined evolution of upper and lower level populations. But they were computationally demanding. However, these initial studies made us understand the complex intricacies by which both level problems can influence each other. Based on this experience, we suggested a less-structural, self-adaptive, computationally fast, and a hybrid evolutionary algorithm coupled with a local search procedure for handling multi-objective bilevel programming problems. This procedure [Deb and Sinha(2010)] and its extension [Sinha(2011)] are discussed in detail in this chapter.

Bilevel programming problems, particularly with multiple conflicting objectives, should have been paid more attention than what has been made so far. As more and more studies are performed, the algorithms must be tested and compared against each other. This process needs an adequate number of test problems with tunable difficulties. In the next section, we describe a generic procedure for developing test problems.

9.3.4 Development of Tunable Test Problems

Bilevel multi-objective optimization problems are different from single-level multi-objective optimization problems. In bilevel multi-objective problems Pareto-optimality of a lower level multi-objective optimization problem is a feasibility requirement to the upper level problem. Thus, while developing a bilevel multi-objective test problem, we should have ways to test an algorithm's ability to handle complexities in both lower and upper level problems independently and additionally their interactions. Further, the test problems should be such that we would have a precise knowledge about the exact location (and relationships) of Pareto-optimal points. Thinking along these lines, we outline a number of desired properties in a bilevel multi-objective test problem:

1. *Exact location of Pareto-optimal solutions in both lower and upper level problems are possible to be established.* This will facilitate a user to evaluate the performance of an algorithm easily by comparing the obtained solutions with the exact Pareto-optimal solutions.
2. *Problems are scalable with respect to number of variables.* This will allow a user to investigate whether the proposed algorithm scales well with number of variables in both lower and upper levels.

3. *Problems are scalable with respect to number of objectives in both lower and upper levels.* This will enable a user to evaluate whether the proposed algorithm scales well with the number of objectives in each level.
4. *Lower level problems are difficult to solve to Pareto-optimality.* If the lower level Pareto-optimal front is not found exactly, the corresponding upper level solution are not feasible. Therefore, these problems will test an algorithm's ability to converge to the correct Pareto-optimal front. Here, ideas can be borrowed from single-level EMO test problems [Deb *et al.*(2005)] to construct difficult lower level optimization problems. The shape (convex, non-convex, disjointedness and multi-modality) of the Pareto-optimal front will also play an important role in this respect.
5. *There exists a conflict between lower and upper level problem solving tasks.* For two solutions $\mathbf{x}$ and $\mathbf{y}$ of which $\mathbf{x}$ is Pareto-optimal and $\mathbf{y}$ is a dominated solution in the lower level, solution $\mathbf{y}$ can be better than solution $\mathbf{x}$ in the upper level. Due to these discrepancies, these problems will cause a conflict in converging to the appropriate Pareto-optimal front in both lower and upper level optimization tasks.
6. *Extension to higher level optimization problems is possible.* Although our focus here is for bilevel problems only, test problems scalable to three or higher levels would be interesting, as there may exist some practical problems formulated in three or higher levels. On the other hand, it will also be ideal to have bilevel test problems which will degenerate to challenging single level test problems, if a single objective function is chosen for each level.
7. *Different lower level problems may contribute differently to the upper level front in terms of their extent of representative solutions on the upper level Pareto-optimal front.* These test problems will test an algorithm's ability to emphasis different lower level problems differently in order to find a well-distributed set of Pareto-optimal solutions at the upper level.
8. *Test problems must include constraints at both levels.* This will allow algorithms to be tested for their ability to handle constraints in both lower and upper level optimization problems.

Different principles are possible to construct test problems following the above guidelines. Here, we present a generalized version of a recently proposed procedure [Deb and Sinha(2009a), Deb and Sinha(2010)].

9.3.4.1 A Multi-objective Bilevel Test Problem Construction Procedure

We suggest a test problem construction procedure for a bilevel problem having M and m objectives in the upper and lower level, respectively. The procedure needs at most three functional forms and is described below:

Step 1: First, a parametric trade-off function $\Phi_U : \mathbb{R}^{M-1} \rightarrow \mathbb{R}^M$ which determines a trade-off frontier $(v_1(\mathbf{u}), \ldots, v_M(\mathbf{u}))$ on the $\mathbf{F}$-space as a function of $(M-1)$ parameters $\mathbf{u}$ (can be considered as a subset of $\mathbf{x}_u$) is chosen. Figure 9.3 shows such a v_1-v_2 relationship on a two-objective bilevel problem.

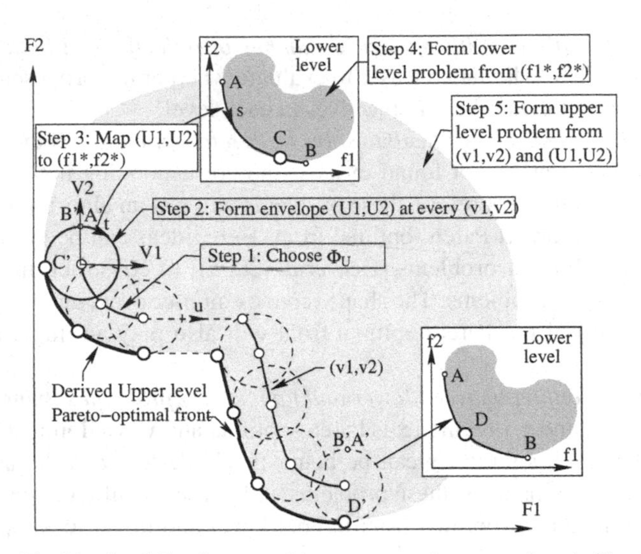

Fig. 9.3 A multi-objective bilevel test problem construction procedure is illustrated through two objectives in both upper and lower levels

Step 2: Next, for every point $\mathbf{v}$ on the Φ_U-frontier, a $(M-1)$-dimensional envelope $(U_1(\mathbf{t}),\ldots,U_M(\mathbf{t}))^{\mathbf{v}})$ on the $\mathbf{F}$-space as a function of $\mathbf{t}$ (having $(M-1)$ parameters) is chosen. The non-dominated part of the agglomerate envelope $\cup_{\mathbf{v}} \cup_{\mathbf{t}} \left[(v_1(\mathbf{u})+U_1(\mathbf{t})^{\mathbf{v}}),\ldots,(v_M(\mathbf{u})+U_M(\mathbf{t})^{\mathbf{v}})\right]$ constitutes the overall upper level Pareto-optimal front. Figure 9.3 indicates this upper level Pareto-optimal front and some specific Pareto-optimal points (marked with bigger circles) derived from specific $\mathbf{v}$-vectors.

Step 3: Next, for every point $\mathbf{v}$ on the Φ_U-frontier, a mapping function Φ_L : $\mathbb{R}^{M-1} \rightarrow \mathbb{R}^{m-1}$ which maps every $\mathbf{v}$-point from the $\mathbf{U}$-frontier to the lower level Pareto-optimal front $(f_1^*(\mathbf{s}),\ldots,f_m^*(\mathbf{s}))^{\mathbf{v}}$ is chosen. Here, $\mathbf{s}$ is a $(m-1)$-dimensional vector and can be considered as a subset of $\mathbf{x}_l$. Figure 9.3 shows this mapping The envelope A′C′B′ (a circle in the figure) is mapped to the lower level Pareto-optimal frontier ACB (inlet figure on top).

Step 4: After these three functions are defined, the lower level problem can be constructed by using a bottom-up procedure adopted in [Deb *et al.*(2005)] through additional terms arising from other lower level decision variables: $f_j(\mathbf{x}_l) = f_j^*(\mathbf{s}) + e_j(\mathbf{x}_l \backslash \mathbf{s})$ with $e_j \geq 0$. The task of the lower level optimization task would be to make the e_j term zero for each objective. The term e_j can be made complex (multi-modal, non-linear, or large-dimensional) to make the convergence to the lower level Pareto-optimal front difficult by an optimization algorithm.

Step 5: Finally, the upper level objectives can be formed from u_j functions by including additional terms from other upper level decision variables. An additive form is as follows: $F_j(\mathbf{x}) = u_j(\mathbf{u}) + E_j(\mathbf{x}_u \backslash \mathbf{u})$ with $E_j \geq 0$. Like the e_j term, the term E_j can also be made complex for an algorithm to properly converge to the upper level Pareto-optimal front.

Step 6: Additionally, a number of linked terms $l_j(\mathbf{x}_u\backslash\mathbf{u},\mathbf{x}_l\backslash\mathbf{s})$ and $L_j(\mathbf{x}_u\backslash\mathbf{u},\mathbf{x}_l\backslash\mathbf{s})$ (non-negative terms) involving remaining $\mathbf{x}_u$ (without $\mathbf{u}$) and $\mathbf{x}_l$ (without $\mathbf{s}$) variables can be added to both lower and upper level problems, respectively, to make sure a proper coordination between lower and upper level optimization tasks is needed to converge to the respective Pareto-optimal fronts.

Another interesting yet a difficult scenario can be created with the linked terms. An identical link term can be added to the lower level problem, but subtracted from the the upper level problem. Thus, an effort to reduce the value of the linked term will make an improvement in the lower level, whereas it will cause a deterioration in the upper level. This will create a conflict in the working of both levels of optimization. Based on this described construction procedure, a set of two-objective test problems is proposed in [Deb and Sinha(2010)]. The test-suite is called the DS test problems and has been used in this chapter to evaluate the solution methodologies. In the next section, we describe the first solution methodology, which can be used to approximate the Pareto-optimal set of a given multi-objective bilevel problem.

9.4 Hybrid Bilevel Evolutionary Multi-Objective Optimization (H-BLEMO) Algorithm

The H-BLEMO [Deb and Sinha(2010)] procedure is an extension of our previously suggested algorithms [Deb and Sinha(2009a), Deb and Sinha(2009b)]. In this section, we provide a step-by-step procedure for the algorithm. A sketch of an iteration of the algorithm and the population structure is shown in Figure 9.4.

An upper level population of size N_u is initialized, which has a subpopulation of lower level variable set $\mathbf{x}_l$ for each upper level variable $\mathbf{x}_u$. Each $\mathbf{x}_u$ is shown in the figure with a dotted line, and its corresponding subpopulation members are shown with continuous lines. To begin with, the subpopulation size ($N_l^{(0)}$) is kept identical for each $\mathbf{x}_u$ variable set, but it is allowed to change adaptively with generation T. At the initialization stage, an empty archive A_0 is created, which is shown in the figure below the upper level population memebers at $T = 0$. For each $\mathbf{x}_u$, we perform a lower level NSGA-II operation on the corresponding subpopulation which contain only lower level variables $\mathbf{x}_l$. The NSGA-II search is not performed till the true lower level Pareto-optimal front is found, rather only until a small number of generations at which the specified lower level termination criterion (discussed in subsection 9.9.2) is satisfied. Thereafter, a local search is performed on a few rank-one lower level solutions until the local search termination criterion is met (discussed in Step 3 in subsection 9.4.3). The archive is maintained at the upper level containing solution vectors $(\mathbf{x}_{u_a},\mathbf{x}_{l_a})$, which are optimal at the lower level and non-dominated at the upper level. The solutions in the archive are updated after every lower level NSGA-II call. The members of the lower level population undergoing a local search are lower level optimal solutions and hence are assigned an 'optimality tag'. These

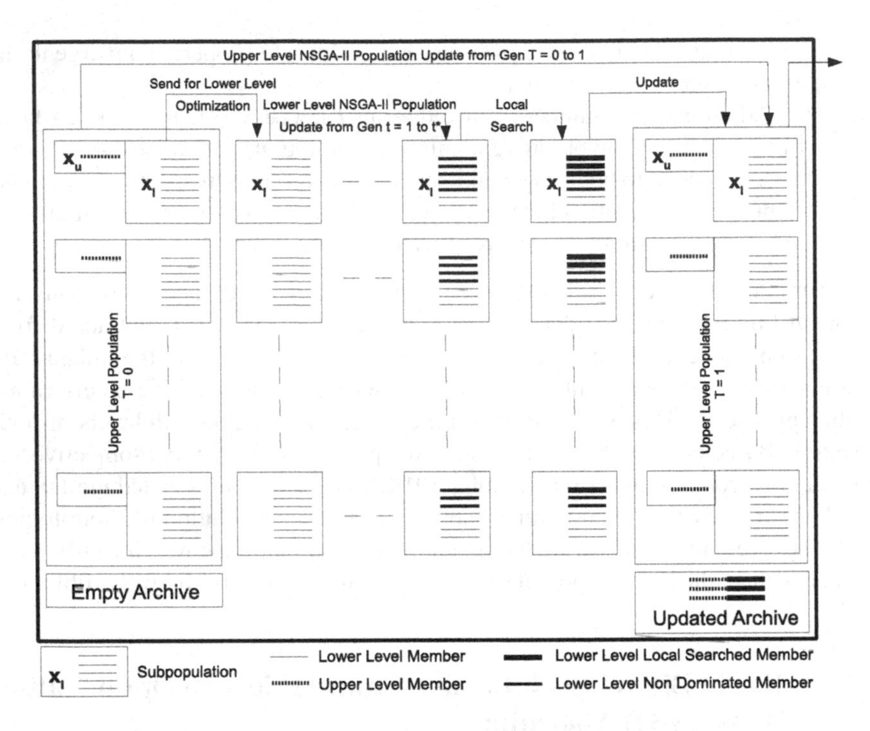

Fig. 9.4 A sketch of an iteration of the H-BLEMO algorithm

local searched solutions ($\mathbf{x}_l$) are then combined with corresponding $\mathbf{x}_u$ variables and become eligible to enter the archive if it is non-dominated when compared to the existing members of the archive. The dominated members in the archive are then eliminated. The solutions obtained from the lower level ($\mathbf{x}_l$) are combined with corresponding $\mathbf{x}_u$ variables and are processed by the upper level NSGA-II operators to create a new upper level population. This process is continued till an upper level termination criterion (described in subsection 9.9.2) is satisfied.

To make the proposed algorithm computationally faster, we have used two different strategies: (i) for every upper level variable vector $\mathbf{x}_u$, we do not completely solve the lower level multi-objective optimization problem, thereby not making our approach a nested procedure, and (ii) the subpopulation size and number of generations for a lower level NSGA-II simulation are computed adaptively based on the relative location of $\mathbf{x}_u$ compared to archive solutions, thereby making the overall procedure less parametric and more computationally efficient in terms of overall function evaluations. However, before we present a detailed step-by-step procedure, we discuss the automatic update procedure of population size and termination criteria of the lower level NSGA-II.

9.4.1 Update of Population Sizes

The upper level population size N_u is kept fixed and is chosen to be proportional to the number of variables. However, the subpopulation size (N_l) for each lower level NSGA-II is sized in a self-adaptive manner. Here we describe the procedure.

In a lower level problem, $\mathbf{x}_l$ is updated by a modified NSGA-II procedure and the corresponding $\mathbf{x}_u$ is kept fixed throughout. Initially, The population size of each lower level NSGA-II ($N_l^{(0)}$) is set depending upon the dimension of lower and upper level variables ($|\mathbf{x}_l|$ and $|\mathbf{x}_u|$, respectively). The number of lower level subpopulations ($n_s^{(0)}$) signifies the number of independent population members for $\mathbf{x}_u$ in a population. Our intention is to set the population sizes ($n_s^{(0)}$ and $N_l^{(0)}$) for $\mathbf{x}_u$ and $\mathbf{x}_l$ proportionately to their dimensions, yielding

$$\frac{n_s^{(0)}}{N_l^{(0)}} = \frac{|\mathbf{x}_u|}{|\mathbf{x}_l|}. \tag{9.2}$$

Noting also that $n_s^{(0)} N_l^{(0)} = N_u$, we obtain the following sizing equations:

$$n_s^{(0)} = \sqrt{\frac{|\mathbf{x}_u|}{|\mathbf{x}_l|} N_u}, \tag{9.3}$$

$$N_l^{(0)} = \sqrt{\frac{|\mathbf{x}_l|}{|\mathbf{x}_u|} N_u}. \tag{9.4}$$

For an equal number of lower and upper level variables, $n_s^{(0)} = N_l^{(0)} = \sqrt{N_u}$. The above values are set for the initial population only, but are allowed to get modified thereafter in a self-adaptive manner by directly relating the location of the corresponding $\mathbf{x}_u$ variable vector from the points in the archive in the variable space. As shown in Figure 9.5, first the maximum Euclidean distance (δ_U) in the $\mathbf{x}_u$-space among the members of the archive is computed. Then, the Euclidean distance (δ_u) between the current $\mathbf{x}_u$ vector and the closest archive member is computed. The subpopulation size N_l is then set proportional to the ratio of δ_u and δ_U as follows:

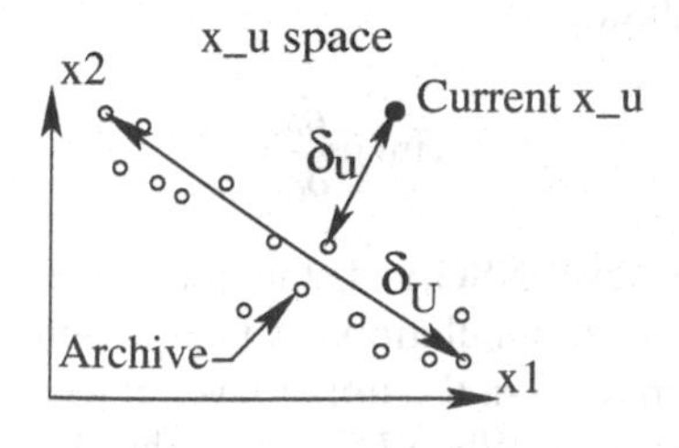

Fig. 9.5 Computation of δ_u and δ_U

$$N_l = (\text{round})\frac{\delta_u}{\delta_U}N_l^{(0)}. \tag{9.5}$$

To eliminate the cases with too low or too large population sizes, N_l is restricted between four (due to the need of two binary tournament selection operations to choose two parent solutions for a single recombination event in the NSGA-II) and $N_l^{(0)}$. If the current $\mathbf{x}_u$ variable vector is far away from the archive members, a large number of generations must have to be spent in the corresponding lower level NSGA-II, as dictated by equation (9.5).

9.4.2 Termination Criteria

In a bilevel optimization, it is clear that the lower level optimization must have to be run more often than the upper level optimization, as the former task acts as a constraint to the upper level task. Thus, any judicial and efficient efforts in terminating a lower level optimization can make a substantial saving in the overall computational effort. For this purpose, we first gauge the difficulty of solving all lower level problems by observing the change in their *hypervolume* measures only in the initial generation ($T = 0$) of the upper level optimization.

The maximum ($H^{\max}$) and minimum ($H^{\min}$) hypervolume is calculated from the lower level non-dominated set (with a reference point constructed from the worst objective values of the set) in every τ generations of a lower level run. The H_l-metric is then computed as follows:

$$H_l = \frac{H_l^{\max} - H_l^{\min}}{H_l^{\max} + H_l^{\min}}. \tag{9.6}$$

If $H_l \leq \varepsilon_l$ (a threshold parameter) is encountered, indicating that an adequate convergence in the hypervolume measure is obtained, the lower level NSGA-II simulation is terminated. The number of lower level generations needed to meet the above criterion is calculated for each subpopulation during the initial generation ($T = 0$) of the upper level NSGA-II and an average (denoted here as $t_l^{\max}$) is computed. Thereafter, no subsequent lower level NSGA-II simulations are allowed to proceed beyond t_l generations (derived from $t_l^{\max}$, as calculated below) or the above $H_l \leq \varepsilon_l$ is satisfied. We bound the limiting generation (t_l) to be proportional to the distance of current $\mathbf{x}_u$ from the archive, as follows:

$$t_l = (\text{int})\ \frac{\delta_u}{\delta_U}t_l^{\max}. \tag{9.7}$$

For terminating the upper level NSGA-II, the normalized change in hypervolume measure H_u of the upper level population (as in equation (9.6) except that the hypervolume measure is computed in the upper level objective space) is computed in every τ consecutive generations. When $H_u \leq \varepsilon_u$ (a threshold parameter) is obtained, the overall algorithm is terminated. We have used $\tau = 10$, $\varepsilon_l = 0.1$ (for a quick termination) and $\varepsilon_u = 0.0001$ (for a reliable convergence of the upper level problem) for all problems in this study.

Now, we are ready to describe the overall algorithm for a typical generation in a step-by-step format.

9.4.3 Step-by-Step Procedure

At the start of the upper level NSGA-II generation T, we have a population P_T of size N_u. Every population member has the following quantities computed from the previous iteration: (i) a non-dominated rank ND_u corresponding to $\mathbf{F}$ and $\mathbf{G}$, (ii) a crowding distance value CD_u corresponding to $\mathbf{F}$, (iii) a non-dominated rank ND_l corresponding to $\mathbf{f}$ and $\mathbf{g}$, and (iv) a crowding distance value CD_l using $\mathbf{f}$. In addition to these quantities, for the members stored in the archive A_T, we have also computed (v) a crowding distance value CD_a corresponding to $\mathbf{F}$ and (vi) a non-dominated rank ND_a corresponding to $\mathbf{F}$ and $\mathbf{G}$.

Step 1a: Creation of new $\mathbf{x}_u$: We apply two binary tournament selection operations on members ($\mathbf{x} = (\mathbf{x}_u, \mathbf{x}_l)$) of P_T using ND_u and CD_u lexicographically. Also, we apply two binary tournament selections on the archive population A_T using ND_a and CD_a lexicographically. Of the four selected members, two participate in the recombination operator based on stochastic events. The members from A_T participate as parents with a probability of $\frac{|A_T|}{|A_T|+|P_T|}$, otherwise the members from P_T become the parents for recombination. The upper level variable vectors $\mathbf{x}_u$ of the two selected parents are then recombined using the SBX operator [Deb and Agrawal(1995)] to obtain two new vectors of which one is chosen for further processing at random. The chosen vector is then mutated by the polynomial mutation operator [Deb(2001)] to obtain a child vector (say, $\mathbf{x}_u^{(1)}$).

Step 1b: Creation of new $\mathbf{x}_l$: First, the population size ($N_l(\mathbf{x}_u^{(1)})$) for the child solution $\mathbf{x}_u^{(1)}$ is determined by equation (9.5). The creation of $\mathbf{x}_l$ depends on how close the new variable set $\mathbf{x}_u^{(1)}$ is compared to the current archive, A_T. If $N_l = N_l^{(0)}$ (indicating that the $\mathbf{x}_u$ is away from the archive members), new lower level variable vectors $\mathbf{x}_l^{(i)}$ (for $i = 1, \ldots, N_l(\mathbf{x}_u^{(1)})$) are created by applying selection-recombination-mutation operations on members of P_T and A_T. Here, a parent member is chosen from A_T with a probability $\frac{|A_T|}{|A_T|+|P_T|}$, otherwise a member from P_T is chosen at random. A total of $N_l(\mathbf{x}_u^{(1)})$ child solutions are created by concatenating upper and lower level variable vectors together, as follows: $c_i = (\mathbf{x}_u^{(1)}, \mathbf{x}_l^{(i)})$ for $i = 1, \ldots, N_l(\mathbf{x}_u^{(1)})$. Thus, for the new upper level variable vector $\mathbf{x}_u^{(1)}$, a subpopulation of $N_l(\mathbf{x}_u^{(1)})$ lower level variable vectors are created by genetic operations from P_T and A_T.

However, if the lower level population size ($N_l(\mathbf{x}_u^{(1)})$) is less than $N_l^{(0)}$ (indicating that the variable set $\mathbf{x}_u$ is close to the archive members), a different strategy is used. First, a specific archive member (say, $\mathbf{x}_u^{(a)}$) closest to $\mathbf{x}_u^{(1)}$ is identified.

Instead of creating new lower level variable vectors, $N_l(\mathbf{x}_u^{(1)})$ vectors are chosen from the subpopulation to which $\mathbf{x}_u^{(a)}$ belongs. Complete child solutions are created by concatenating upper and lower level variables vectors together. If however the previous subpopulation does not have $N_l(\mathbf{x}_u^{(1)})$ members, the remaining slots are filled by creating new child solutions by the procedure of the previous paragraph.

Step 2: Lower level NSGA-II: For each subpopulation, we now perform a NSGA-II procedure using lower level objectives ($\mathbf{f}$) and constraints ($\mathbf{g}$) for t_l generations (equation (9.7)). It is important to reiterate that in each lower level NSGA-II run, the upper level variable vector $\mathbf{x}_u$ is not changed. The selection process is different from that in the usual NSGA-II procedure. If the subpopulation has no member in the current archive A_T, the parent solutions are chosen as usual by the binary tournament selection using ND_l and CD_l lexicographically. If, however, the subpopulation has a member or members which already exist in the archive, only these solutions are used in the binary tournament selection. This is done to emphasize already-found good solutions. The mutation operator is applied as usual. After the lower level NSGA-II simulation is performed on a subpopulation, the members are sorted according to the constrained non-domination level [Deb *et al.*(2002)] and are assigned their non-dominated rank (ND_l) and crowding distance value (CD_l) based on lower level objectives ($\mathbf{f}$) and lower level constraints ($\mathbf{g}$).

Step 3: Local search: The local search operator is employed next to provide us with a solution which is guaranteed to be on a locally Pareto-optimal front. Since the local search operator can be expensive, we use this operator sparingly. We apply the local search operator to good solutions having the following properties: (i) it is a non-dominated solution in the lower level having $ND_l = 1$, (ii) it is a non-dominated solution in the upper level having $ND_u = 1$, and (iii) it does not get dominated by any current archive member, or it is located at a distance less than $\delta_U N_l / N_l^{(0)}$ from any of the current archive members. In the local search procedure, the achievement scalarizing function problem [Wierzbicki(1980)] formulated at the current NSGA-II solution ($\mathbf{x}_l$) with $z_j = f_j(\mathbf{x}_l)$ is solved:

$$\begin{aligned} &\text{Minimize}_{\mathbf{p}} \max_{j=1}^{m} \frac{f_j(\mathbf{p})-z_j}{f_j^{\max}-f_j^{\min}} + \rho \textstyle\sum_{j=1}^{m} \frac{f_j(\mathbf{p})-z_j}{f_j^{\max}-f_j^{\min}}, \\ &\text{subject to } \mathbf{p} \in \mathscr{S}_l, \end{aligned} \tag{9.8}$$

where $\mathscr{S}_l$ is the feasible search space for the lower level problem. The minimum and maximum function values are taken from the NSGA-II minimum and maximum function values of the current generation. The optimal solution $\mathbf{p}^*$ to the above problem is guaranteed to be a Pareto-optimal solution to the lower level problem [Miettinen(1999)]. Here, we use $\rho = 10^{-6}$, which prohibits the local search to converge to a weak Pareto-optimal solution. We use a popular software KNITRO [Byrd *et al.*(2006)] (which employs a sequential quadratic programming (SQP) algorithm) to solve the above single objective optimization problem. The KNITRO software terminates when a solution satisfies the

Karush-Kuhn-Tucker (KKT) conditions [Reklaitis *et al.*(1983)] with a pre-specified error limit. We fix this error limit to 10^{-2} in all problems of this study here. The solutions which meet this KKT satisfaction criterion are assigned an 'optimal tag' for further processing. For handling non-differentiable problems, a non-gradient, adaptive step-size based hill-climbing procedure [Nolle(2006)] can be used.

Step 4: Updating the archive: The optimally tagged members, if feasible with respect to the upper level constraints (**G**), are then compared with the current archive members. If these members are non-dominated when compared to the members of the archive, they become eligible to be added into the archive. The dominated members in the archive are also eliminated, thus the archive always keeps non-dominated solutions. We limit the size of archive to $10N_u$. If and when more members are to be entered in the archive, the archive size is maintained to the above limit by eliminating extra members using the crowding distance (CD_a) measure.

Step 5: Formation of the combined population: Steps 1 to 4 are repeated until the population Q_T is filled with newly created solutions. Each member of Q_T is now evaluated with **F** and **G**. Populations P_T and Q_T are combined together to form R_T. The combined population R_T is then ranked according to constrained non-domination [Deb *et al.*(2002)] based on upper level objectives (**F**) and upper level constraints (**G**). Solutions are thus, assigned a non-dominated rank (ND_u) and members within an identical non-dominated rank are assigned a crowding distance (CD_u) computed in the **F**-space.

Step 7: Choosing half the population: From the combined population R_T of size $2N_u$, half of its members are retained in this step. First, the members of rank $ND_u = 1$ are considered. From them, solutions having $ND_l = 1$ are noted one by one in the order of reducing crowding distance CD_u. For each such solution, the entire N_l subpopulation from its source population (either P_T or Q_T) are copied in an intermediate population S_T. If a subpopulation is already copied in S_T and a future solution from the same subpopulation is found to have $ND_u = ND_l = 1$, the subpopulation is not copied again. When all members of $ND_u = 1$ are considered, a similar consideration is continued with $ND_u = 2$ and so on till S_T has N_u population members.

Step 6: Upgrading old lower level subpopulations: Each subpopulation of S_T which are not created in the current generation are modified using the lower level NSGA-II procedure (Step 2) applied with **f** and **g**. This step helps progress each lower level population towards their individual Pareto-optimal frontiers.

The final population is renamed as P_{T+1}. This marks the end of one generation of the overall H-BLEMO algorithm.

9.4.4 Algorithmic Complexity

With self-adaptive operations to update population sizes and number of generations, it becomes difficult to compute an exact number of function evaluations (FE) needed in the proposed H-BLEMO algorithm. However, using the maximum allowable values of these parameters, we estimate that the worst case function evaluations is $N_u(2T_u+1)(t_l^{\max}+1)+FE_{\text{LS}}$. Here T_u is the number of upper level generations and FE_{LS} is the total function evaluations used by the local search (LS) algorithm. Lower level NSGA-II is able to quickly bring the members close to the Pareto-optimal front and then local search operator is used on few members. Moreover, towards the end of a simulation, most upper level solutions are close to the archive, thereby requiring a much smaller number of function evaluations than that used in the above expression.

However, it is important to note that the computations needed in the local search may be substantial and any effort to reduce the computational effort will be useful. In this regard, the choice of the local search algorithm and the chosen KKT error limit for terminating the local search will play an important role. Also, the termination parameter for lower level NSGA-II run (parameter ε_l) may also make an important contribution.

9.5 Results on Test Problems

To evaluate the H-BLEMO solution methodology we use the DS test suite, and provide the results in this section. We use the following standard NSGA-II parameter values in both lower and upper levels on all problems considered in this section: Crossover probability of 0.9, distribution index for SBX of 15, mutation probability of 0.1, distribution index for the polynomial mutation of 20. The upper level population size is set proportional to the total number of variables (n): $N_u = 20n$. As described in the algorithm, all other parameters including the lower level population size, termination criteria are all set in a self-adaptive manner during the optimization run. The approximate Pareto-optimal fronts obtained from H-BLEMO procedure for

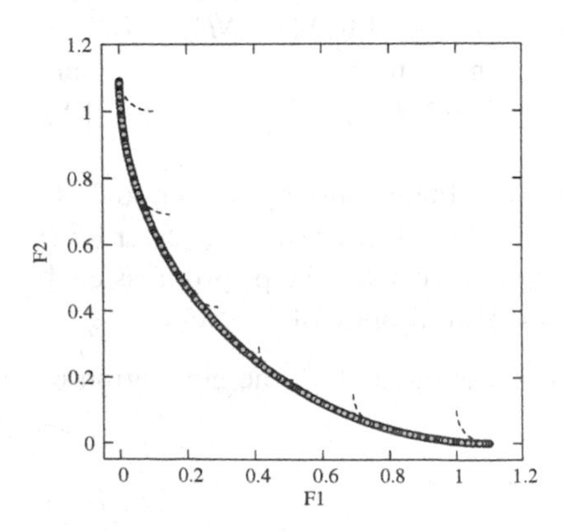

Fig. 9.6 Final archive solutions for problem DS1

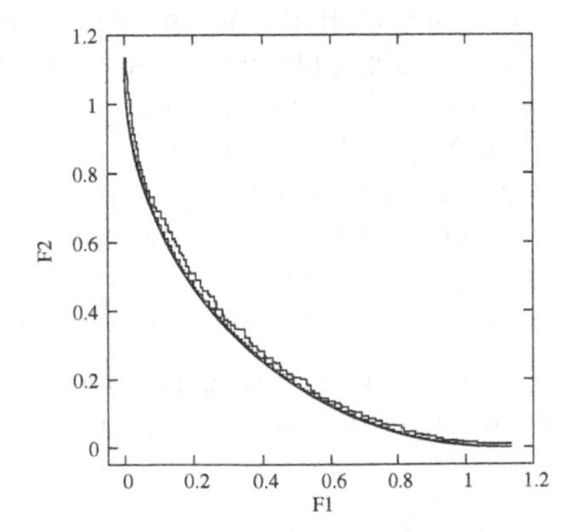

Fig. 9.7 Attainment surfaces (0%, 50% and 100%) for problem DS1 from 21 runs

each of the test problems are shown in Figures 9.6, 9.8, 9.10, 9.12 and 9.14. Each figure represents the approximation of the Pareto-frontier obtained from a particular run of the H-BLEMO procedure. Moreover, in all cases, we have used 21 different simulations starting from different initial populations and show the 0, 50, and 100% attainment surfaces [Fonseca and Fleming(1996)] in Figures 9.7, 9.9, 9.11, 9.13 and 9.15 to describe the robustness of the proposed procedure.

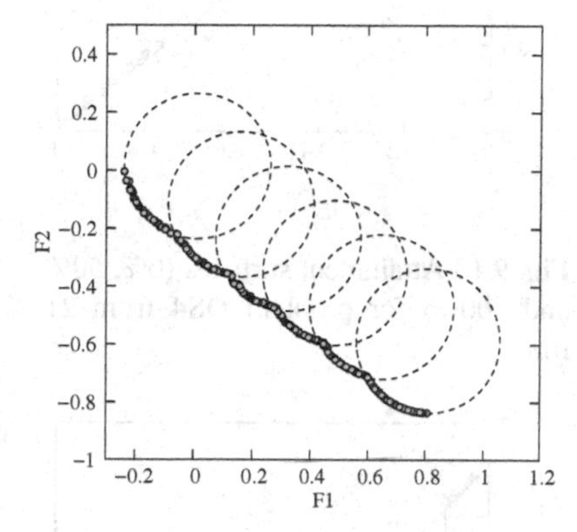

Fig. 9.8 Final archive solutions for problem DS2

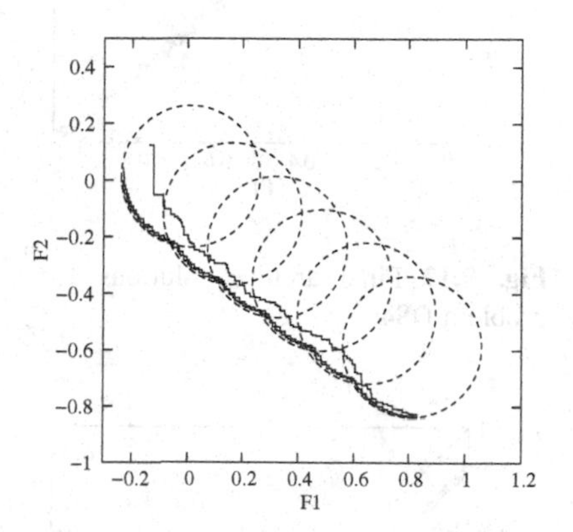

Fig. 9.9 Attainment surfaces (0%, 50%, 75% and 100%) for problem DS2 from 21 runs

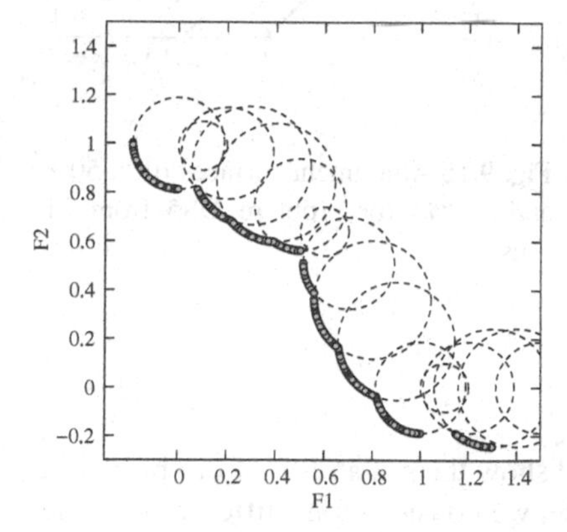

Fig. 9.10 Final archive solutions for problem DS3

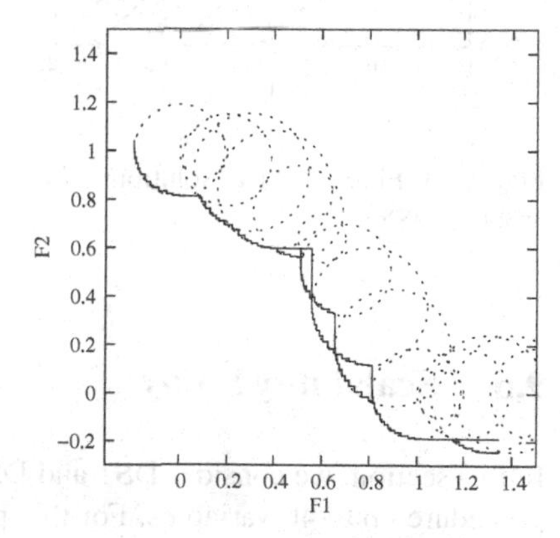

Fig. 9.11 Attainment surfaces (0%, 50% and 100%) for problem DS3 from 21 runs

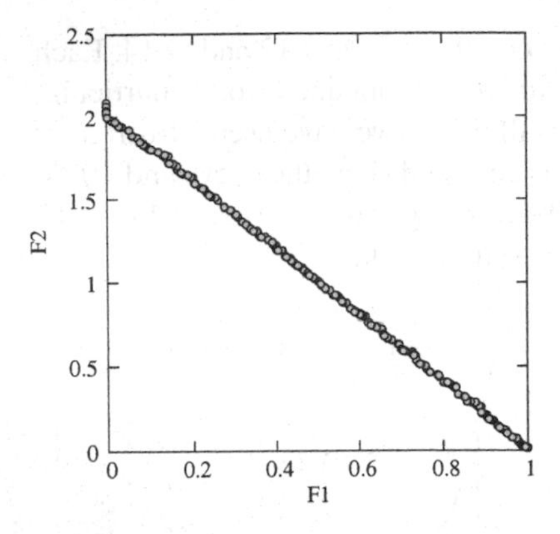

Fig. 9.12 Final archive solutions for problem DS4

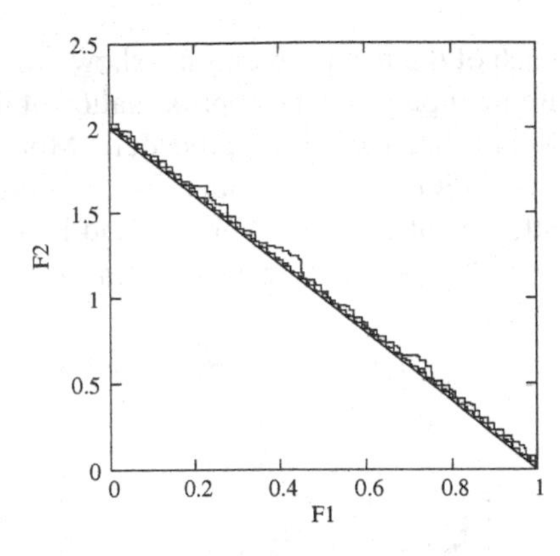

Fig. 9.13 Attainment surfaces (0%, 50% and 100%) for problem DS4 from 21 runs

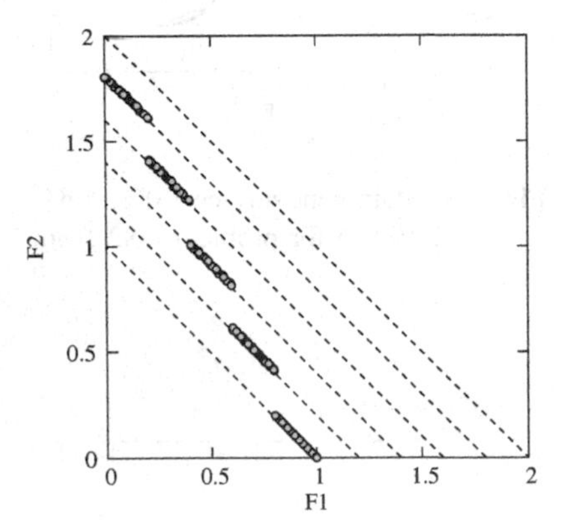

Fig. 9.14 Final archive solutions for problem DS5

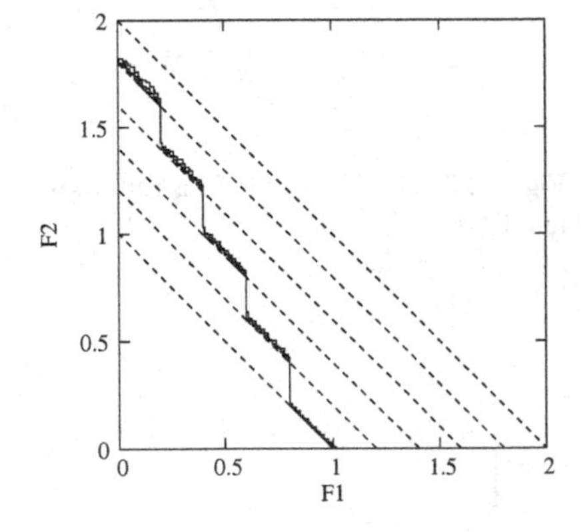

Fig. 9.15 Attainment surfaces (0%, 50% and 100%) for problem DS5 from 21 runs

9.6 Scalability Study

In this section, we consider DS1 and DS2 and show the scalability of our proposed procedure up to 40 variables. For this purpose, we consider four different variable sizes: $n = 10$, 20, 30 and 40. Based on parametric studies performed on these problems in section 9.5, we set $N_u = 20n$. All other parameters are automatically set in a self-adaptive manner during the course of a simulation, as before.

Figure 9.16 shows the variation of function evaluations for obtaining a fixed termination criterion on normalized hypervolume measure ($H_u < 0.0001$) calculated using the upper level objective values for problem DS1.

Since the vertical axis is plotted in a logarithmic scale and the relationship is found to be sub-linear, the hybrid methodology performs better than an exponential algorithm. The break-up of computations needed in the local search, lower level NSGA-II and upper level NSGA-II indicate that majority of the computations is spent in the lower level optimization task. This is an important insight to the working of the proposed H-BLEMO algorithm and suggests that further efforts must be put in making the lower level optimization more computationally efficient.

Figure 9.17 shows the similar outcome for problem DS2, but a comparison with that for problem DS1 indicates that DS2 is more difficult to be solved with an increase in problem size than DS1.

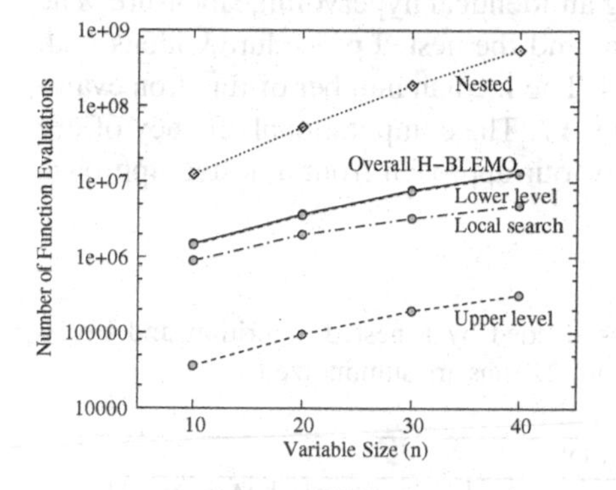

Fig. 9.16 Variation of function evaluations with problem size n for DS1

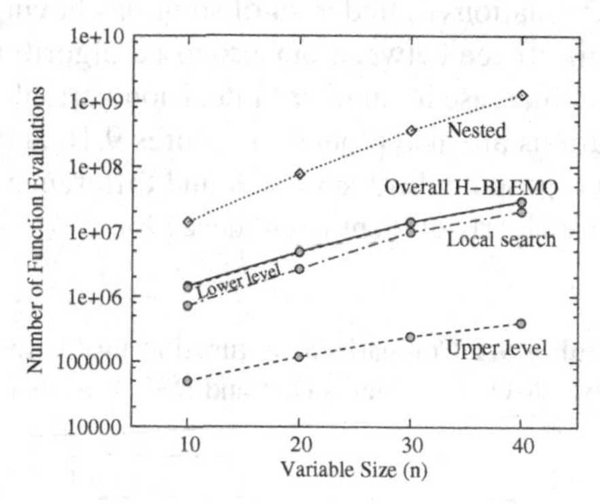

Fig. 9.17 Variation of function evaluations with problem size n for DS2

9.7 Comparison with a Nested Algorithm

We have argued before that by allowing lower level and upper level NSGA-IIs to proceed partially in tandem, we have created a computationally efficient and accurate algorithm which progresses towards the true Pareto-optimal front on a number of difficult problems (Section 9.5). The algorithm is even found to converge in problems in which there is a conflict between upper and lower level problems [Deb and Sinha(2010)]. The proposed algorithm is also found to solve scaled-up problems up to 40 real-parameter variables (Section 9.6). In this section, we compare the proposed H-BLEMO algorithm with an efficient yet nested bilevel optimization algorithm using the NSGA-II-cum-local-search procedure. This algorithm uses a fixed population structure, but for every $\mathbf{x}_u$, the lower level optimization is terminated by performing a local search to all non-dominated solutions of the final lower level NSGA-II population. The termination criterion for lower and upper level NSGA-IIs and that for the local search procedure are identical to that in H-BLEMO algorithm. Since for every $\mathbf{x}_u$, we find a set of well-converged and well-distributed lower level Pareto-optimal solutions, this approach is truly a nested bilevel optimization procedure.

For the simulation with this nested algorithm on DS1 and DS2 problems, we use $N_u = 400$. To make a fair comparison, we use the same subpopulation size N_l as that was used in the very first iteration of our H-BLEMO algorithm using equation (9.4). The number of generations for the lower level NSGA-II is kept fixed for all upper level generations to that computed by equation (9.3) in the initial generation. Similar archiving strategy and other NSGA-II parameter values are used as before. Table 9.1 shows the comparison of overall function evaluations needed by the nested algorithm and by the hybrid BLEMO algorithm. The table shows that for both problems, the nested algorithm takes at least one order of magnitude of more function evaluations to find a set of solutions having an identical hypervolume measure. The difference between our proposed algorithm and the nested procedure widens with an increase in number of decision variables. The median number of function evaluations are also plotted in Figures 9.16 and 9.17. The computational efficacy of our proposed hybrid approach and difference of our approach from a nested approach are clearly evident from these plots.

Table 9.1 Comparison of function evaluations needed by a nested algorithm and by H-BLEMO on problems DS1 and DS2. Results from 21 runs are summarized.

Problem DS1						
n	Algo.	Median			Min. overall FE	Max. overall FE
		Lower FE	Upper FE	Overall FE		
10	Nested	12,124,083	354,114	12,478,197	11,733,871	14,547,725
10	Hybrid	1,454,194	36,315	1,490,509	1,437,038	1,535,329
20	Nested	51,142,994	1,349,335	52,492,329	42,291,810	62,525,401
20	Hybrid	3,612,711	94,409	3,707,120	2,907,352	3,937,471
30	Nested	182,881,535	4,727,534	187,609,069	184,128,609	218,164,646
30	Hybrid	7,527,677	194,324	7,722,001	6,458,856	8,726,543
40	Nested	538,064,283	13,397,967	551,462,250	445,897,063	587,385,335
40	Hybrid	12,744,092	313,861	13,057,953	10,666,017	15,146,652

Problem DS2						
n	Algo.	Median			Min. overall FE	Max. overall FE
		Lower FE	Upper FE	Overall FE		
10	Nested	13,408,837	473,208	13,882,045	11,952,650	15,550,144
10	Hybrid	1,386,258	50,122	1,436,380	1,152,015	1,655,821
20	Nested	74,016,721	1,780,882	75,797,603	71,988,726	90,575,216
20	Hybrid	4,716,205	117,632	4,833,837	4,590,019	5,605,740
30	Nested	349,242,956	5,973,849	355,216,805	316,279,784	391,648,693
30	Hybrid	13,770,098	241,474	14,011,572	14,000,057	15,385,316
40	Nested	1,248,848,767	17,046,212	1,265,894,979	1,102,945,724	1,366,734,137
40	Hybrid	28,870,856	399,316	29,270,172	24,725,683	30,135,983

9.8 Incorporating Decision Maker Preferences in H-BLEMO

From the results presented in the previous sections, we observe that the H-BLEMO procedure is able to perform significantly better than a nested approach. However, the number of function evaluations required by the H-BLEMO approach are still high, and we seek to further reduce the evaluations. In order to achieve this, we make a shift from an a posteriori approach to a progressively interactive approach, where the upper level decision maker interacts with the H-BLEMO algorithm, and a single point on the Pareto-optimal front is the target. Concepts from a Progressively Interactive Evolutionary Multi-objective Optimization algorithm (PI-EMO-VF) [Deb *et al.*(2010)] have been integrated with the Hybrid Bilevel Evolutionary Multi-objective Optimization algorithm (H-BLEMO). In the suggested methodology, preference information from the decision maker at the upper level is used to direct the search towards the most preferred solution. Incorporating preferences from the decision maker in the optimization run makes the search process more efficient in terms of function evaluations as well as accuracy. The integrated methodology, interacts with the decision maker after every few generations of an evolutionary algorithm and is different from an a posteriori approach, as it explores only the most preferred point. An a posteriori approach like the H-BLEMO and other evolutionary multi-objective optimization algorithms [Deb *et al.*(2002), Zitzler *et al.*(2001)] produce the entire efficient frontier as the final solution and then a decision maker is asked to pick up the most preferred point. However, an a posteriori approach is not a viable methodology for problems which are computationally expensive and/or involve high number of objectives (more than three) where EMOs tend to suffer in convergence as well as maintaining diversity. Figure 9.18 shows the interaction of the upper level decision maker at various stages of a progressively interactive algo-

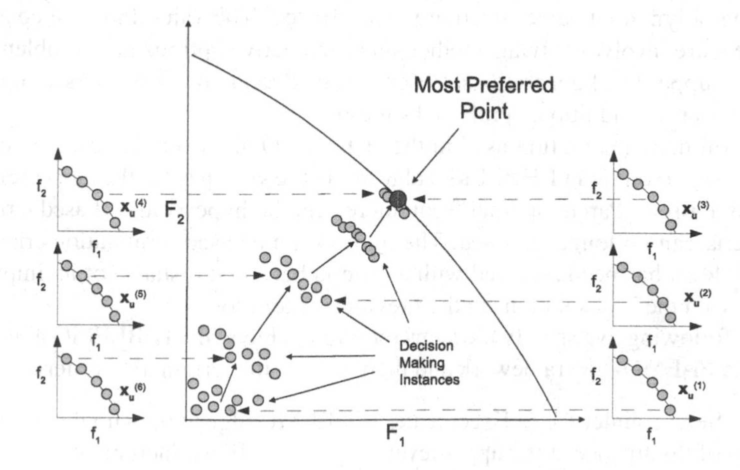

Fig. 9.18 Progressive interaction with the decision maker at the upper level of multi-objective bilevel problem. Upper level frontier and lower level frontiers for different $\mathbf{x}_u$ are shown for a maximization problem at both levels.

rithm. Each broken line shows a Pareto-optimal point in the lower level objective space for a particular $\mathbf{x}_u$, and the same point in the upper level objective space.

The integration procedure, Progressively Interactive Hybrid Bilevel Evolutionary Multi-objective Optimization (PI-HBLEMO) algorithm, has been discussed in the next section, and thereafter, the performance of the algorithm has been evaluated on the DS test problems [Deb and Sinha(2010)]. A comparison for the savings in computational cost has also been done with a posteriori H-BLEMO approach.

9.9 Progressively Interactive Hybrid Bilevel Evolutionary Multi-objective Optimization Algorithm (PI-HBLEMO)

In this section, the changes made to the Hybrid Bilevel Evolutionary Multi-objective Optimization (H-BLEMO) [Deb and Sinha(2010)] algorithm have been stated. The major change made to the H-BLEMO algorithm is in the domination criteria. The other change which has been made is in the termination criteria. The Progressively Interactive EMO using Value Function (PI-EMO-VF) [Deb *et al.*(2010)] is a generic procedure which can be integrated with any Evolutionary Multi-objective Optimization (EMO) algorithm. Here, we integrate the procedure at the upper level execution of the H-BLEMO algorithm.

After every χ upper level generations of the H-BLEMO algorithm, the decision-maker is provided with η (≥ 2) well-sparse non-dominated solutions from the upper level set of non-dominated points. The decision-maker is expected to provide a complete or partial preference information about superiority of one solution over the other, or indifference towards the two solutions. In an ideal situation, the DM can provide a complete ranking (from best to worst) of these solutions, but partial preference information is also allowed. With the given preference information, a strictly increasing polynomial value function is constructed. The value function construction procedure involves solving another single-objective optimization problem. Till the next χ upper level generations, the constructed value function is used to direct the search towards additional preferred solutions.

The termination condition used in the H-BLEMO algorithm is based on hypervolume. In the modified PI-HBLEMO algorithm, the search is for the most preferred point and not for a Pareto optimal front, therefore, the hypervolume based termination criteria can no longer be used. The hypervolume based termination criteria at the upper level has been replaced with a criteria based on distance of an improved solution from the best solutions in the previous generations.

In the following, we specify the steps required to blend the H-BLEMO algorithm within the PI-EMO-VF framework and then discuss the termination criteria.

Step 1: Set a counter $t = 0$. Execute the H-BLEMO algorithm with the usual definition of dominance at the upper level for χ generations. Increment the value of t by one after each generation.

Step 2: If t is perfectly divisible by χ, then use the k-mean clustering approach [Deb(2001), Zitzler *et al.*(2001)] to choose η diversified points from the non-dominated solutions in the archive; otherwise, proceed to Step 5.
Step 3: Elicit the preferences of the decision-maker on the chosen η points. Construct a value function $V(\mathbf{f})$, emulating the decision maker preferences, from the information. The value function is constructed by solving an optimization problem (VFOP), described in Section 9.9.1. If a feasible value function is not found which satisfies all DM's preferences then proceed to Step 5 and use the usual domination principle in H-BLEMO operators.
Step 4: Check for termination. The termination check (described in Section 9.9.2) is based on the distance of the current best solution from the previous best solutions and requires a parameter Δ_u. If the algorithm terminates, the current best point is chosen as the final solution.
Step 5: An offspring population at the upper level is produced from the parent population at the upper level using a modified domination principle (discussed in Section 9.9.3) and H-BLEMO algorithm's search operators.
Step 6: The parent and the offspring populations are used to create a new parent population for the next generation using the modified domination based on the current value function and other H-BLEMO algorithm's operators. The iteration counter is incremented as $t \leftarrow t+1$ and the algorithm proceeds to Step 2.

The parameters used in the PI-HBLEMO algorithm are χ, η and Δ_u.

9.9.1 Step 3: Preference Elicitation and Construction of a Value Function

Whenever a DM call is made, a set of η points are presented to the decision maker (DM). The preference information from the decision maker is accepted in the form of pairwise comparisons for each pair in the set of η points. A pairwise comparison of a give pair could lead to three possibilities, the first being that one solution is preferred over the other, the second being that the decision maker is indifferent to both the solutions and the third being that the two solutions are incomparable. Based on such preference information from a decision maker, for a given pair (i,j), if i-th point is preferred over j-th point, then $P_i \succ P_j$, if the decision maker is indifferent to the two solutions then it establishes that $P_i \equiv P_j$. There can be situations such that the decision maker finds a given pair of points as incomparable and in such a case the incomparable points are dropped from the list of η points. If the decision maker is not able to provide preference information for any of the given solution points then algorithm moves back to the previous population where the decision maker was able to take a decisive action, and uses the usual domination instead of modified domination principle to proceed the search process. But such a scenario where no preference is established by a decision maker is rare, and it is likely to have at least one point which is better than another point. Once preference information is available, the task is to construct a polynomial value function which satisfies the preference statements of the decision maker.

9.9.1.1 Polynomial Value Function for Two Objectives

A polynomial value function is constructed based on the preference information provided by the decision maker. The parameters of the polynomial value function are optimally adjusted such that the preference statements of the decision maker are satisfied. We describe the procedure for two objectives as all the test problems considered in this chapter have two objectives. The value function procedure described below is valid for a maximization problem therefore we convert the test problems into a maximization problem while implementing the value function procedure. However, while reporting the results for the test problems they are converted back to minimization problems.

$$\begin{aligned} &V(F_1,F_2) = (F_1 + k_1F_2 + l_1)(F_2 + k_2F_1 + l_2), \\ &\text{where } F_1,F_2 \quad \text{are the objective values} \\ &\text{and} \quad k_1,k_2,l_1,l_2 \text{ are the value function parameters} \end{aligned} \tag{9.9}$$

The description of the two objective value function has been taken from [Deb *et al.*(2010)]. In the above equations it can been seen that the value function V, for two objectives, is represented as a product of two linear functions $S_1 : \mathbb{R}^2 \rightarrow \mathbb{R}$ and $S_2 : \mathbb{R}^2 \rightarrow \mathbb{R}$. [1] The parameters in this value function which are required to be determined optimally from the preference statements of the decision maker are k_1, k_2, l_1 and l_2. Following is the value function optimization problem (VFOP) which should be solved with the value function parameters (k_1, k_2, l_1 and l_2) as variables. The optimal solution to the VFOP assigns optimal values to the value function parameters. The above problem is a simple single objective optimization problem which can be solved using any single objective optimizer. Here, the problem has been solved using a sequential quadratic programming (SQP) procedure from the KNITRO [Byrd *et al.*(2006)] software.

$$\begin{aligned} &\text{Maximize } \varepsilon, \\ &\text{subject to } V \text{ is non-negative at every point } P_i, \\ &\qquad V \text{ is strictly increasing at every point } P_i, \\ &\qquad V(P_i) - V(P_j) \geq \varepsilon, \quad \text{for all } (i,j) \text{ pairs} \\ &\qquad\qquad \text{satisfying } P_i \succ P_j, \\ &\qquad \left|V(P_i) - V(P_j)\right| \leq \delta_V, \quad \text{for all } (i,j) \text{ pairs} \\ &\qquad\qquad \text{satisfying } P_i \equiv P_j. \end{aligned} \tag{9.10}$$

The above optimization problem adjusts the value function parameters in such a way that the minimum difference in the value function values for the ordered pairs of points is maximum.

[1] A generalized version of the polynomial value function can be found in [Sinha *et al.*(2010)].

9.9.2 Termination Criterion

Distance of the current best point is computed from the best points in the previous generations. In the simulations performed, the distance is computed from the current best point to the best points in the previous 10 generations and if each of the computed distances $\delta_u(i), i \in \{1,2,\ldots,10\}$ is found to be less than Δ_u then the algorithm is terminated. A value of $\Delta_u = 0.1$ has been chosen for the simulations.

9.9.3 Modified Domination Principle

In this sub-section we define the modified domination principle proposed in [Deb *et al.*(2010)]. The value function V is used to modify the usual domination principle so that more focussed search can be performed in the region of interest to the decision maker. Let $V(F_1,F_2)$ be the value function for a two objective case. The parameters for this value function are optimally determined from the VFOP. For the given η points, the value function assigns a value to each point. Let the values be $V_1, V_2, \ldots, V_\eta$ in the descending order. Now any two feasible solutions ($\mathbf{x}^{(1)}$ and $\mathbf{x}^{(2)}$) can be compared with their objective function values by using the following modified domination criteria:

1. If both points have a value function value *less* than V_2, then the two points are compared based on the usual dominance principle.
2. If both points have a value function value *more* than V_2, then the two points are compared based on the usual dominance principle.
3. If one point has a value function value more than V_2 and the other point has a value function value less than V_2, then the former dominates the latter.

The modified domination principle has been explained through Figure 9.19 which illustrates regions dominated by two points A and B. Let us consider that the second best point from a given set of η points has a value V_2. The function $V(F) = V_2$ represents a contour which has been shown by a curved line [2]. The first point A has a value V_A which is smaller than V_2 and the region dominated by A is shaded in the figure. The region dominated by A is identical to what can be obtained using the usual domination principle. The second point B has a value V_B which is larger than V_2, and, the region dominated by this point is once again shaded. It can be observed that this point no longer follows the usual domination principle. In addition to usual region of dominance this point dominates all the points having a smaller value function value than V_2.

The above modified domination principle can easily be extended for handling constraints as in [Deb *et al.*(2002)]. When two solutions under consideration for a dominance check are feasible, then the above modified domination principle should be used. If one solution is feasible and the other is infeasible, then the feasible

[2] The reason for using the contour corresponding to the second best point can be found in [Deb *et al.*(2010)]

solution is considered as dominating the other. If both the solutions are found to be infeasible then the one with smaller overall feasibility violation (sum of all constraint violations) is considered to be dominating the other solution.

9.10 Results

In this section, results have been presented on the set of 5 DS test problems. In all simulations, the crossover and mutation parameters of the NSGA-II algorithm are kept same as before. Population size $N_u = 40$, number of points given to the DM for preference information $\eta = 5$, and number of generations between two consecutive DM calls: $\chi = 5$ has been used. A point, $(F_1^{(b)}, F_2^{(b)})$, on the Pareto-front of the upper level is assumed as the most preferred point and then a DM emulated value function is selected which assigns a maximum value to the most preferred point. The value function selected is $V(F_1, F_2) = \frac{1}{1+(F_1-F_1^{(b)})^2+(F_2-F_2^{(b)})^2}$. It is noteworthy that the value function selected to emulate a decision maker is a simple distance function and therefore has circles as indifference curves which is not a true representative of a rational decision maker. A circular indifference curve may lead to assignment of equal values to a pair of points where one dominates the other. For a pair of points it may also lead assignment of higher value to a point dominated by the other. However, only non-dominated set of points are presented to a decision maker, therefore, such discrepancies are avoided and the chosen value function is able to emulate a decision maker by assigning higher value to the point closest to the most preferred point and lower value to others.

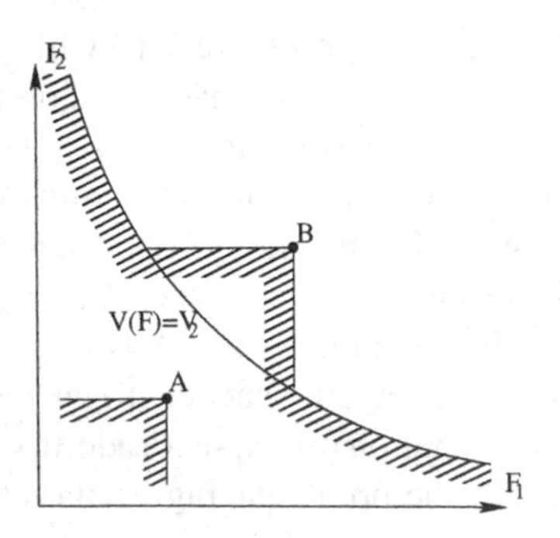

Fig. 9.19 Dominated regions in case of two points A and B using the modified definition

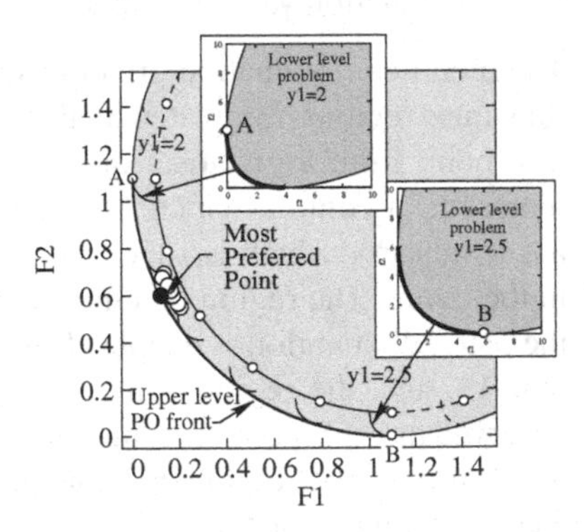

Fig. 9.20 Pareto-optimal front for problem DS1. Final parent population members have been shown close to the most preferred point

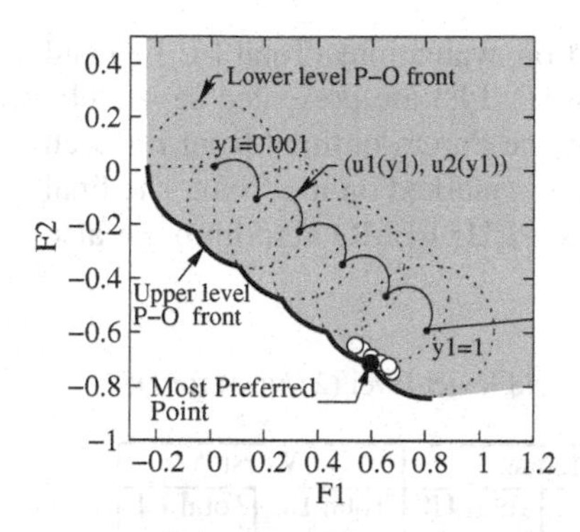

Fig. 9.21 Pareto-optimal front for problem DS2. Final parent population members have been shown close to the most preferred point.

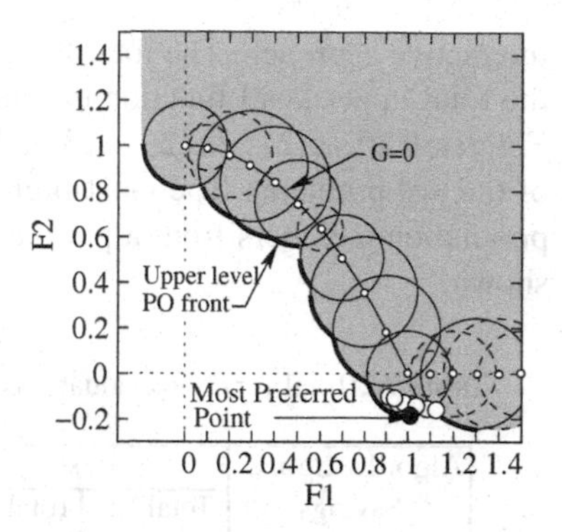

Fig. 9.22 Pareto-optimal front for problem DS3. Final parent population members have been shown close to the most preferred point.

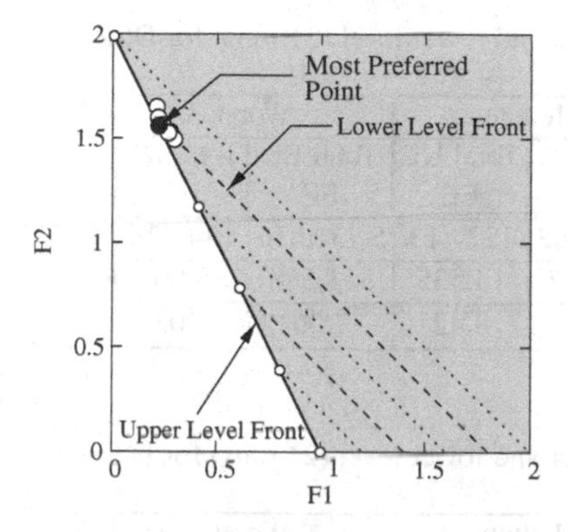

Fig. 9.23 Pareto-optimal front for problem DS4. Final parent population members have been shown close to the most preferred point.

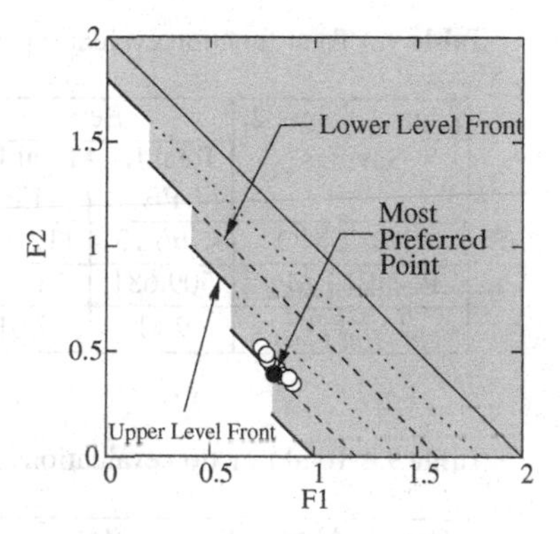

Fig. 9.24 Pareto-optimal front for problem DS5. Final parent population members have been shown close to the most preferred point.

The DS test problems are minimization problems and the progressively interactive procedure using value function works only on problems to be maximized. Therefore, the procedure has been executed by converting the test problems into a maximization problem by putting a negative sign before each of the objectives. However, the final results have once again been converted and the solution to the minimization problem has been presented. The upper level and lower level function evaluations have been reported for each of the test problems. A comparison has been made between the H-BLEMO algorithm and PI-HBLEMO procedure in the tables 9.2, 9.3, 9.4, 9.5 and 9.6. The tables show the savings in function evaluations which could be achieved moving from an a posteriori approach to a progressively interactive approach. The total lower level function evaluations (Total LL FE) and the total upper level function evaluations (Total UL FE) are presented separately. Figures 9.20, 9.21, 9.22, 9.23 and 9.24 show the Pareto-optimal front for each of the test problems. The most-preferred solution is marked on the front. The final population members from a particular run of the PI-HBLEMO algorithm are also shown.

Table 9.2 Total function evaluations for the upper and lower level (21 runs) for DS1

Algo. 1, Algo. 2, Savings	Best		Median		Worst	
	Total LL FE	Total UL FE	Total LL FE	Total UL FE	Total LL FE	Total UL FE
H-BLEMO	2,819,770	87,582	3,423,544	91,852	3,829,812	107,659
PI-HBLEMO	329,412	12,509	383,720	12,791	430,273	10,907
$\frac{H-BLEMO}{PI-HBLEMO}$	8.56	7.00	8.92	7.18	8.90	9.87

Table 9.3 Total function evaluations for the upper and lower level (21 runs) for DS2

Algo. 1, Algo. 2, Savings	Best		Median		Worst	
	Total LL FE	Total UL FE	Total LL FE	Total UL FE	Total LL FE	Total UL FE
H-BLEMO	4,796,131	112,563	4,958,593	122,413	5,731,016	144,428
PI-HBLEMO	509,681	14,785	640,857	14,535	811,588	15,967
$\frac{H-BLEMO}{PI-HBLEMO}$	9.41	7.61	7.74	8.42	7.06	9.05

Table 9.4 Total function evaluations for the upper and lower level (21 runs) for DS3

Algo. 1, Algo. 2, Savings	Best		Median		Worst	
	Total LL FE	Total UL FE	Total LL FE	Total UL FE	Total LL FE	Total UL FE
H-BLEMO	3,970,411	112,560	4,725,596	118,848	5,265,074	125,438
PI-HBLEMO	475,600	11,412	595,609	16,693	759,040	16,637
$\frac{H-BLEMO}{PI-HBLEMO}$	8.35	9.86	7.93	7.12	6.94	7.54

9.11 Accuracy and DM calls

Table 9.7 represents the accuracy achieved and the number of decision maker calls required while using the PI-HBLEMO procedure. In the above test problems the most preferred point which the algorithm is seeking is pre-decided and the value function emulating the decision maker is constructed. When the algorithm terminates it provides the best achieved point. The accuracy measure is the Euclidean distance between the best point achieved and the most preferred point. It can be observed from the results of the PI-HBLEMO procedure that preference information

Table 9.5 Total function evaluations for the upper and lower level (21 runs) for DS4

Algo. 1, Algo. 2, Savings	Best		Median		Worst	
	Total LL FE	Total UL FE	Total LL FE	Total UL FE	Total LL FE	Total UL FE
H-BLEMO	1,356,598	38,127	1,435,344	53,548	1,675,422	59,047
PI-HBLEMO	149,214	5,038	161,463	8,123	199,880	8,712
$\frac{H-BLEMO}{PI-HBLEMO}$	9.09	7.57	8.89	6.59	8.38	6.78

Table 9.6 Total function evaluations for the upper and lower level (21 runs) for DS5

Algo. 1, Algo. 2, Savings	Best		Median		Worst	
	Total LL FE	Total UL FE	Total LL FE	Total UL FE	Total LL FE	Total UL FE
H-BLEMO	1,666,953	47,127	1,791,511	56,725	2,197,470	71,246
PI-HBLEMO	168,670	5,105	279,568	6,269	304,243	9,114
$\frac{H-BLEMO}{PI-HBLEMO}$	9.88	9.23	6.41	9.05	7.22	7.82

Table 9.7 Accuracy and the number of decision maker calls for the PI-HBLEMO runs (21 runs). The distance of the closest point to the most preferred point achieved from the H-BLEMO algorithm has been provided in the brackets.

		Best	Median	Worst
DS1	Accuracy	0.0426 (0.1203)	0.0888 (0.2788)	0.1188 (0.4162)
	DM Calls	12	13	29
DS2	Accuracy	0.0281 (0.0729)	0.0804 (0.4289)	0.1405 (0.7997)
	DM Calls	12	15	25
DS3	Accuracy	0.0498 (0.0968)	0.0918 (0.3169)	0.1789 (0.6609)
	DM Calls	7	17	22
DS4	Accuracy	0.0282 (0.0621)	0.0968 (0.0981)	0.1992 (0.5667)
	DM Calls	7	15	23
DS5	Accuracy	0.0233 (0.1023)	0.0994 (0.1877)	0.1946 (0.8946)
	DM Calls	7	14	22

from the decision maker leads to a high accuracy (Table 9.7) as well as huge savings (Table 9.2,9.3,9.4,9.5,9.6) in function evaluations. Producing the entire front using the H-BLEMO procedure has its own merits but it comes with a cost of huge function evaluations and there can be instances when the entire set of close Pareto-optimal solutions will be difficult to achieve even after high number of evaluations. The accuracy achieved using the H-BLEMO procedure has been reported in the brackets; the final choice made from a set of close Pareto-optimal solutions will lead to a poorer accuracy than a progressively interactive approach.

9.12 Conclusions

In the realm of research in optimization, multi-objective bilevel problems provide significant challenges to researchers by elevating every aspect of an optimization effort at a higher level. This makes the problem demanding as well as intriguing to pursue. Extensive attempts have been made to develop solution methodologies for bilevel problems, however, the instance of multiple objectives at both levels has received less attention from researchers. This chapter provides an insight into multi-objective bilevel programming and highlights some of the past efforts made towards handling these problems. It discusses a generic procedure for test problem development by featuring various intricacies which may exist in a practical multi-objective bilevel problem. Further, the chapter describes a couple of recently proposed solution methodologies for multi-objective bilevel problems and evaluates its performance on the DS test suite.

The first methodology (H-BLEMO) is a hybrid evolutionary-cum-local-search algorithm, where the evolutionary part contributes towards handling the intertwined and multi-solution aspect of the problem and the local search part contributes towards ensuring lower level optimality. The algorithm is also self adaptive, allowing an automatic update of the key parameters during an optimization run without any user intervention. The procedure approximates the entire Pareto-optimal front for the bilevel problem allowing the decision maker to choose the most preferred point a posteriori. Simulation results on the test suite amply demonstrate the effectiveness of evolutionary algorithms in solving such complex problems. The second methodology (PI-HBLEMO) extends the first methodology by incorporating preference information from the decision maker at the upper level. It makes the procedure interactive and produces the most preferred point as the final solution. The methodology showcases the power of amalgamating evolutionary algorithm's parallel search with principles from the field of multi-criteria decision making. Incorporating preferences from the decision maker in the intermediate steps of the algorithm offers a dual advantage of reduced computational expense and an enhanced accuracy.

To conclude, further investigation in the direction of bilevel programming is required urgently. Theoretical research in bilevel optimization has led to the formulation of optimality conditions, however, viable methodologies to implement them in practice are challenging. There is still a lack of a generic and efficient procedure which guarantees convergence for such problems. Although a nested procedure could be used to tackle the problem, the enormous computational expense required would render the procedure redundant for problems with high number of variables. Under such circumstances, developing bilevel optimization algorithms by hybridizing principles from evolutionary and classical research provide immense implementation opportunities.

Acknowledgements. Large parts of the chapter, including text, figures and tables, have been taken from the articles [Deb and Sinha(2010)] and [Sinha(2011)]. Authors acknowledge the support from the Academy of Finland under research project number 133387.

References

[Abass(2005)] Abass, S.A.: Bilevel programming approach applied to the flow shop scheduling problem under fuzziness. Computational Management Science 4(4), 279–293 (2005)

[Alexandrov and Dennis(1994)] Alexandrov, N., Dennis, J.E.: Algorithms for bilevel optimization. In: AIAA/NASA/USAF/ISSMO Symposium on Multidisciplinary Analyis and Optimization, pp. 810–816 (1994)

[Bard(1998)] Bard, J.F.: Practical Bilevel Optimization: Algorithms and Applications. Kluwer, The Netherlands (1998)

[Bianco *et al.*(2009)] Bianco, L., Caramia, M., Giordani, S.: A bilevel flow model for hazmat transportation network design. Transportation Research. Part C: Emerging technologies 17(2), 175–196 (2009)

[Byrd *et al.*(2006)] Byrd, R.H., Nocedal, J., Waltz, R.A.: KNITRO: An integrated package for nonlinear optimization, pp. 35–59. Springer (2006)

[Calamai and Vicente(1994)] Calamai, P.H., Vicente, L.N.: Generating quadratic bilevel programming test problems. ACM Trans. Math. Software 20(1), 103–119 (1994)

[Colson *et al.*(2007)] Colson, B., Marcotte, P., Savard, G.: An overview of bilevel optimization. Annals of Operational Research 153, 235–256 (2007)

[Deb(2001)] Deb, K.: Multi-objective optimization using evolutionary algorithms. Wiley, Chichester (2001)

[Deb and Agrawal(1995)] Deb, K., Agrawal, R.B.: Simulated binary crossover for continuous search space. Complex Systems 9(2), 115–148 (1995)

[Deb *et al.*(2010)] Deb, K., Sinha, A., Korhonen, P., Wallenius, J.: An interactive evolutionary multi-objective optimization method based on progressively approximated value functions. IEEE Transactions on Evolutionary Computation 14(5), 723–739 (2010)

[Deb and Sinha(2010)] Deb, K., Sinha, A.: An efficient and accurate solution methodology for bilevel multi-objective programming problems using a hybrid evolutionary-local-search algorithm. Evolutionary Computation Journal 18(3), 403–449 (2010)

[Deb and Sinha(2009a)] Deb, K., Sinha, A.: Constructing test problems for bilevel evolutionary multi-objective optimization. In: Proceedings of the Congress on Evolutionary Computation (CEC 2009). IEEE Press, Piscataway (2009a); Also KanGAL Report No. 2008010

[Deb and Sinha(2009b)] Deb, K., Sinha, A.: Solving bilevel multi-objective optimization problems using evolutionary algorithms. In: Ehrgott, M., Fonseca, C.M., Gandibleux, X., Hao, J.-K., Sevaux, M. (eds.) EMO 2009. LNCS, vol. 5467, pp. 110–124. Springer, Heidelberg (2009b)

[Deb *et al.*(2002)] Deb, K., Agrawal, S., Pratap, A., Meyarivan, T.: A fast and elitist multi-objective genetic algorithm: NSGA-II. IEEE Transactions on Evolutionary Computation 6(2), 182–197 (2002)

[Deb *et al.*(2005)] Deb, K., Thiele, L., Laumanns, M., Zitzler, E.: Scalable test problems for evolutionary multi-objective optimization. In: Abraham, A., Jain, L., Goldberg, R. (eds.) Evolutionary Multiobjective Optimization, pp. 105–145. Springer, London (2005)

[Dempe(2002)] Dempe, S.: Foundations of bilevel programming. Kluwer, Dordrecht (2002)

[Dempe *et al.*(2006)] Dempe, S., Dutta, J., Lohse, S.: Optimality conditions for bilevel programming problems. Optimization 55(5-6), 505–524 (2006)

[Dimitriou *et al.*(2008)] Dimitriou, L., Tsekeris, T., Stathopoulos, A.: Genetic computation of road network design and pricing stackelberg games with multi-class users. In: Giacobini, M., Brabazon, A., Cagnoni, S., Di Caro, G.A., Drechsler, R., Ekárt, A., Esparcia-Alcázar, A.I., Farooq, M., Fink, A., McCormack, J., O'Neill, M., Romero, J., Rothlauf, F., Squillero, G., Uyar, A.Ş., Yang, S. (eds.) EvoWorkshops 2008. LNCS, vol. 4974, pp. 669–678. Springer, Heidelberg (2008)

[Eichfelder(2007)] Eichfelder, G.: Solving nonlinear multiobjective bilevel optimization problems with coupled upper level constraints. Technical Report Preprint No.320, Preprint-Series of the Institute of Applied Mathematics, Univ. Erlangen-Nürnberg, Germany (2007)

[Eichfelder(2008)] Eichfelder, G.: Multiobjective bilevel optimization. Mathematical Programming (2008), doi: 10.1007/s10107-008-0259-0

[Fampa *et al.*(2008)] Fampa, M., Barroso, L.A., Candal, D., Simonetti, L.: Bilevel optimization applied to strategic pricing in competitive electricity markets. Comput. Optim. Appl. 39, 121–142 (2008)

[Fliege and Vicente(2006)] Fliege, J., Vicente, L.N.: Multicriteria approach to bilevel optimization. Journal of Optimization Theory and Applications 131(2), 209–225 (2006)

[Fonseca and Fleming(1996)] Fonseca, C.M., Fleming, P.J.: On the performance assessment and comparison of stochastic multiobjective optimizers. In: Ebeling, W., Rechenberg, I., Voigt, H.-M., Schwefel, H.-P. (eds.) PPSN 1996. LNCS, vol. 1141, pp. 584–593. Springer, Heidelberg (1996)

[Fudenberg and Tirole(1993)] Fudenberg, D., Tirole, J.: Game theory. MIT Press (1993)

[Gaur and Arora(2008)] Gaur, A., Arora, S.R.: Multi-level multi-attribute multi-objective integer linear programming problem. AMO-Advanced Modeling and Optimization 10(2), 297–322 (2008)

[Halter and Mostaghim(2006)] Halter, W., Mostaghim, S.: Bilevel optimization of multi-component chemical systems using particle swarm optimization. In: Proceedings of World Congress on Computational Intelligence (WCCI 2006), pp. 1240–1247 (2006)

[Hecheng and Wang(2007)] Li, H., Wang, Y.: A genetic algorithm for solving a special class of nonlinear bilevel programming problems. In: Shi, Y., van Albada, G.D., Dongarra, J., Sloot, P.M.A. (eds.) ICCS 2007, Part IV. LNCS, vol. 4490, pp. 1159–1162. Springer, Heidelberg (2007)

[Herskovits *et al.*(2000)] Herskovits, J., Leontiev, A., Dias, G., Santos, G.: Contact shape optimization: A bilevel programming approach. Struct. Multidisc. Optimization 20, 214–221 (2000)

[Koh(2007)] Koh, A.: Solving transportation bi-level programs with differential evolution. In: 2007 IEEE Congress on Evolutionary Computation (CEC 2007), pp. 2243–2250. IEEE Press (2007)

[Li and Wang(2007)] Li, H., Wang, Y.: A hybrid genetic algorithm for solving nonlinear bilevel programming problems based on the simplex method. In: Third International Conference on Natural Computation (ICNC 2007), pp. 91–95 (2007)

[Li *et al.*(2006)] Li, X., Tian, P., Min, X.: A hierarchical particle swarm optimization for solving bilevel programming problems. In: Rutkowski, L., Tadeusiewicz, R., Zadeh, L.A., Żurada, J.M. (eds.) ICAISC 2006. LNCS (LNAI), vol. 4029, pp. 1169–1178. Springer, Heidelberg (2006)

[Mathieu *et al.*(1994)] Mathieu, R., Pittard, L., Anandalingam, G.: Genetic algorithm based approach to bi-level linear programming. Operations Research 28(1), 1–21 (1994)

[Miettinen(1999)] Miettinen, K.: Nonlinear Multiobjective Optimization. Kluwer, Boston (1999)

[Nolle(2006)] Nolle, L.: On a hill-climbing algorithm with adaptive step size: Towards a control parameter-less black-box optimization algorithm. In: Reusch, B. (ed.) Computational Intelligence, Theory and Applications, pp. 587–596. Springer, Berlin (2006)

[Oduguwa and Roy(2002)] Oduguwa, V., Roy, R.: Bi-level optimisation using genetic algorithm. In: Proceedings of the 2002 IEEE International Conference on Artificial Intelligence Systems (ICAIS 2002), pp. 322–327 (2002)

[Pakala(1993)] Pakala, R.R.: A Study on Applications of Stackelberg Game Strategies in Concurrent Design Models. Master's thesis, Department of Mechanical Engineering: University of Houston (1993)

[Paruchuri *et al.*(2008)] Paruchuri, P., Pearce, J.P., Marecki, J., Tambe, M., Ordonez, F., Kraus, S.: Efficient algorithms to solve bayesian Stackelberg games for security applications. In: Proceedings of the Twenty-Third AAAI Conference on Artificial Intelligence, pp. 1559–1562 (2008)

[Rao(1984)] Rao, S.S.: Optimization: Theory and Applications. Wiley, New York (1984)

[Reklaitis *et al.*(1983)] Reklaitis, G.V., Ravindran, A., Ragsdell, K.M.: Engineering Optimization Methods and Applications. Wiley, New York (1983)

[Shi and Xia(2001)] Shi, X., Xia, H.S.: Model and interactive algorithm of bi-level multi-objective decision-making with multiple interconnected decision makers. Journal of Multi-Criteria Decision Analysis 10(1), 27–34 (2001)

[Sinha and Deb(2009)] Sinha, A., Deb, K.: Towards understanding evolutionary bilevel multi-objective optimization algorithm. In: Proceedings of the IFAC Workshop on Control Applications of Optimization, Jyväskylä, Finland, May 6-8, pp. 6–8 (2009); Also KanGAL Report No. 2008006, Indian Institute of Technology Kanpur, India

[Sinha *et al.*(2010)] Sinha, A., Deb, K., Korhonen, P., Wallenius, J.: Progressively interactive evolutionary multi-objective optimization method using generalized polynomial value functions. In: 2010 IEEE Congress on Evolutionary Computation (CEC 2010), pp. 1–8. IEEE Press (2010)

[Sinha(2011)] Sinha, A.: Bilevel multi-objective optimization problem solving using progressively interactive EMO. In: Takahashi, R.H.C., Deb, K., Wanner, E.F., Greco, S. (eds.) EMO 2011. LNCS, vol. 6576, pp. 269–284. Springer, Heidelberg (2011)

[Sun *et al.*(2006)] Sun, D., Benekohal, R.F., Waller, S.T.: Bi-level programming formulation and heuristic solution approach for dynamic traffic signal optimization. Computer-Aided Civil and Infrastructure Engineering 21(5), 321–333 (2006)

[Vicente and Calamai(2004)] Vicente, L.N., Calamai, P.H.: Bilevel and multilevel programming: A bibliography review. Journal of Global Optimization 5(3), 291–306 (2004)

[Wang *et al.*(2008)] Wang, G., Wan, Z., Wang, X., Lv, Y.: Genetic algorithm based on simplex method for solving linear-quadratic bilevel programming problem. Comput. Math. Appl. 56(10), 2550–2555 (2008)

[Wang *et al.*(2007)] Wang, G.-M., Wang, X.-J., Wan, Z.-P., Jia, S.-H.: An adaptive genetic algorithm for solving bilevel linear programming problem. Applied Mathematics and Mechanics 28(12), 1605–1612 (2007)

[Wang and Periaux(2001)] Wang, J.F., Periaux, J.: Multi-point optimization using gas and Nash/Stackelberg games for high lift multi-airfoil design in aerodynamics. In: Proceedings of the 2001 Congress on Evolutionary Computation (CEC 2001), pp. 552–559 (2001)

[Wang and Ehrgott(2011)] Wang, J.Y.T., Ehrgott, M.: Transport Sustainability Analysis with a Bilevel Biobjective Optimisation Model. In: Proceedings of the 2011 Conference on Multi-Criteria Decision Making, MCDM 2011 (2011)

[Wang *et al.*(2005)] Wang, Y., Jiao, Y.-C., Li, H.: An evolutionary algorithm for solving nonlinear bilevel programming based on a new constraint-handling scheme. IEEE Transactions on Systems, Man, and Cybernetics, Part C: Applications and Reviews 35(2), 221–232 (2005)

[Wierzbicki(1980)] Wierzbicki, A.P.: The use of reference objectives in multiobjective optimization. In: Fandel, G., Gal, T. (eds.) Multiple Criteria Decision Making Theory and Applications, pp. 468–486. Springer (1980)

[Yin(2000)] Yin, Y.: Genetic algorithm based approach for bilevel programming models. Journal of Transportation Engineering 126(2), 115–120 (2000)

[Zhang *et al.*(2007)] Zhang, G., Liu, J., Dillon, T.: Decntralized multi-objective bilevel decision making with fuzzy demands. Knowledge-Based Systems 20, 495–507 (2007)

[Zitzler *et al.*(2001)] Zitzler, E., Laumanns, M., Thiele, L.: SPEA2: Improving the strength pareto evolutionary algorithm for multiobjective optimization. In: Giannakoglou, K.C., Tsahalis, D.T., Périaux, J., Papailiou, K.D., Fogarty, T. (eds.) Evolutionary Methods for Design Optimization and Control with Applications to Industrial Problems. International Center for Numerical Methods in Engineering (CMINE), Athens, Greece, pp. 95–100 (2001)

Index

Printed in the United States
By Bookmasters